Inorganic
Mass Spectrometry

CHEMICAL ANALYSIS

A SERIES OF MONOGRAPHS ON ANALYTICAL CHEMISTRY AND ITS APPLICATIONS

VOLUME 95

A WILEY-INTERSCIENCE PUBLICATION

JOHN WILEY & SONS

New York / Chichester / Brisbane / Toronto / Singapore

Inorganic
Mass Spectrometry

Edited by

F. Adams
R. Gijbels
R. Van Grieken

University of Antwerp
Department of Chemistry
Antwerpen-Wilrijk, Belgium

A WILEY-INTERSCIENCE PUBLICATION

JOHN WILEY & SONS

New York / Chichester / Brisbane / Toronto / Singapore

Copyright © 1988 by John Wiley & Sons, Inc.

Library of Congress Cataloging in Publication Data:

Inorganic mass spectrometry.

(Chemical analysis ; v. 95)
"A Wiley-Interscience publication."
Includes index.
1. Mass spectrometry. 2. Chemistry, Inorganic.
I. Adams, F. (Freddy) II. Gijbels, R. (Renaat) III. Grieken,
R. Van (René) IV. Series.
QD96.M3I55 1988   543'.0873   88-21561
ISBN 0-471-82364-3

Printed in the United States of America

10 9 8 7 6 5 4 3 2 1

CONTRIBUTORS

F. Adams, University of Antwerp, Antwerpen-Wilrijk, Belgium

F.J. Bruynseels, University of Antwerp, Antwerpen-Wilrijk, Belgium

I. Cornides, Budapest, Hungary

R. Gijbels, University of Antwerp, Antwerpen-Wilrijk, Belgium

A. L. Gray, University of Surrey, Guilford, Surrey, United Kingdom

W. W. Harrison, University of Virginia, Charlottesville, Virginia

Klaus G. Heumann, University of Regensburg, Regensburg, Federal Republic of Germany

Alexander Lodding, Chalmers University of Technology, Gothenburg, Sweden

G. Ramendik, Vernadsky Institute of Geochemistry and Analytical Chemistry, Moscow, USSR

R. Van Grieken, University of Antwerp, Antwerpen-Wilrijk, Belgium

A. H. Verbueken, University of Antwerp, Antwerpen-Wilrijk, Belgium

J. Verlinden, University of Antwerp, Antwerpen-Wilrijk, Belgium

v

PREFACE

This book is published at a moment when there is, again, a sense of real excitement at the forefront of inorganic mass spectrometry. Indeed, in this field, spectacular advances have been made recently in the diversification of powerful ionization sources, the commercial availability of the instruments, and the fields of application. It is to be expected that in the future these significant new developments will further enhance the position of mass spectrometry relative to the other analysis methods available for inorganic characterization, multielement trace analysis, and micro- and surface analysis.

The scientific literature and conferences prove that mass spectrometry is most widely employed in the field of organic analysis, but this was not always the case. Mass spectrometry was, in fact, born when J. J. Thompson was exploring pathways of "positive electricity" around 1910. Although he did not set out deliberately to invent mass spectrometry, he quickly realized the implications for analysis of the experiments in which he passed his "positive rays" through combined magnetic and electric fields to produce streaks on fluorescent screens or photographic plates, and he wrote that "there are many problems in chemistry which could be solved much better by this than by any other method." Using mass spectrographs, Aston measured the isotopic composition of most elements, and Dempster finished the job in 1933. It was only in the 1940s that mass spectrometry began to be used in industrial analytical applications, and it was found to be particularly valuable in the petroleum industry for the analysis of mixtures of hydrocarbons. Yet an early bibliography on mass spectrometry covering 1938–1950 listed 700 papers on topics ranging from instrument design through operating characteristics and applications; of these, only 72 dealt with organic analysis. The rise in organic mass spectrometry began in the 1950s, mostly in connection with the qualitative analysis of organic compounds. It had been demonstrated that accurate measurements can tie down the molecular formula corresponding to a particular mass peak and that knowledge of fragmentation patterns can lead to identification of complex molecular structures. The coupling of a quadrupole mass spectrometer to a gas chromatograph introduced mass spectrometry to countless applications in industrial biochemical and medical laboratories. Tandem mass spectrometers and Fourier transform instruments constitute some other, more recent, novel developments in organic mass spectrometry.

Several reasons can be suggested to explain why, in the last 30 years, mass spectrometry has been invoked relatively less frequently for inorganic analysis than for applications in the organic field. The most significant reasons are perhaps the following: (a) In contrast to organic matrices, which are often volatile, it is usually not a simple matter to vaporize and ionize solid inorganic samples for introduction into the vaccum of a mass spectrometer; a very powerful ion source is mandatory and, in addition, some previous sample preparation might be required; (b) even if successfully introduced into the high-vacuum environment of the mass spectrometer, many complex inorganic ions are dissociated only with difficulty into their constituent atoms, and sometimes multiple ionization can complicate the interpretation of the mass spectra; and (c) multielement analysis is often required at ultratrace levels, and since mass spectrometry is essentially a multielement technique, extreme sensitivity is normally expected of it.

Recent developments in inorganic mass spectrometry—inductively coupled plasma mass spectrometry, glow discharge mass spectrometry, and new developments in spark source mass spectrometry—have all implied a remedy, at least partially, to these limitations. Other novelties are the advantageous microanalysis capabilities of laser microprobe mass analysis and of the modern instruments for secondary ion mass spectrometry. The principles, instrumental features, characteristics, and new application fields of spark source, glow discharge, inductively coupled plasma, and isotope dilution mass spectrometry and of micro- and surface analysis via laser microprobe mass analysis and secondary ion mass spectrometry versions constitute the core of this monograph.

Chapter 1 reviews the historical background of inorganic solids mass spectrometry. Chapter 2 accounts for spark source mass spectrometry, which was introduced in 1934. The principles and instrumentation of spark source mass spectrometry are described, but even more emphasis is placed on the physics of the vacuum spark discharge and the energy approach to the major steps of the mass spectrum formation. Apart from the conventional multielement bulk analysis applications of spark source mass spectrometry, the capabilities for semimicro, local, and in-depth analysis are outlined. Chapter 3 discusses a new approach to inorganic mass spectrometry using a glow discharge ion source, a technique for which commercial equipment is now available. Secondary ion mass spectrometry has, in principle, been available as a sensitive surface analysis technique for several decades. A discussion of the new commercial instrumentation developments and of the inherent sensitivity and quantitative character, the spatial analysis, and imaging capabilities of secondary ion mass spectrometry constitutes Chapter 4 of this book.

The youngest of all mass spectrometry techniques are undoubtedly the ones using lasers as ion sources, such as the "laser microprobe mass analyzer," in which a focused and pulsed laser is combined with a time-of-flight mass spectrometer in a commercial instrument for trace analysis with micrometer size spatial resolution; most of Chapter 5 is devoted to this technique. The instrumentation and its analytical characteristics are outlined, as well as the fundamentals of laser–solid interactions and applications in biology, medicine, and environmental research. Inductively coupled plasma source mass spectrometry is another new and very exciting development in inorganic mass spectrometry. In Chapter 6, much emphasis is put on the ion source, ion extraction, and mass analysis systems, but also on the diverse applications of the technique. Chapter 7 deals with isotope dilution mass spectrometry, probably the most accurate inorganic trace analysis yet developed. Again, both the principles and ionization, analysis, and detection methods are described, and much emphasis is given to the applications of isotope dilution mass spectrometry to metals, nonmetals, and gases. Finally, Chapter 8 discusses ongoing developments and the probable future of inorganic mass spectrometry.

This book was deliberately made a multiauthor effort. We believe that having scientists actively engaged in a particular technique cover those areas for which they are particularly qualified, all presenting their own points of view and general approach, certainly outweighs the advantage of the uniformity characteristic of a single-author book. Two of the chapters were written by the editors and their coworkers. For all the other chapters, we have had the cooperation of eminent and renowned specialists. The editors wish to thank F. J. Bruynseels, I. Cornides, A. L. Gray, W. W. Harrison, K. G. Heumann, A. Lodding, G. Ramendik, A. H. Verbueken and J. Verlinden for their hard work in providing material for this book.

The editors and authors hope that novices in inorganic mass spectrometry will find this book useful and instructive, and that our fellow mass spectrometrists will benefit from the large amount of information available in this compact form.

F. ADAMS
R. GIJBELS
R. VAN GRIEKEN

Antwerp, Belgium
March 1987

CONTENTS

CHAPTER

1

MASS SPECTROMETRIC ANALYSIS OF INORGANIC SOLIDS—THE HISTORICAL BACKGROUND

I. CORNIDES

Budapest, Hungary

Looking at the exceedingly great diversity of the applications of mass spectrometry, not only in almost all branches of pure science but also in the applied sciences, such as technology, agriculture, and medicine, it is admirable how appropriately *J. J. Thomson* was able to outline the potentialities of the method he used to investigate the nature of the rays of "positive" electricity. In fact, the activities of mass spectrometrists are still—after more than 70 years—within the scope of Thomson's suggestions to use the method of analysis of "positive rays" (a) for detection and abundance measurement of isotopes, (b) for chemical analysis, and (c) for investigations in chemical physics.

According to general opinion, Thomson was one of those who opened the gate to the modern physics of the atoms, by making it possible to give an exact experimental answer to one of the most important questions of the science of that time, the question of the existence of isotopes. Nevertheless, and though he was a physicist, he seems to have been more interested in paving the way to chemical applications and left the exploitation of the more immediately rewarding isotope research to his students. (*Aston* became the Nobel laureate for obtaining mass spectra of isotopes of most elements.)

In the preface to his famous book, *Rays of Positive Electricity and their Application to Chemical Analysis* (82), Thomson says:

One of the main reasons for writing this book was the hope that it might induce others, and especially chemists, to try this method of analysis. I feel sure that there are many problems in chemistry which could be solved with far greater ease by this than any other method. The method is surprisingly sensitive—more so even than that of spectrum analysis, requires an infinitesimal amount of material, and does not require this to be specially purified; the technique is not difficult if appliances for producing high vacua are available.

1

This was written in 1913, and the same reasoning is often used even now by analysts who have never read Thomson's book.

Of course, time was needed for the necessary technical development (and not only in vacuum technique) before mass spectometry began to be acknowledged as a method of analytical chemistry in the early 1940s. *Hoover and Washburn* (53) published the first paper of this kind (mass spectrometric hydrocarbon analyses) in 1940. It is interesting to note that just about five years earlier, when the isotope analysis of the last elements—Pd, Pt, Au, and Ir—had been carried out, Aston had stated that mass spectrometry had served its purpose and would die away as a field of research. As we know, this short-term prediction was completely erroneous; the results of the chemical applications of mass spectrometry over the past four decades have splendidly justified the wisdom of Thomson's long-term prophecy and have fulfilled his expectations.

Fortunately, during the two decades devoted to the investigation of "isotope mass spectrometry," a substantial improvement of the technical background has taken place to meet the increasing demands in this field. Detection of low-abundance isotopes required both much higher sensitivity and lower detection limits. Moreover, reliable calculation of the chemical "atomic weights" necessitated isotope abundance measurements of high precision and accuracy. The pioneers in the design and construction of mass spectrometric equipment had also realized the possibility of producing commercial mass spectrometers, which was a precondition for using mass spectrometry as a method of analytical chemistry. Let us now briefly summarize the main steps of this development.

Analyzers with much better focusing were designed: the first velocity-focusing analyzer of Aston (1) and its improved versions built by Costa (19) and Aston (3, 4), on the one hand, and the direction-focusing type of *Dempster* (20) and *Swann* (79) using only magnetic deflection of 180°, on the other hand. Proper focusing increased the sensitivity of mass spectrometry and also improved the separation of isotopes in the higher mass range, and finally, which was so important for the measurement of accurate mass values, the resolution was improved sufficiently to separate doublets of ions of nearly the same mass number. The degree of improvement can be judged by the resolving power of the instruments developed in the period in question:

Thomson's parabolic instrument	15
Dempster's magnetic spectrometer	100
Aston's first mass spectrograph	130
Costa's velocity-focusing instrument	600
Aston's 3rd mass spectrograph	2000

This list suggests that magnetic direction focusing is inferior to velocity focusing in terms of resolving power. It must, however, be kept in mind that in velocity-focusing instruments the direction spread of the ion beam is carefully suppressed by using a collimating system of narrow slits. When Bainbridge (6) added a velocity filter to Swann's magnetic analyzer to suppress the velocity spread, he obtained a resolving power of 500 for this magnetic instrument.

Furthermore, the results of theoretical ion optics, attained by a great number of specialists (e.g., Barber, Stephens, Henneberg, Herzog, Mattauch, and many others, including mass spectrometrists) had already shown in the early 1930s the potential of the magnetic instruments for further development.

It has been found, for instance, that both direction and velocity focusing can be realized, within limits, in one instrument, thereby reducing the loss of ions, that is, the loss of sensitivity caused by the suppression of either the direction or the velocity (energy) spread. The first such double-focusing mass spectrometer was built by Dempster in 1935 (22), and another by Bainbridge and Jordan in 1936 (7). At about the same time, *Mattauch* (60) constructed an instrument based on the special ion optics developed by *Herzog* (43) and Mattauch and Herzog (61). This ion optical system is unique in that it provides double focusing simultaneously for all masses, and the entire mass spectrum can be detected simultaneously on a plane detector such as a photoplate. Because of these properties, this instrument is still extremely important in the field of inorganic solid analysis.

Double-focusing mass spectrometers are rather bulky and expensive instruments. For this reason another line in the development of magnetic mass spectrometers was especially significant in making mass spectrometry available to analytical chemists. At a given degree of focusing, the resolution of the magnetic analyzer is dependent on the radius of the path of the ions deflected in the magnetic field, a larger radius providing a proportionally higher value of the resolving power. However, the ion path in the analyzers built in the manner of Dempster is a semicircle (the deflection is 180°), and a large radius would require an impracticably large magnet. For this reason it was of very great importance that in 1940 *Nier* (67) built a mass spectrometer with a "sector" magnetic field (of 60° ion deflection), the focusing properties of which had been recognized some years earlier. It was found, as Nier remembers (69),

that a mass spectrometer using a 60° wedge magnet had essentially as good focusing properties as the earlier 180° instruments. The saving in weight and power consumption was substantial, and the mass spectrometer became a tool within the reach of many more researchers in fields where gas or isotope analyses were required.

Although Nier, as a physicist, was interested mainly in isotope research (in fact, he belonged to those who completed the list of the stable isotopes existing in nature and also pioneered isotope studies in the earth and planetary sciences), he was aware of the importance of mass spectrometry for chemical analysis, especially for quick and reliable analyses of complex hydrocarbon mixtures in oil refineries.

The financial support for the production of commercial mass spectrometers was indeed provided by the oil industry. The first analytical instruments were of the 180° deflection type, such as the Consolidated Engineering Corporation's Model 21-101 instruments, the first of which was installed in 1942 for the Atlantic Refining Company. This was the starting point for the large-scale application of analytical mass spectrometry. Very shortly, however, the sector-type instruments became overwhelmingly dominant. This was due also to a series of physical and technical (including vacuum technical and electronic) improvements implemented by Nier (68) over the next few years. The ion optical refinements and the "Nier-type" electron impact ion source of remarkably homogeneous ion energy made possible a very considerable increase in resolving power, the value of which can reach a few thousands.

The further development of mass analyzers was governed mainly by chemical requirements. The so-called nonmagnetic or dynamic mass spectrometers, the development of which started in the 1950s, were expected to be relatively small and inexpensive instruments of high analytical performance. The elimination of the magnetic field (i.e., the magnet and its power supply) promised not only a significant decrease in weight and size, but also considerably faster scanning of the spectra, that is, very short response times.

As is well known, quadrupole mass spectrometers, the development of which was initiated by Paul and coworkers (70, 71), have become very important instruments of modern mass spectrometry. In the quadrupole "mass filter," the beam of ions is sent axially between four parallel rods, and, in the quadrupole electric field of the direct voltages between neighboring pairs of rods, the ions are forced to oscillate perpendicularly to the axis by a superimposed radiofrequency field. The amplitudes of the oscillations are dependent on mass, and only ions of a definite mass-to-charge ratio will not leave the beam and will therefore be able to emerge in the direction of the axis. By changing the combination of the dc and rf voltages, it is possible to select another mass and thus another kind of ion to pass the "mass filter"—in other words, to scan the mass spectrum.

Quadrupole mass spectrometry (QMS) has had a wonderful career. Combined with a gas chromatograph for sample introduction, the *GC-QMS system* has become the most powerful analytical tool of organic chemistry, including biochemistry, pharmaceutical research, and industry, environmental studies, and many other applications. The triple quadrupole system is an important tool of gas-phase ion chemistry. And quadrupoles are usually the analyzers in the most recently developed instruments for the analysis of inorganic solids, the glow discharge (GD) and inductively coupled (ICP) mass spectrometers (see Chapters 3 and 6).

Another type of dynamic mass spectrometer, also used for solid analysis, applies the time-of-flight principle to separate ions according to their masses, which is the simplest and most direct method of separation. The ions acted upon by the same electric impulse acquire velocities inversely proportional to their mass, and accordingly they pass the same distance through a field-free space to a detector in different times. Obviously, this principle can be realized only with pulses of ions that separate during the time of flight to the detector into smaller bunches according to their mass and are detected in the time sequence of their arrival. The first time-of-flight (TOF) instruments were built in the late 1940s [e.g., by Cameron and Eggers (11)], and the resolving power was gradually increased from below 100 to about 1000 as the electronic methods of the pulse technique improved. This is completely satisfactory for inorganic analysis. Time-of-flight analyzers are often operated with laser ion sources; the pulse character of both units can advantageously be matched.

Other than the analyzer, the ion source is the most important part of any mass spectrometer, and the development of ion sources closely followed that of analyzers.

The first main problem to be solved was the ionization of the atoms of solid samples. Thomson analyzed the ions of positive rays emerging from gas discharges. However, this type of ion source could only be used to ionize gases and materials of relatively high vapor pressure from which a sufficient number of atoms or molecules could enter the discharge to be ionized by fast particles and provide detectable ion currents. For this reason, when Aston started to carry out his program of detecting all existing isotopes of all known elements, and to measure their masses and abundances, it was necessary for him to devise appropriate ion sources for the great majority of elements, which were available only as solids of low volatility—refractory metals, metallic oxides, and so on (4, 5).

The design of the first ion sources for solids could not break away from the gas discharge type. Still using a gas discharge to ionize the samples' atoms and/or molecules they were, in fact, modifications of Thomson's ion source. In the case of the *composite anode* method, even the evaporation of the solid material into the discharge was brought about by the discharge itself. The powdered sample mixed with graphite powder was pressed into a hole drilled in the end of a cylindrical anode and heated by the impinging cathode rays. This type of ion source was first used for isotope analysis by G. P. Thomson in 1921 (80).

In the *hot anode* method, the previously dissolved sample was deposited on a strip of platinum foil from which it was evaporated into the discharge by an electric current passing through the platinum strip. Aston (2) made use of this type of source for the isotope analysis of alkali metals, but a similar heated anode method had been used somewhat earlier by Dempster (20) when his magnetic mass spectrometer was built. Dempster tried several modifications, including the direct use of thermally evaporated ions and ionizing vapors by electron bombardment (electron impact) as early as 1922.

Looking back even more into the history of solid ion sources, we find that both ways of producing "anode rays" were discovered by Gehrcke and Reichenheim in 1906–1908 (36); however, at that time the means for mass analysis were not yet available.

In spite of some difficulties and limitations, the ion sources dealt with above made possible the isotope analysis of all elements of the periodic system, except for four noble metals, within about 15 years. The work was, however, subsequently completed in a very short time by Dempster, who was able to evaporate and ionize the four elements in his vacuum spark ion source, later of so much importance in mass spectrometric solid analysis, even for the most refractory materials (23). By the isotope analysis of platinum, palladium, and gold, and finally by the determination of the $^{193}Ir/^{191}Ir$ isotope ratio (24), the first great undertaking of mass spectrometry of basic importance for both physics and chemistry was accomplished.

And though some very rare isotopes were discovered after 1935 and the isotope abundance data have been undergoing refinement ever since, in the late 1930s a new period of mass spectrometry began: the development of mass spectrometric chemical analysis.

As we have already seen, commercial mass spectrometers, necessary to make mass spectrometry acceptable as a method of analytical chemistry, started to appear in the early 1940s. The real breakthrough came with Nier's famous instrument (68), the peak product of the isotope period but also the ideal model for all modern commercial mass spectrometers of the new era, including, of course, analytical mass spectrometers.

The Nier-type instruments, with their high-performance gas ion source, quickly became very successful in the field of gas analysis, including the analysis of hydrocarbons and a great many organic compounds of sufficient volatility.

The very important results attained by solid-state physics after the World War II and the surprisingly quick response of the industry initiated the search for new methods for the analysis of purified inorganic solid materials. Semiconductor physics and technology required full coverage of the periodic system at part-per-million and sub-ppm concentration levels, and this analytical problem was as difficult for conventional analytical chemistry as full analysis of complex hydrocarbon mixtures within a time short enough for process control. Mass spectrometry was tried again and was again successful.

The main problem is to achieve the evaporation and ionization of atoms of all elements in a nonselective way, so that the ion beam properly represents the composition of the solid sample.

Simple thermal evaporation from a small heated crucible, and subsequent ionization of the vapor, cannot meet this requirement because of the inherent selectivity of the first process. Though by using total vaporization this method could be made quantitative [see, e.g., Hickam (45)], for a single element this approach was by no means satisfactory.

The high energy concentrated into a very small volume by a vacuum spark promised total evaporation of the material in these small volumes and, accordingly, a reasonably representative ion beam. On the basis of results achieved during World War II, Dempster (25) suggested that the vacuum spark discharge, brought about by high-frequency high voltage applied to the sample electrodes and found indispensable in isotope analysis, should be tried also as a general-purpose analytical ion source.

After two earlier attempts to build instruments utilizing the spark ion source [Shaw and Rall (77) and Gorman et al. (37)], the first modern spark source mass spectrometer with a Mattauch–Herzog analyzer was built by Hannay (40), and its analytical capabilities, including impurity analysis down to 0.1 ppm, were demonstrated by Hannay and Ahearn (41). The Mattauch–Herzog ion optics have been proven to be of important practical advantage (simultaneous recording of all components on the photoplate); no manufacturer has ever chosen another type of analyzer. The first commercial instrument, the Metropolitan-Vickers MS-7, was available as early as 1958; the presently best (and only) spark source mass spectrometer is the JMS-O1BM type of JEOL.

Since the appearance of the first commercial instruments, spark source mass spectrometry (SSMS) has become a popular general-purpose method of inorganic solid analysis. Its homogeneously high sensitivity throughout the periodic system (i.e., full elemental coverage even at low concentrations in the sub-ppm range) plus the relatively simple sample preparation procedures it requires make SSMS unsurpassable by any other method (including ICP optical emission spectrometry and neutron activation analysis) in the field of full and survey (panoramic) analysis, purity control of materials, comparative analysis, and so on.

As a matter of fact, its moderately unsatisfactory precision and accuracy, caused by the erratic behavior of the vacuum spark, is considered the only weak point of SSMS. And quite reasonably, this was and is the starting point of most of the development work in inorganic MS analytics. Improvement was looked for in several ways:

1. By controlling the spark process and/or using isotopic calibration.
2. By testing other more predictable and stable methods of evaporation and ionization.

Since most of this work has been done during the last decade and will be dealt with in detail in the subsequent chapters of this book, just a brief overview will be given here to make clear the logic and linkage of the developments in question.

An improvement of the SSMS method was attempted by measures aimed at making the sparking process more stable and reproducible by controlling the mechanical and electrical spark parameters, including the physical-chemical properties of the electrodes (i.e., by eliminating any kind of matrix effect insofar as possible). If the spark source phenomena (including ion extraction) are made sufficiently reproducible, correct calibration is possible, for example, by means of an internal standard and relative sensitivity factors, and the accuracy is limited only by the photoplate detection-evaluation errors to about 4–6%.

Even with poor reproducibility, the above limits of accuracy can be attained by using isotope standards for calibration, by fully simultaneous measurement and calibration. This isotope dilution (ID)–SSMS procedure is the most advanced multielement method of inorganic solid analysis. It was first suggested and tried by Leipziger in 1965 (57).

As for SSMS, two more points remain to be made. First, it should be observed that although SSMS provides bulk concentration data, the consumed amount of sample per analysis is very small, roughly 100 times less than in optical emission spectroscopy. Accordingly, it is essentially a micro method, and for macrosamples reproducible results can be obtained only if the sample is sufficiently homogeneous or homogenized.

Second, we have to realize that the molecular ions formed in the spark plasma (17) may seriously jeopardize the reliability of the quantitative and even the qualitative analytical results by interfering with analyte atomic ions. In fact, molecular interferences constitute the main obstacle to lowering the detection limits into the sub-ppb region. In some cases this supersensitivity was achieved by using ion beams of only multiply charged ions A^{n+} with $n = 3$. It was, however, recently proved by Morvay and Cornides (66) that some molecules may exist even in the triply charged state.

With its capability for renewal a full 50 years after Dempster's first measurements, SSMS is still in the forefront of the analytical methods. Its development over the past 20 years was considerably helped by some basic studies of the fundamental processes taking place in the vacuum spark carried out in several laboratories. This work was started in Hintenberger's laboratory (Mainz), mainly by Franzen (33–35), and in the United States by Morrison and coworkers (63–65). Important contributions to the understanding of the physics and chemistry of the vacuum spark have been made in the past decade by Chupakhin, Derzhiev, and Ramendik in Moscow (13, 14, 26–28, 72–74); Stefani's group in Grenoble (10); the Antwerp group of Adams, Gijbels, and Van Grieken (83–85); Cornides, and coworkers in Budapest and Nitra (16–18); Dietze, and Becker in Leipzig (30–32).

In another approach to improve the performance of mass spectrometric analysis of inorganic solids, alternative ion sources have been tried, of which the triggered low-voltage dc arc stands quite close to the spark. Furthermore it is quite reasonable to expect that the use of a steady current of ions or photons, the intensity of which can easily be controlled, to bombard the solid sample instead of the particles of an erratic discharge current might result in a much more stable ion source.

Secondary ions were first observed by Thomson. I quote from his work in 1910 (81): "I had occasion . . . to investigate the secondary Canalstrahlen produced when primary Canalstrahlen strike against a metal plate." The first instrumental setup for secondary ion mass spectrometric measurements was, however, reported much later by Herzog and Viehböck (44), and the clear idea to apply the secondary ions for surface analysis was born in 1950 in the RCA Laboratories, Princeton, where the first results were attained in the years 1954–1955 by Honig (47).

The beginnings and further development of secondary ion mass spectrometry (SIMS) are beautifully described by Honig himself (50), who also presented the first report on mass spectrometric studies of solid surfaces (48). SIMS indeed developed as, and inherently is, a surface analytical method (including depth profiling); that is, it is a special-purpose method, and therefore replacing SSMS by SIMS is out of the question. Even apart of this, the quantitation of SIMS always presented similar or more serious problems.

Early SIMS studies used primary beams of millimeter dimensions. The reduction in beam size began in 1963 with an improved source design by Liebl and Herzog. In the early 1970s SIMS work differentiated and developed along two separate routes. Low primary current density, or static, SIMS using broad beams was introduced by Benninghoven and co-workers (8, 9) for the surface analysis of organic samples. High primary current density, or dynamics, SIMS was used by Wittmaack et al. (86) and by Magee et al. (58), among others, to obtain concentration profiles in depth on inorganic samples and to identify trace impurities.

High lateral resolution instrumentation developed along two parallel paths: ion microscopes and ion microprobes. In the ion microscope the sample is bombarded with a broad beam of primary ions. High resolution, of the order of micrometers, is achieved by forming on a screen a magnified image of the object with a narrow beam of secondary particles that have passed through appropriate ion optics and a mass filter. The ion microprobe uses a primary beam, focused to micrometer dimensions, that can be rastered over the sample; a selected portion of the secondary ions is analyzed in a suitable mass filter.

The prototype of the ion microscope was developed by Castaing and Slodzian (12) and served as the basis for three generations of sophisticated and successful commercial instruments: the Cameca IMS-300 described by Rouberol et al. (75); the Cameca IMS-3F (76), which contains a triple focusing mass spectrometer of high mass resolution housed in an ultrahigh vacuum system; and the Cameca IMS-4F (62), which can be considered as a combined ion microscope/ion microprobe. For the latter application, a finely focused Cs^+ beam allows a lateral resolution of 0.2 μm.

The first commercial ion microprobe was the IMMA of Applied Research Laboratories, developed by Liebl in 1967, allowing a lateral resolution of about 1 μm (87). Substantial improvements have been made recently by using a liquid metal ion source, as will be discussed in more detail in Chapter 4.

Due to their full chemical and electrical neutrality, the use of photons to supply energy for the evaporation and/or ionization of solids may seem to be even more attractive. Exploratory experiments in this field have also been carried out in the RCA Laboratories, where Honig and co-workers (52) produced positive ions by laser irradiation of metals, semiconductors, and insulators more than 20 years ago. Since then predictions have been made from time to time that the role of the spark source will soon be taken over by laser ion sources. But even as recently as 1984 Svec (78) could still only promise that laser ionization would make SSMS obsolete.

The coupling of lasers (mostly Nd:YAG lasers) to double-focusing mass spectrometers is still being actively pursued [see, e.g., Jansen and

Witmer (55) and Dietze and Becker (31)], primarily for the analysis of bulk samples.

One important aim of this book is to present a clear account of the most recent developments and to evaluate the relative prospects of SSMS, laser-induced mass spectrometry (LIMS), and other alternative MS methods for the near future. As for the present, it is obvious that laser ionization mass spectrometry has proved its capabilities for local microanalysis, including the trace element analysis of biological microsamples. To obtain good lateral resolution, the laser pulse length was reduced, the laser beam was focused, its wavelength was decreased, and a transmission configuration was chosen for ion production and extraction. Additional improvements were obtained by using a light microscope for further focusing of the laser beam, a time-of-flight mass spectrometer for high transmission, and a time-focusing ion reflector for better mass resolution (46, 56). Laser microprobe mass spectrometers are commercially available from Leybold-Heraeus and Cambridge Mass Spectrometry Ltd.

The development of a second group of alternative ion sources originated from the fact that the pioneers of mass spectrometry used as ion sources phenomena—gas discharges, sparks—that had been the main light sources of optical emission spectroscopy (OES) since its very beginning. It was therefore like carrying on a tradition when the more modern optical sources of considerably higher stability were tested also as ion sources.

A glow discharge ion source was first used by Coburn (15), who applied a simple planar diode geometry. The high-performance hollow-cathode light source was developed into an ion source by Harrison and Daughtrey (42), and, as another variant, a coaxial cathode geometry was also tested successfully by the same group in 1976 (59).

Besides the relatively low-pressure (from about 10 to some hundreds of pascals) GD sources, mass spectrometrists have become interested also in the atmospheric pressure gas discharges used for optical emission. At first Gray extracted ions from the dc plasma of the capillary arc (38), but later he turned to the utilization of inductively coupled plasma (ICP), the light source of the most advanced OES method, as a possible ion source (39).

The development and the expected wide application of GD, ICP, and other possible plasma sources is in fact a revival of the gas discharge ion source. The GD source is a modern form of Aston's anode ray source. In the case of the ICP source the excitation of the plasma and the sample introduction (by nebulizing a solution) are carried out in new ways.

Finally, it is to be mentioned that the most accurate method of solids analysis, isotope dilution thermal ionization mass spectrometry, makes use of an ion source that is the modern version [developed basically by Inghram and Chupka (54)] of the hot anode of earlier times.

This historical introduction may seem unbalanced to the reader. I hope, however, that the reason for this is also obvious. The optimal choice(s) from the possible values of the numerous (geometrical, gas, and electric) discharge parameters and from the possible inlet-system–ion-source–analyzer–detector combinations and, finally, from the almost limitless number of operation modes of the mass spectrometers built from these units could not be made as yet.

For this reason the ending to this chapter is inevitably incomplete. Instead of presenting a full and clear picture of the present capabilities of mass spectrometric solid analysis, I have to confine myself to giving only my idea of what might be optimal at present and probably in the near future. Some years ago (18) I suggested that inorganic mass spectrometrists should follow the organic mass spectrometrists—they should have three or four interchangeable ion sources, a double-focusing analyzer (with both photo- and electronic detectors), and possibly a quadrupole, from which an instrument of the capability required can be easily and quickly built up for any analytical problem.

As an aid in making proper selections, readers will find all presently available knowledge in the subsequent chapters, written by specialists.

REFERENCES

1. F. W. Aston, *Phil. Mag.*, **38**, 707 (1919).

2. F. W. Aston, *Phil. Mag.*, **42**, 436 (1921).

3. F. W. Aston, *Proc. Roy. Soc. A.*, **115**, 487 (1927).

4. F. W. Aston, *Proc. Roy. Soc. A.*, **163**, 391 (1937).

5. F. W. Aston, *Mass Spectra and Isotopes*, Arnold, London, (1942).

6. K. T. Bainbridge, *Phys. Rev.*, **42**, 1 (1932).

7. K. T. Bainbridge and E. B. Jordan, *Phys. Rev.*, **50**, 282 (1936).

8. A. Benninghoven, *Z. Phys.*, **230**, 403 (1970).

9. A. Benninghoven, D. Jaspers, and W. Sichtermann, *Appl. Phys.*, **11**, 35 (1976).

10. J. Berthod, A. M. Andreani, and R. Stefani, *Int. J. Mass Spectrom. Ion Phys.*, **27**, 305 (1978).

11. A. E. Cameron and D. F. Eggers, *Rev. Sci. Instrum.*, **19**, 605 (1948).

12. R. Castaing and G. Slodzian, *J. Microsc.*, **1**, 395 (1962).

13. M. S. Chupakhin, O. I. Kryuchkova, and G. I. Ramendik, *Analytical Po-*

tentialities of Spark Source Mass Spectrometry (in Russian), Atomizdat, Moscow, 1972.

14. M. S. Chupakhin, G. I. Ramendik, V. I. Derzhiev, Yu. G. Tatsii, and M. A. Potapov, *Dokl. Akad. Nauk USSR,* **210,** 1074 (1973).

15. J. W. Coburn, *Rev. Sci. Instrum.,* **41,** 1219 (1970).

16. I. Cornides, *Adv. Mass Spectrom.,* **8A,** 452 (1980).

17. I. Cornides and T. Gál, *High Temp. Sci.,* **10,** 171 (1978).

18. I. Cornides, (1981) in L. Niinistö, Ed., *Euroanalysis IV, Reviews on Analytical Chemistry,* Akadémiai Kiadó, Budapest, 1982, p. 105.

19. J. L. Costa, *Ann. Phys.* **4,** 425 (1925).

20. A. J. Dempster, *Phys. Rev.,* **11,** 316 (1918).

21. A. J. Dempster, *Phys. Rev.,* **20,** 631 (1922).

22. A. J. Dempster, *Proc. Am. Phil. Soc.,* **75,** 755 (1935).

23. A. J. Dempster, *Nature,* **135,** 993; **136,** 65 (1935).

24. A. J. Dempster, *Nature,* **136,** 909 (1935).

25. A. J. Dempster, MDDC 370, U.S. Dept. of Commerce, Washington, DC, 1946.

26. V. I. Derzhiev, V. Liebich, H. Mai, and G. I. Ramendik, *Int. J. Mass Spectrom. Ion Phys.,* **32,** 345 (1980).

27. V. I. Derzhiev, A. Yu. Zakharov, and G. I. Ramendik, *J. Tech. Phys. USSR,* **48,** 1877 (1978).

28. V. I. Derzhiev and G. I. Ramendik, *J. Tech. Phys. USSR,* **48,** 312 (1978).

29. H. J. Dietze, *Massenspektroskopische Spurenanalyse,* Akad. Verlagsges. Geest und Portig KG, Leipzig, 1975.

30. H. J. Dietze and S. Becker, *ZFI Mitt.,* **57,** 1 (1982).

31. H. J. Dietze and S. Becker, *ZFI Mitt.,* **101,** 5 (1985).

32. H. J. Dietze and S. Becker, *Fresenius Z. Anal. Chem.,* **321,** 490 (1985).

33. J. Franzen, in A. J. Ahearn, Ed., *Trace Analysis by Mass Spectrometry,* Academic, New York, 1972, p. 11.

34. J. Franzen and H. Hintenberger, *Z. Naturforsch.,* **18a,** 397 (1963).

35. J. Franzen and K. D. Schuy, *Z. Naturforsch.,* **20a,** 176 (1965).

36. E. Gehrcke and O. Reichenheim, *Verh. Phys. Ges.,* **8,** 559 (1906).

37. J. G. Gorman, E. J. Jones, and J. A. Hipple, *Anal. Chem.,* **23,** 438 (1951).

38. A. L. Gray, *Proc. Soc. Anal. Chem.,* **11,** 182 (1974).

39. A. L. Gray, in D. Price and J. F. J. Todd, Eds., *Dynamic Mass Spectrometry,* Vol. 5, Heyden, London, 1978, p. 106.

40. N. B. Hannay, *Rev. Sci. Instrum.,* **25,** 644 (1954).

41. N. B. Hannay and A. J Ahearn, *Anal. Chem.,* **26,** 1056 (1954).

42. W. W. Harrison and E. H. Daughtrey, *Anal. Chem. Acta,* **65,** 35 (1973).

43. R. Herzog, *Z. Physik.,* **89,** 447 (1934).

44. R. Herzog and F. P. Viehböck, *Phys. Rev.*, **76**, 855L (1949).

45. W. M. Hickam, *ASTM Spec. Tech. Pub.*, **149**, 7 (1951).

46. F. Hillenkamp, E. Unsöld, R. Kaufmann, and R. Nitsche, *Appl. Phys.*, **8**, 341 (1975).

47. R. E. Honig, *J. Appl. Phys.*, **29**, 549 (1958).

48. R. E. Honig, *Adv. Mass Spectrom.*, **2**, 25 (1962).

49. R. E. Honig, *Appl. Phys. Lett.*, **3**, 8 (1963).

50. R. E. Honig, *32nd Annual Conf. Mass Spectrom. Allied Topics*, Retrospective Lectures, Am. Soc. Mass Spectrometry, 1984, p. 19.

51. R. E. Honig, *Int. J. Mass Spectrom. Ion Processes*, **66**, 31 (1985).

52. R. E. Honig and J. R. Wolston, *Appl. Phys. Lett.*, **2**, 138 (1963).

53. H. Hoover and H. W. Washburn, Amer. Inst. Mining Metallurgical Engrs. Tech. Publ. No. 1205, 1940.

54. M. G. Inghram and W. A. Chupka, *Rev. Sci. Instrum.*, **24**, 518 (1953).

55. J. A. J. Jansen and A. W. Witmer, *Spectrochim. Acta*, **37B**, 483 (1982).

56. R. L. Kaufmann, H. J. Heinen, M. W. Schürmann, and R. M. Wechsung, in D. E. Newburg, Ed., *Microbeam Analysis—1979*, San Francisco Press, San Francisco, 1979, p. 63.

57. F. D. Leipziger, *Anal. Chem.*, **37**, 171 (1965).

58. C. W. Magee, W. L. Harrington, and R. E. Honig, *Rev. Sci. Instrum.*, **49**, 477 (1978).

59. W. A. Mattson, B. L. Bentz, and W. W. Harrison, *Anal. Chem.*, **48**, 489 (1976).

60. J. Mattauch, *Phys. Rev.*, **50**, 617 (1936).

61. J. Mattauch and R. Herzog, *Z. Physik.*, **89**, 786 (1934).

62. H. N. Migeon, C. Le Pipec, and J. J. Le Goux, in A. Benninghoven et al., Eds., *SIMS-V*, Springer-Verlag, Heidelberg, 1986, pp. 155–157.

63. G. H. Morrison, *Crit. Rev. Anal. Chem.*, **8**, 287 (1979).

64. G. H. Morrison and A. T. Kashuba, *Anal. Chem.*, **41**, 1842 (1969).

65. G. H. Morrison and R. K. Skogerboe, in G. H. Morrison, Ed., *Trace Analysis: Physical Methods*, Wiley, New York, 1965, p. 16.

66. L. Morvay and I. Cornides, *Int. J. Mass Spectrom. Ion Processes*, **62**, 263 (1984).

67. A. O. Nier, *Rev. Sci. Instrum.*, **11**, 252 (1940).

68. A. O. Nier, *Rev. Sci. Instrum.*, **18**, 398 (1947).

69. A. O. Nier, *Ann. Rev. Earth Planet. Sci.*, **9**, 1 (1981).

70. W. Paul and H. Steinwedel, *Z. Naturforsch.*, **8a**, 448 (1953).

71. W. Paul and M. Raether, *Z. Physik.*, **140**, 262 (1955).

72. G. I. Ramendik and V. I. Derzhiev, *J. Anal. Chem. USSR*, **32**, 1197, 1204 (1977).

73. G. I. Ramendik and V. I. Derzhiev, *J. Anal. Chem. USSR,* **34,** 647 (1979).

74. G. I. Ramendik, in L. Niinistö, Ed., *Euroanalysis IV, Reviews on Analytical Chemistry,* Akadémiai Kiado, Budapest, 1982, p. 57.

75. J. M. Rouberol, J. Guernet, P. Deschamps, J. P. Dagnot, and J. M. Guyon de la Berge, *Proc. 5th Int. Conf. X-Ray Opt. Microanal.,* Springer-Verlag, Heidelberg, 1969, pp. 311–318.

76. J. M. Rouberol, M. Lepareur, B. Autier, and J. M. Gourgout, in R. E. Ogilvie and D. B. Wittry, Eds., *X-ray Optical Microanalysis,* Pendell, Midland, MI, 1980.

77. A. E. Shaw and W. Rall, *Rev. Sci. Instrum.,* **18,** 278 (1947).

78. H. J. Svec, 32nd Annual Conf. MS and Allied Topics, Retrospect. lectures, *Am. Soc. Mass Spectrom.,* 1984, p. 61.

79. W. F. G. Swann, *J. Franklin Inst.,* **210,** 751 (1930).

80. G. P. Thomson, *Phil. Mag.,* **42,** 857 (1921).

81. J. J. Thomson, *Phil. Mag.,* **20,** 752 (1910).

82. J. J. Thomson, *Rays of Positive Electricity and Their Application to Chemical Analyses,* Longmans, Green, London, 1913.

83. E. Van Hoye, F. Adams, and R. Gijbels, *Talanta,* **26,** 285 (1979).

84. J. Van Puymbroeck, R. Gijbels, M. Viczián, and I. Cornides, *Int. J. Mass Spectrom. Ion Processes,* **56,** 269 (1984).

85. L. Vos and R. Van Grieken, *Int. J. Mass Spectrom. Ion Phys.,* **51,** 63 (1983).

86. K. Wittmaack, J. Maul, and F. Schulz, *Int. J. Mass Spectrom. Ion Phys.,* **11,** 35 (1973).

87. H. J. Liebl, *J. Appl. Phys.,* **38,** 5277 (1967).

CHAPTER

2

SPARK SOURCE MASS SPECTROMETRY

G. RAMENDIK

Vernadsky Institute of Geochemistry and Analytical Chemistry,
Moscow, USSR

J. VERLINDEN and R. GIJBELS

University of Antwerp, Antwerpen-Wilrijk, Belgium

1. INTRODUCTION

The spark source was introduced in mass spectroscopy in 1934 by Dempster (1, 2), who immediately realized the possibilities of this ion source for the analysis of solids. During World War II Dempster and his group worked on the further development and application of the method and indicated that spark source mass spectroscopy (SSMS) was not only a very sensitive technique but also one of broad applicability. The construction of an instrument designed for analytical purposes was described by Shaw and Rall in 1947 (3), but no analytical results were reported. In 1951, Gorman, Jones, and Hipple, utilizing standard samples and electrical detection, demonstrated that quantitative results could be obtained (4). Three years later, Hannay (5) described a double-focusing mass spectrograph utilizing the Mattauch–Herzog geometry with a pulsed radiofrequency spark source and photographic plate detection. Hannay and Ahearn (6) demonstrated that impurities down to 0.1 ppma could be detected and that the mass spectrograph response was directly proportional to the concentration over a large range. Finally, in 1958, the first commercial instruments became available (7). These greatly improved versions can, under favorable conditions, achieve sensitivities at the ppba level. After a period of rapid development and applications in the 1960s, without significant improvement of the instrument itself, activity has declined. In fact, the introduction of SSMS in practice was steady but slow, owing to some rather specific analytical characteristics of the technique, which are discussed in many original papers as well as in some reviews (8–13). Spark source mass spectrometers have been built by English, French, German, Russian, American, Japanese, and Chinese manufacturers. The only manufacturer at present, however, appears to be JEOL (Tokyo), which so far has built over 70 instruments. The number of instruments in actual use is limited; it has been estimated at ~60 in Europe, excluding the Soviet Union (14) and ~40 in Japan (Fig. 2.1); no reliable estimates are available for the United States and other countries.

The main reasons that SSMS has not yet become a routine analytical technique lie in the difficulty of obtaining reliable quantitative results. Other disadvantages are:

1. The analysis with photographic detection is time-consuming, the main inconvenience being the necessity to interrupt the analysis process for the development and further processing of the photoplate. The time spent directly on the analysis is, however, compensated for by fast sample preparation and the high volume of information provided by the method.

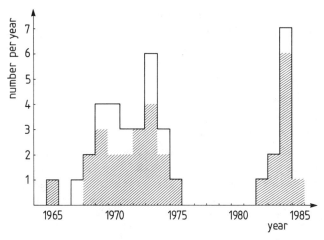

Figure 2.1. JEOL JMS 01-BM spark source spectrometers installed in Japan since 1965, showing a renewed interest in this technique since 1982. The shaded area refers to instruments installed in industrial laboratories (data from JEOL, Tokyo, March 1985).

2. The equipment is complex and costly, and it needs highly trained personnel. The price of an instrument is now more than $300,000 (U.S.), and only well-equipped laboratories have this type of instrument (12–14).

On the other hand, SSMS has some unique features that makes it the method of choice for a variety of applications. Almost every material can be readily analyzed. Chemical processing of the sample is not necessary, and just a few milligrams of the material is sufficient for analysis. The simultaneous detection capability of all elements in a wide concentration range is an important advantage of SSMS over other analytical techniques. There are no "white spots" in the periodic system of the elements. With detection limits of 1–100 ppba, one can determine about 70 elements in high-purity substances and about 40 in multicomponent natural objects (13). The analytical signal is linearly proportional to the element concentration over at least five orders of magnitude (15). Hence, it is usually sufficient to have one standard sample to carry out the analysis. Even if a standard is not available, results can be obtained that are accurate within the range of a factor 0.3–3. Thus, SSMS has no proven competitors as a method of survey ("panoramic") semiquantitative analysis. For these reasons it is expected that SSMS will remain very useful and even become more popular in the near future, since there is increasing interest in the analysis of high-purity materials, especially for the semiconductor industry, as reflected in Fig. 2.1. In recent years research in the field of spark source mass spectrometry has been carried out in the following major directions:

1. Expansion of the field of application, although this is already rather broad (see Section 4).
2. Elaboration of a theoretical basis in order to reach a more profound understanding of ion-formation mechanisms (see Section 3).
3. Improvement of the analytical characteristics, mainly precision and accuracy. Most papers are devoted to this problem. This has led to evident success, though largely due to procedure improvement, in particular the stabilization of ion formation and detection conditions, but also by employing multielement isotope dilution.

In contrast to the majority of other analytical instruments, spark source mass spectrometry design has not essentially changed in the course of more than 25 years. The main improvement, so far not fully completed, is connected with ion current electrical detection (see Section 2).

2. PRINCIPLES OF SPARK SOURCE MASS SPECTROMETRY AND INSTRUMENTATION

Spark source mass spectrometry is based on the formation of ions from a conducting sample brought about by an electrical discharge. The positive ions are accelerated and, after energy selection, separated according to their mass-to-charge ratio in a magnetic field. Detection of the ions can be achieved simultaneously on an ion-sensitive emulsion (usually called a photoplate) or sequentially by means of an electron multiplier. Various reviews on the instrumentation of SSMS have appeared in the literature. [Ahearn (8) has reviewed all developments in SSMS up to 1972, and the paper of Bacon and Ure (16) covers the period since 1972.] Essentially an SSMS contains an ion source, an electrostatic sector, a magnetic sector, and a detection system. Figure 2.2 shows the essential parts of the JEOL instrument.

2.1. Ion Source

Two kinds of ion sources are used to produce ions in SSMS: the radiofrequency spark source and the low-voltage dc arc source. All commercial instruments, except one, have used the spark source; therefore, only it

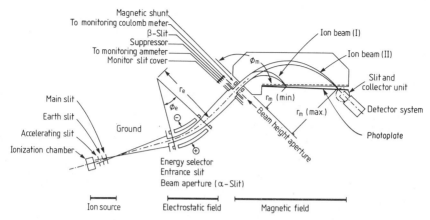

Figure 2.2. Double-focusing mass spectrometer with Mattauch–Herzog geometry.

will be discussed. This ion source consists of a vacuum chamber in which electrodes are mounted (Fig. 2.3). A pulsed 1-MHz radiofrequency voltage of several tens of kilovolts is applied in short pulse trains across a small gap between the electrodes. The pulse length and repetition rate are easily variable between wide limits, thus allowing the adjustment of spark parameters according to the characteristics of the sample and also providing a convenient means of selecting a wide range of exposures.

The electrodes are usually encapsulated by a spark housing, which, together with the electrodes, is connected to the accelerating voltage potential. The spark housing defines a potential around the spark and prevents most of the ions formed in the spark from being accelerated toward

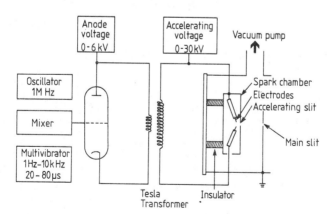

Figure 2.3. The spark ion source.

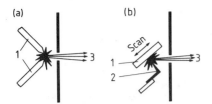

Figure 2.4. Arrangement of electrodes in (a) traditional and (b) counterprobe analysis technique: 1, sample; 2, counter-probe electrode; 3, ion beam.

the walls of the vacuum system (source housing), which would cause sputtering of the source housing material and heavy loading of high-voltage connectors, insulators, and so on.

In most applications samples are sparked directly in an electrode configuration like the one shown in Fig. 2.4a or in a point-to-plane geometry (Fig. 2.4b) where the sample is sparked against a pointed counter probe, usually of a refractory metal. In some instruments devices are incorporated for rotation of cylindrical electrodes (17) and for automatic scanning of surfaces (18, 19). Such systems are very useful in the analysis of layers and surfaces (see Section 4), and furthermore, analytical errors due to sample inhomogeneities can be reduced. For the latter purpose also, ion beam gating techniques have been designed that increase the sample consumption for a given collected charge. Two types are in use: the synchronized chopper (20–22), which allows only preselected time intervals of the whole discharge to pass the analyzer, and the random ion beam chopper (23), in which only a certain number of whole pulses are accepted.

Many papers have shown that changes in the spark gap width adversely affect the analytical results. Therefore most SSMS instruments are equipped with automatic spark gap control devices (24–26) that regulate the mean distance between the electrodes and cause one electrode to oscillate slowly about a reference position, thus averaging the electrode separations. For hard-to-spark samples, a system for the self-vibration of electrodes is sometimes used (27). Also various modifications in the spark circuit are described that aim at improving the sparking procedure by making the discharge unipolar (21, 28–30) and by reducing its duration (31, 32), and intensity (33, 34).

To make the instrument even more versatile, some spark source units have been redesigned to make them interchangeable with other types of ion sources. The incorporation of a laser source (35–37) seems very attractive, for example, since this allows the analysis of insulators without any treatment. According to Dietze (38), the sensitivity is, however, worse than after mixing with a conductor because of the lower ionization probability in insulating samples. Since the energy density of both sources is of the same order of magnitude (see Section 3), the mass spectra are similar, although molecular ions, such as carbides and oxides, tend to be more prominent with the laser source (36). Other sources that have been attached to the double-focusing mass spectrometer for inorganic applications are the Knudsen cell (39), ion microprobe source (40, 41), dc arc discharge source (42), and hollow cathode gas discharge source (43).

2.2. Analyzers

The spark source, being a high-voltage device, does not produce monoenergetic ions. In fact, the ions have an energy spread of 2–3 kV. This characteristic imposes the necessity for velocity focusing in addition to directional focusing. The Mattauch–Herzog geometry (44) copes with this problem: By a suitable combination of electric and magnetic sector fields, the energy dispersion of the magnetic field is compensated by the equal but opposite dispersion of the electric field. The ion optics produce an image of the entrance slit for each ionic species, and all these images are situated on a planar locus. A single photoplate can be used to detect ions in a wide mass range $m_1 < m < m_2$ where $m_2 \approx 36m_1$ (e.g., from lithium to uranium).

When the ions leave the ion source, their kinetic energy is the product of their charge z and the accelerating voltage V_{acc}:

$$\frac{mv^2}{2} = V_{acc}z \qquad (2.1)$$

where m is the ion mass and v its velocity.

The electrostatic sector analyzer consists of two spherical, cylindrical, or toroidal electrodes of equal but opposite voltage (Fig. 2.2). The radius of curvature through the electrostatic sector depends on the kinetic energy of the ion: the higher its energy, the greater the radius. Ions of very high energy will be deflected so little that they impinge on the positive electrode, whereas low-energy ions are discharged at the negative electrode. Ions will follow an exactly circular path of radius r_e if their energy is such that exact compensation occurs between the centripetal and electric forces:

$$\frac{mv^2}{r_e} = zE \qquad (2.2)$$

where v is the velocity of the injected ions with mass m and charge z, E

the electric field strength, and r_e the radius of the electrostatic sector. With Eq. (2.1), Eq. (2.2) can be rewritten as

$$r_e = \frac{2V_{acc}}{E} \tag{2.3}$$

The electrostatic sector thus serves as a kinetic energy analyzer. Only ions of a given energy (V_{acc}) describe a circular path with radius r_e for a given field strength E.

More specifically, the electrostatic sector analyzer serves to sort out ions of nearly equal kinetic energy and bring them to a common focus. Thus an ion beam emanating from a single point (source) is brought to a common focus at many points, each representing a common kinetic energy. An exit slit between the electric and magnetic fields (β slit) allows only a narrow energy band of ions to pass through to the magnetic sector. Usually ions with a spread in initial energy well over 100 eV are admitted.

In the Mattauch–Herzog ion optics, the magnetic field shape is designed so as to refocus the mass-separated beam at one point for each m/z ratio. Furthermore, all points lie in one plane.

In the magnetic field the ions experience a centripetal force Hzv, where H is the field strength that causes deflection. This force must be balanced by the centrifugal force of the ions mv^2/r_m, where r_m is the radius of curvature in the magnetic field. Hence,

$$\frac{mv^2}{r_m} = Hzv \tag{2.4}$$

or

$$mv = Hzr_m \tag{2.5}$$

The magnetic field thus acts as a momentum analyzer; for example, all singly charged ions of the same mv follow the same trajectory. If the ion velocity v derived from Eq. (2.5) is substituted in Eq. (2.1), one obtains

$$m/z = \frac{H^2 r_m^2}{2V_{acc}} \tag{2.6a}$$

For a given magnetic field strength H and a given accelerating potential V_{acc}, ions of a given m/z will follow a path with radius r_m:

$$r_m = \sqrt{\frac{2V_{acc}(m/z)}{H}} \tag{2.6b}$$

The magnetic sector thus acts as a mass analyzer if preceded by a velocity analyzer. Equation (2.6b) also shows that the mass scale on the photoplate varies with $\sqrt{m/z}$.

In electrical detection the detector is placed at one point of the plane. A scan of the mass spectrum can be accomplished either by keeping H constant and changing the accelerating voltage V_{acc} or vice versa.

In mass spectrometry, mass resolution R is defined as

$$R = m/\Delta m \qquad (2.7)$$

where Δm is the mass difference between two resolved peaks and m is the nominal mass at which the peak occurs. A resolution of 5000 would indicate that $m/z = 50.00$ would be resolved from $m/z = 50.01$. Resolution of 150,000 can be achieved with one commercial double-focusing spectrometer, and a resolution of 20,000–50,000 is not uncommon. The higher the mass resolution, the lower the ion transmission for a given instrument, however. In some applications, time-of-flight (45, 46) and quadrupole analyzers (45, 47) have been used; however, the mass resolution was very poor.

2.3.   Detectors

The ions may be recorded either photographically by means of an ion-sensitive emulsion or electrically with a Faraday cup or a multiplier, the plate detection method being generally preferred.

With a photoplate, the magnetic field H is held constant and each analysis gives information for the whole spectrum with good detection power and excellent mass resolution. The photoplate has the advantage of being an integrating detector that can give a reliable spectrum even when the evaporation of the sample is discontinuous in time or even sporadic. It is also useful for small samples because the entire mass spectrum is exposed at the same time. A disadvantage of the plate is that it is necessary to develop and identify the lines in order to obtain the mass spectrum. Also, intensity data suffer in accuracy and precision (see Section 4.1.4) due to plate nonuniformity and nonlinear response and low dynamic range.

In electrical detection two modes can be applied: peak scanning and peak switching. In the peak-scanning mode a whole spectrum can be registered by continuously changing the magnetic field H, allowing the different ion species to reach the detector in succession. Since each species is registered for a very short time, the effect of sample inhomogeneity and the unstable character of the spark make scanning unreliable for quantitative work and suitable only for survey analysis.

Most electrical detection systems therefore rely on peak switching. Here a number of mass lines can be preselected by an appropriate choice of magnetic field strength (magnetic peak switching) or acceleration voltage [and condensor voltage, see Eq. (2.3)] (electrostatic peak switching). The ion current of each mass line is integrated over a preset period of time, and the total ion charge measured by a monitor prior to mass separation is taken as a measure of exposure. The use of an electron multiplier has some advantages over photoplate detection; it provides a direct output linearly related to the ion intensity and can detect single ions. Peak switching has the highest detection power and accuracy, while peak scanning is the fastest method for survey analysis. The disadvantages of electrical detection are the restriction to a preselected number of mass lines in switching and the low detection power for equal analysis time in scanning. Also, most systems operate at a mass resolution of ~500. Such a low resolving power limits the practical application of electrical detection because of mass spectral interferences. The short exposures, which are typical for electrical detection, are also subject to a considerable sampling error (see Section 4.1.3).

Development work on electrical detection is still going on. Dale et al. (48) succeeded in magnetic step scanning with an online microcomputer via a digital-to-analog converter. A mass resolution of around 3000 was achieved with limits of detection of 5 ppba. Several papers describe the use of position-sensitive electro-optical ion detectors (49–51). These detectors are the electronic equivalent of photoplates, and they incorporate many desirable features of the ion-sensitive emulsion and the multiplier (e.g., increased detection sensitivity). Currently available detectors are small, mass resolution is limited, and there are still a number of less desirable features, especially the complexity. There is still a need for the development of new simultaneous electrical detection systems. Possible approaches have been suggested by Ure (52): channel electron multiplier, charge-coupled devices, electrostatic buildup on insulated plates, and Fourier transform mass spectrometry.

2.4. Spectral Information

Only a few reports deal with the formation of negative ions in the spark (45, 53–56). For analytical purposes, only positive ions are normally measured. Various types of positive ions are produced in a spark discharge: singly and multiply charged atomic ions, charge-exchange ions, dimer and polymer ions, and heterogeneous compound ions.

A spark source mass spectrum is always dominated by singly and multiply charged matrix ions. Very high charge levels have been observed (higher than 10 charge units) (57). The intensity usually decreases by a factor of 3 for each charge increment, with shell effects occurring for a charge state corresponding to the first ionization of a closed shell (58). Also, for impurity elements the singly charged lines are the most intense and are usually the only ones used for analytical purposes. Other line groups are important mainly because of possible mass spectral interferences.

At a mass resolution of ~10,000, most interferences except those from isobaric species can be resolved. If such high mass resolution is not available, analysis can be based on the doubly charged ions or, when the spectrum is known, the degree of interference can be estimated (59, 60). Time-dependent sampling of the ions may also considerably reduce the importance of some spectral interferences, as shown by the work of Viczian et al. (61) and Van Puymbroeck et al. (62).

A minor feature of the mass spectrum is the charge-exchange spectrum. When an ion collides with another particle during its flight through the instrument, there is a finite probability that it will exchange charge during the collision. If the collision takes place between the electrostatic sector and the magnetic analyzer, the result will be a relatively well defined line at mass $(p/q^2)m$, where p is the charge of the ion before collision, q is the charge after collision, and m is the mass of the ion. The line will not be quite as sharp as other mass lines because of an uncertainty in the amount of momentum transferred during the collision, and it will be slightly displaced from the location calculated.

With the large amount of energy available in the spark, it would be expected that few molecular ions would be produced. Yet there is some tendency for compounds at high concentration, such as the matrix, to form molecular ions. Cluster ions of the type M_n^+ were studied by Franzen and Hintenberger (63), and a value of n up to 25 could be observed for a beryllium matrix. A comparative study on the building of molecular ions in laser and spark plasmas showed that the formation probability of molecular ions is similar in both plasmas in the range $5 \times 10^8 - 1 \times 10^{10}$ W/cm^2 (64). For a detailed study on the occurrence of molecular ions in various matrices, see Becker and Dietze (60).

The mass spectrometer source region can give rise to background lines, due to the presence of residual gases, that consist of singly and multiply charged carbon, nitrogen, and oxygen ions; hydrogen can be recorded if the magnetic field is sufficiently lowered to bring it into position on the detector. Lines due to hydrocarbons are also frequently observed. These background lines usually determine the detection limit for the analysis of hydrogen, carbon, nitrogen, and oxygen. Various methods exist, however, to reduce the ion source pressure, and they have proved to be successful in the analysis of these elements to the sub-ppm level (65–69), especially cryopumping and cryosorption (see Section 4.1.2).

Another important class of lines are those of the heterogeneous compound ions formed by association of the matrix with hydrogen, carbon, nitrogen, and oxygen. Such ions may constitute a relatively large percentage of the total number of ion lines observed. Figure 2.5 shows part of the mass spectrum of a rhenium matrix, illustrating the abundance of this class of ions. When complex matrices are studied, complex compound ions may be observed containing various species (59, 70–72).

The existence of multiply charged molecular ions is not generally accepted. Various authors have, however, actually observed doubly charged molecular ions (73–76), and in one report even the detection of triply charged molecular ions was mentioned (77). The apparent concentrations of these ions are usually very low, from 10 ppm to the sub-ppm level (57).

3.   ION FORMATION IN THE VACUUM DISCHARGE

Many experiments and calculations in the field of spark discharge and plasma physics (see, e.g., reviews 78–81) and concerning the mechanisms of ion formation in the plasma of a spark discharge have been carried out in recent years (see reviews 11–13, 57, 82, 83). In this section the existing information on the physics of the vacuum spark discharge will be presented, and some concepts connected with ion formation mechanisms in mass spectrometry will be worked out on the basis of these data.

3.1.   Physics of the Vacuum Spark Discharge

The discharge parameters are mainly determined by the spark voltage generator circuit (Fig. 2.6), by the electrode arrangement within the ion source, and by the electrode material and form. A sinusoidal spark voltage V of frequency $\nu \approx 0.5$–1 MHz, lying within the radio range, is commonly used in spark source mass spectrometry. The electrodes are supplied with voltage pulses having a length $T = 20$–300 μs and with a repetition frequency of 1–10,000 Hz. Due to a weak coupling between L_1 and L_2 and the high Q factor of the electrode circuit, the voltage in this circuit increases during 5–25 cycles until it reaches the breakdown voltage V_0. The value of V_0 is 5–25 kV for an interelectrode gap width $d = 10$–250

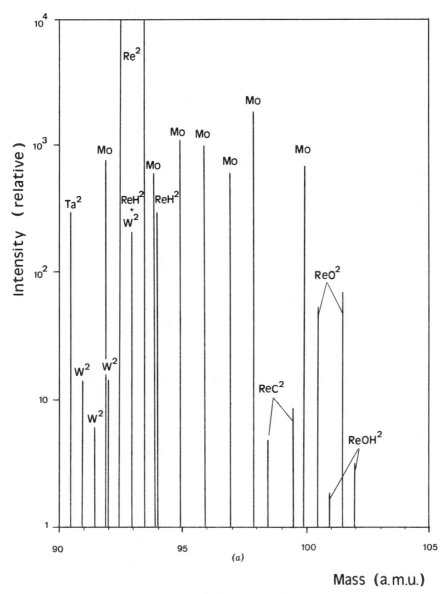

Figure 2.5. Spark source mass spectrum of rhenium in the mass range (*a*) 90–105 and (*b*) 170–250 amu (57).

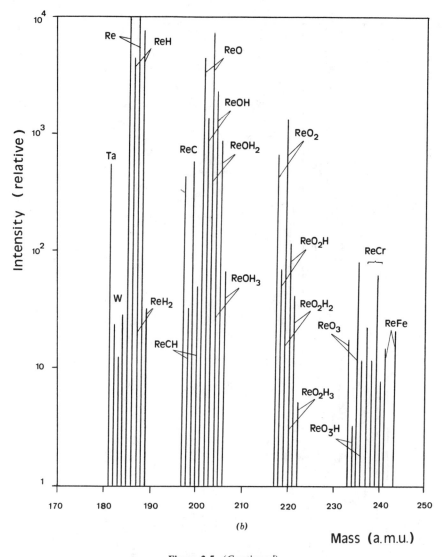

Figure 2.5. (*Continued*)

μm (84) typical of spark source mass spectrometry. If the radiofrequency oscillation pulse (total duration *T*) has not finished, the voltage starts increasing again, and so the breakdown can be repeated several times (Fig. 2.7).

Since the term "spark" can be used in more than one sense, it seems

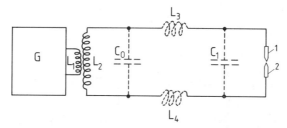

Figure 2.6. Spark circuit. G, radiofrequency generator; L_1, L_2, Tesla transformer; L_3, L_4, leakage inductances; C_0, C_1, effective capacity of master oscillator circuit and electrodes, respectively; 1, 2, electrodes.

worthwhile to define the main stages of a discharge. They are qualitatively shown in Fig. 2.8 (82) on a voltage–current diagram.

Prebreakdown Stage. When the potential applied to the electrodes is increased, a prebreakdown current reaching a value of 10^{-4}–10^{-3} A appears. This stage is fully reversible, that is, the vacuum insulation properties are restored at a decrease in the voltage. Its duration depends on the voltage growth rate dV/dt and typically reaches 100–250 ns.

Breakdown Stage. The current grows, and the voltage rapidly drops by a factor of 100–1000. This irreversible stage, usually lasting no longer than 100–1000 ns, in its turn consists of two parts:

(a) *Breakdown initiation*, the time interval between the beginning of the irreversible growth and the moment of the interelectrode gap occupation by a plasma. Since processes taking place during breakdown initiation are very short, the main energy stored in the oscillation circuit L_2–C_0 (Fig. 2.6) does not play an important

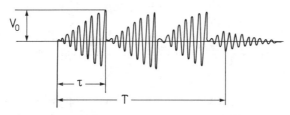

Figure 2.7. Schematic view of oscillograms of rf voltage on the electrodes. V_0 breakdown voltage; T, duration of the rf pulse; τ, time interval from the beginning of the pulse to the first breakdown.

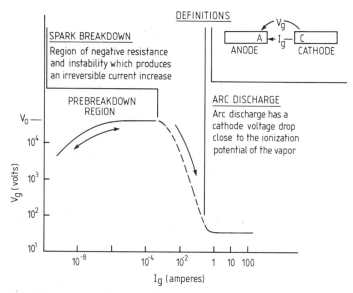

Figure 2.8. Voltage–current diagram of a complete electrical discharge cycle in vacuum.

role. The gap gets energy supplies mainly from the effective capacity of the electrodes and the adjacent source details (C_1).

(b) *Spark discharge*, a high-current and high-voltage stage of the discharge. It is connected with the development of the plasma filament and is characterized by preferable destruction of the anode (bombarded by an electron beam), by the nonstationary character of the major processes taking place in the interelectrode gap, and by a relatively high voltage between the electrodes (from hundreds of volts to a kilovolt).

It will be shown later that a distinction is necessary between parts (a) and (b) of the breakdown. Otherwise, mass spectrometrists can misinterpret the data taken from papers on breakdown physics.

If the power source is able to supply the gap with current exceeding some definite value (usually hundreds of amperes), the discharge enters the final stage:

Arc Stage. At this stage the current flow through the plasma filament and other physical processes become quasi-stationary, and only the cathode is being destroyed. The main voltage drop is concentrated near the cathode. Its value is comparable with the first ionization potential of cathode material atoms. In commercial spark source mass spectrometers this stage of discharge is not reached.

The application of special low-voltage discharge sources for mass spectrometric analysis of solids has been studied in some papers (85). However, this trend was not developed mainly because of difficulties arising during analysis of dielectrics and semiconductors by means of a low-voltage source. That is why we are not going to discuss the discharge arc stage, the more so as it has been described in detail by Franzen (82).

We will now discuss the main stages of the spark discharge in more detail.

3.1.1. Prebreakdown Stage

The breakdown process depends largely on the vacuum in the source chamber and on the material, form, microstructure, and surface condition of the electrodes. Most information in the literature concerns metallic electrodes.

The prebreakdown current is emitted mainly by sharp microprotrusions ("whiskers") (86, 87) that exist on the electrode surfaces or appear under the influence of an electric field. The density of the field emission current is described by the Fowler–Nordheim equation (88), which takes into account that the real field strength near microprotrusions is a factor $\beta = 10$–1000 times higher than the medium field strength within the gap. The area S_1 of the spot on the anode surface bombarded by the field electron beam is given by (84):

$$S_1 = 2\pi l d \qquad (3.1)$$

where l is the height of the microprotrusion on the cathode surface and d is the interelectrode distance, often called the gap width. Thus, for $l = 1$ μm and $d = 40$ μm, $S_1 = 250$ μm^2.

If the electrode surface is contaminated, ions formed from the desorbed atoms and molecules can make a definite contribution to the prebreakdown current. However, their share in the total ion current as a result of a discharge is usually minor.

When the field strength near the microprotrusions reaches a value of $E_0 = (5$–$11) \times 10^7$ V/cm (82, 83), the breakdown starts.

3.1.2. Breakdown Initiation

The dependence of breakdown voltage V_0 on d has the form

$$V_0 = a d^{\alpha} \qquad (3.2)$$

the most probable value of α being 0.6–0.7 (85). Verlinden et al. (89) obtained a value $\alpha = 0.61$ for aluminum, zirconium, molybdenum, and gold electrodes. At constant breakdown voltage, d was shown to be inversely proportional to the melting point of the metal (90).

The conducting medium that is necessary for breakdown development can only be created by the electrode material. Most authors share the opinion that the breakdown is caused by field emission from microprotrusions on the cathode surface. That is why in the point-to-plane electrode configuration (28, 70, 91, 92) the breakdown occurs more often when the negative half-period of the radiofrequency voltage is applied to the sharp counterelectrode (93).

Several models, still disputed in the literature, have been proposed to describe the mechanism of breakdown initiation.

According to the *cathode mechanism*, a plasma cloud is initially formed near the cathode surface due to an avalanche process. The mechanism has been consistently specified. Dyke et al. (94) assumed that field emission turns into thermofield emission as a result of the current heating the top of the protrusion, resulting in evaporation and ionization of cathode material. Sokol'skaya and Fursey (95) considered the breakdown the result of a process similar to the explosion of a thin wire heated by current. Mesyats and Proskurovsky (79, 81, 96) showed that the explosive transition of the cathode material into a dense plasma was due to a high energy concentration in the microscopic volume of the cathode: the current density reaches a value of 10^8–10^9 A/cm^2 (81). The so-called cathode flares are formed, and the electron beam current increases about a hundredfold, mainly due to an increase in the emission surface. This phenomenon is called *explosive electron emission* (97).

The plasma dispersal rate of the cathode flare is $(1–2) \times 10^4$ m/s (79).

According to the *anode mechanism* (80, 82, 98), bombardment of the anode surface by an energetic beam of field electrons causes desorption of gases and material evaporation from the anode surface. The existence of a powerful electron beam influencing the anode during the growth of the discharge current is confirmed by the detection of X rays during the initial stages of the breakdown (9, 99).

Ions formed during breakdown initiation near the anode surface were detected by means of a mass spectrometer (100). These ions are accelerated toward the cathode and create a positive space charge that increases the field near the cathode and the electron emission (101). The process rapidly (0.1–10 ns) becomes critical, and the field emission increases 10^2–10^4 times (102). The anode material is ionized and creates the anode flare, whose spreading rate is $(4–10) \times 10^3$ m/s (80, 103), a bit less than that of the cathode flare.

Earlier the "clump" or Cranberg model of the vacuum breakdown initiation by the impact of a microparticle on an electrode (80, 104) was the most popular. Such a particle can be desorbed or torn away from the opposite electrode surface by the electric field. A number of statements connected with this model, however, received sound criticism (78).

The Davies and Biondi "drop" model can be considered to be a combination of the anode and Cranberg mechanisms (105–107). These authors assume that field electrons emitted by cathode protrusions heat the anode surface locally and cause the formation of a protrusion. It melts and is torn off from the anode, and a drop of positively charged molten material moves rapidly toward the cathode. Since the drop is still under electron bombardment but has lost its thermal contact with the bulk material, it is heated up rapidly to high temperatures and is evaporated on its way to the cathode. The vapor is ionized according to the Paschen mechanism, that is, as a result of electron avalanche formation. Alternatively the drop may survive complete evaporation, hit the cathode surface, and explode with further vapor ionization according to the Paschen mechanism.

The final choice of breakdown initiation model still remains one of the main problems of vacuum spark discharge physics. Attempts have been made to find conditions when either the cathode or anode mechanism works (87, 108).

As a matter of fact, supporters of the cathode mechanism do not refute the fact that material enters the gap mostly from the anode. However, they attribute the main role in the breakdown initiation to the cathode flares, which appear a little earlier (109). In the case of the narrow vacuum gaps characteristic of spark source mass spectrometry, this time difference must not exceed several nanoseconds. At the same time, the anode flare formation process can start before the cathode plasma reaches the anode (110). Anyway, plasma fills the interelectrode gap because of the destruction and ionization of electrode material, and at the same time the electric current that has initiated the discharge increases.

The duration of breakdown initiation—the filling of the interelectrode gap with the initial plasma (leading to a sharp drop in the resistance of the interelectrode gap)—is equal to

$$t_{init} = d/v \tag{3.3}$$

Assuming $v = 2 \times 10^4$ m/s, we obtain for the case of mass spectrometry values of t_{init} ranging from 0.5 ns (a minimum gap) to 12 ns (a maximum gap). Both the duration and ion contribution to the total ion current lead to quite a small contribution of the initiation stage to the whole process (as will be shown in Section 3.2).

3.1.3. Spark Discharge

Vapors that have filled the interelectrode gap quickly become ionized due to the formation of a self-dependent discharge (80) as well as, perhaps, the collective interaction of the electron beam with plasma (111), causing electrostatic oscillations with a frequency close to the plasma Langmuir frequency. As a result, the plasma filament is formed; the time of its formation is determined by the plasma filament resistance, inductivity, and effective capacity (83). For mass spectrometers this time interval can be very short, of the order of 1 ns. The amount of material entering the plasma sharply increases, mainly because of anode processes, the share of the cathode material dropping to $\leq 1\%$ (28). At the same time, the plasma filament broadens, which leads to an increase in the anode spot area. The final area S_2 for $d = 40$ μm is equal to 4000 μm^2 (84), which is 20 times higher than the value of S_1 (see Section 3.1.2).

The current through the gap increases up to 10–30 A (112–114), and the gap resistance drops to ≤ 100 ohms (113), while the interelectrode voltage drops correspondingly.

Damping oscillations can arise in the discharge circuit (112, 115); their damping increment is a function of the value of the ohmic resistance in the electrode circuit, and their frequency (1–100 MHz) also depends on the values and the distribution of leakage capacities and inductivities (Fig. 2.6). It has been assumed that these oscillations are also connected with plasma oscillations at the Langmuir frequency (116). The total duration of the discharge in commercial instruments is 500–3000 ns. It can be decreased if either an additional capacity (117) or an ohmic resistance (112) is introduced into the discharge circuit.

After the discharge is over, the radiofrequency oscillation energy starts passing on from the primary to the secondary circuit (Fig. 2.6). If the duration of the oscillations is long enough, the discharge can be repeated one or several times (Fig. 2.7).

3.2. Energy Approach to the Major Steps of Mass Spectrum Formation

A number of nonstationary processes take place simultaneously during short time intervals (1–10 ns) in regions having a small characteristic size (10–250 μm), while the discharge is going on. This strongly complicates theoretical and experimental research.

Ramendik (13, 118–121) has proposed a model in which ion formation

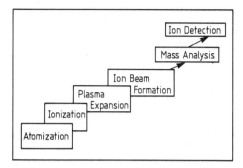

Figure 2.9. The main steps of mass spectrum formation.

is considered on a microscopic level, allowing mass spectrum formation to be described as a series of steps rapidly changing from one into another (Fig. 2.9). Such a stepwise approach allows us to simplify the overall discussion.

Solids consist of strongly bound atoms and molecules. To carry out the analysis, we must detect lines in the mass spectrum belonging to singly or multiply charged ions. Thus, first it is necessary to evaporate atoms (or molecules, in which case it is necessary later to break intermolecular bonds). This step is called atomization.

Then, as a second step, atom ionization occurs in the discharge plasma, and after that other steps follow.

To describe the mass spectrum formation process it is necessary to consider the distribution of the energy introduced during the various steps of the discharge. In general, in methods of physical analysis, the energy necessary to generate the analytical signal is imparted to the sample by some type of radiation (122); in spark source mass spectrometry the energy is transmitted by electrons emitted from the cathode.

The total energy flux Q_Σ of an electron beam passing through a plasma cross section with area S per unit time is given by the expression

$$Q_\Sigma(t) = \frac{dE(t)}{dt} = V(t)I(t), \qquad (3.4)$$

where $V(t)$ and $I(t)$ are instantaneous values of the interelectrode gap voltage and the current through it.

Assuming the energy flux to be uniformly distributed over the area, one can obtain the expression for the total energy flux density q_Σ (in laser source mass spectrometry this value is often called power density):

$$q_\Sigma(t) = \frac{Q_\Sigma(t)}{S} \qquad (3.5)$$

Assuming that S is equal to S_1 at the breakdown initiation stage and to S_2 at the spark stage, one can evaluate $q_\Sigma(t)$ according to expressions (3.4) and (3.5) on the basis of values $V(t)$ and $I(t)$ obtained from synchronous oscilloscopic observation (113, 123, 124).

When a sharp-pointed cathode is used, the peak value of $q_\Sigma(t)$ can reach 10^{11} W/cm^2 (113, 124). Measurements carried out by Ramendik and coworkers (11) on their instrument showed that $(q_\Sigma)_{max} \approx 10^9$ W/cm^2 during approximately the first 10 ns of the spark discharge, while the voltage on the electrodes is still high; during the remaining time of discharge $q_\Sigma(t)$ drops by two orders of magnitude, without, however, becoming constant, because of oscillations appearing in the electrode circuit.

Thus, from the point of view of energetics, the discharge can be subdivided into:

1. A high-energy phase including the end of the breakdown initiation stage and the beginning of the spark stage. Its total duration does not usually exceed 10–30 ns.
2. A low-energy stage during the remaining discharge (Fig. 2.10).

The mass spectrum is the overall result of all ion-formation mechanisms acting during the entire discharge. On the average, the process can be characterized by evaluating the q_Σ value averaged over the discharge duration, after substituting in expression (3.4) the total energy released during the discharge (84):

$$E = \frac{C_0 V_0{}^2}{2} \tag{3.6}$$

For commercial mass spectrometers, E lies within the range 10^{-4}–10^{-2} J (118). Assuming that the energy is uniformly released in the process of discharge, one obtains for standard conditions $q_\Sigma = n \times 10^8$ W/cm^2 (84). This is close to radiation parameters of a laser working in the Q-switched mode (125). Radiation with q values of 10^7–10^9 W/cm^2 is typical for "moderate" lasers (126). This consideration allows the use of results of laser research for spark source mass spectrometry, and illustrates the similarity between ion formation mechanisms in spark and laser ion sources (120, 127).

In spite of a relatively large number of papers devoted to the physics of the vacuum spark discharge and to the effects of laser beams on solids, processes of atomization, ionization, and plasma dispersal cannot yet be

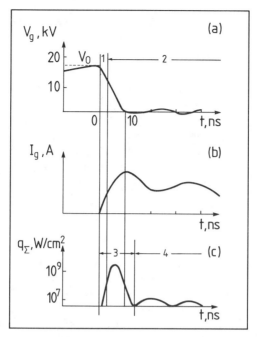

Figure 2.10. Schematic views of oscillograms of the voltage on (*a*) the interelectrode gap, (*b*) the current through the gap, and (*c*) the approximate diagram of the energy release in the spark discharge for a gap width $d = 100$ μm. 1, Breakdown initiation stage; 2, spark stage; 3, discharge high-energy phase; 4, low-energy phase.

described by means of simple equations. One has to consider them at a qualitative level or study them on the basis of computer simulation, solving systems comprising many equations.

3.2.1. Material Atomization

Atomization at the high-energy discharge phase is accomplished by direct heating of the anode with a beam of fast electrons. Almost all energy expenditure of q_Σ is transmitted to the solid sample. The regime of heat release and material evaporation for this stage is discussed by Mesyats and Proskurovsky (81). Sample destruction during this period must resemble an explosion, sputtering, or sublimation. It seems that at this stage no discriminations occur between volatile and barely volatile elements upon entry into the plasma (118).

It is not yet clear what fraction of the material is being destroyed according to this mechanism. At the breakdown initiation stage the anode destruction depth does not exceed 10^{-2} μm, which is only 1–10% of the final crater depth.

After the plasma cloud forms, it starts absorbing most of the total energy flux. As the low-energy phase of the discharge begins, q_Σ also sharply decreases. As a result, the energy flux density spent on sample atomization (q_{atom}) drops to $\leq 10^6$ W/cm^2.

During the low-energy phase of the spark discharge, the near-surface layers of the sample are heated by the electron component of the discharge current (128), by the heat conducted from the discharge zone (121), and by a directed flow of relatively hot (50,000–150,000 K; see Section 3.3.2) plasma electrons (83). In the course of one discharge of duration t_d, the sample can be melted to a depth

$$X_{lp} \approx (\gamma t_d)^{1/2} \tag{3.7}$$

where γ is the thermal conductivity coefficient (13). At $t_d \approx 1000$ ns, the liquid-phase layer thickness reaches $X_{lp} \geq 1$ μm (13). The effect becomes more important at high pulse repetition frequency (1000–3000 Hz). Results of such melting (57, 90, 129) as well as traces of vapor bubbles formed in the liquid layer (129) have been detected by means of scanning electron microscopy (Fig. 2.11). Atoms (molecules) of different elements (compounds) may evaporate from the melt into the vacuum to different extents. The evaporation process can endure for one to several microseconds after the discharge is finished (130). Analysis carried out by means of secondary ion mass spectrometry has shown that the composition of the electrode surface layers that have undergone remelting differs from the bulk composition (131, 132) (see Fig. 2.12). This can be explained by processes similar to zone refining as well as by diffusion in the liquid phase or by partitioning of elements between the solid and liquid phases (132) (see also Section 4.1).

Part of the material can be transported into the interelectrode gap in the form of drops (114, 133), evaporating in the plasma according to a mechanism similar to that proposed by Davies and Biondi (105, 106). In some cases, particularly during the analysis of pressed fine powders, solid particles can enter the plasma: this resembles the Cranberg mechanism (104). It should be realized, however, that at the spark stage drops or particles enter a plasma, whereas at the breakdown initiation stage (to which the Davies–Biondi and Cranberg models refer) they enter vacuum. They no longer appear to be the cause of the breakdown but supply material for its further development and maintenance.

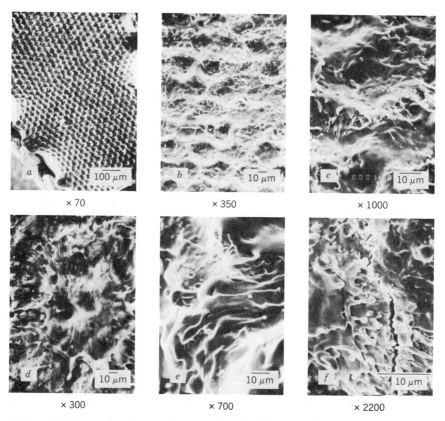

Figure 2.11. Secondary electron images of sparked Co (a–c), Ge (d), Pb (e), and Re (f) electrode surfaces (57).

Melting and evaporation processes can lead to differences between plasma and solid elemental composition (13, 118, 127, 129).

Thus, the destruction of electrode material during the spark discharge is not a uniform process. The energy density flux spent on atomization is, on the average, $q_{atom} \approx 0.01 \, q_{\Sigma}$, or 10^6–10^7 W/cm^2 under standard experimental conditions during the discharge (84). The average rate of sample destruction is close to 1 m/s real sparking time (13).

If the substance evaporates in the form of molecules, the interatomic bonds are completely broken later on. Each particle receives an energy of about 3–4 eV, which is more than sufficient to destroy virtually every chemical compound. That is why the term "atomization" has a real physical meaning: solid matter is being turned into atomic vapor.

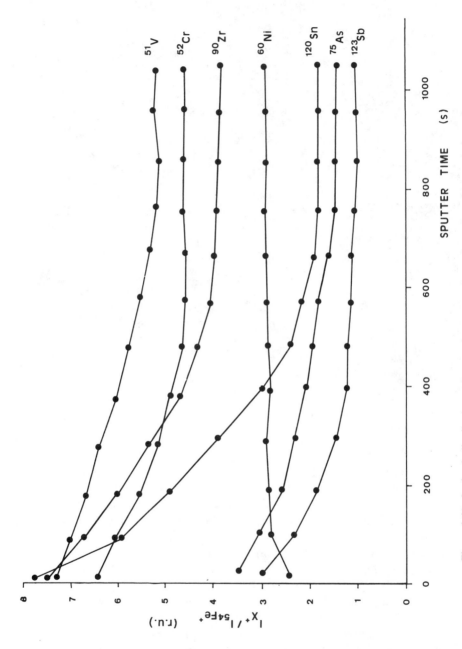

Figure 2.12. In-depth distributions of some elements at the surface of sparked steel electrodes, as measured by secondary ion mass spectometry (132).

42

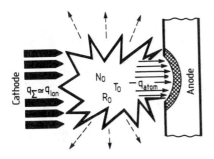

Figure 2.13. The distribution of the energy flux density during plasma heating by the spark discharge.

3.2.2. Ionization of the Atoms

At the beginning of the high-voltage phase, plasma filament formation is complete, and the plasma effectively starts to absorb the electron beam energy. The average flux density spent on atom ionization in the plasma over the whole discharge is $q_{ion} \geq 0.99 \, q_{\Sigma}$ (Fig. 2.13).

Ionization of atoms entering the plasma is due mainly to plasma heating by electrons accelerated by the electric field applied to the plasma. When the gap current is maximum, the voltage drop over the plasma reaches a value of 500 V (113). The energy of the fast electrons is transmitted to plasma electrons, ions, and atoms. Ionization occurs according to a thermal mechanism—it is carried out by electrons that can be characterized by a definite temperature:

$$A^{n+} + e \rightleftarrows A^{(n+1)+} + 2e \tag{3.8}$$

where n changes from zero (neutral atoms) to n_{max}.

If the particle density N in the plasma does not exceed about $10^{17} \, cm^{-3}$, electrons emitted by the cathode lose their energy (are thermalized) more slowly. In this case a beam of high-energy electrons with current density $j \geq 10^5 \, A/cm^2$ can arise in the plasma. Such electrons mainly create multiply charged ions. If the rate of ionization by the beam exceeds the thermal ionization rate, a large number of multiply charged ions can be detected in the mass spectrum (134).

The theory of the heating of a plasma with a complex chemical composition has been discussed on the basis of computer simulation by Derzhiev et al. (135).

At the discharge initiation stage, the cathode flare plasma consists mainly of singly or doubly charged ions. The density of the electrons in the flare N_e is 10^{17}–$10^{18} \, cm^{-3}$ (81), and their temperature $T_e \approx 5 \, eV$ (136). The anode flare density is a little higher: $N_e = 10^{18}$–$10^{19} \, cm^{-3}$, but T_e

≈ 0.5 eV (110, 137). That is why the degree of ionization in the anode flare plasma does not exceed 10%; it is much lower than that of the cathode flare.

The total number of ions formed at the discharge initiation stage is about 10^7; this is much smaller than the total number of ions (10^{12}–10^{14}) formed during one discharge lasting 100 ns (138). Thus, the discharge initiation stage contributes negligibly to the total ion current.

Multiply charged ions are formed mainly during the high-energy phase of the spark discharge at the maximum of q_Σ (61, 62, 114, 118). Experiments have been carried out with a "nanosecond" generator providing a minimum achievable duration of discharge of about 100 ns (139). It has been shown that at the initial stage of the spark discharge, there are approximately equal amounts of cathode and anode material in the plasma and that the degree of ionization is rather high. The concentration of multiply charged ions in the mass spectrum is high; in some cases the peak of the charge distribution falls on the triply charged ions. When the discharge duration is increased to 150 ns the mass spectrum contains primarily ions from the anode; the fraction of the multiply charged ions decreases by a factor of 3 and more. This agrees with the conclusion drawn by Galburt et al. (114) and Viczian et al. (61, 62) that during the initial phase of the discharge mainly multiply charged ions are formed but that after a few tens of nanoseconds singly charged ions start to dominate.

On the average for the whole discharge, the initial plasma density (i.e., before dispersal) depends on the electrode material and on the conditions of the discharge, $N_0 = 10^{17}$ to $n \times 10^{19}$ cm^{-3} (57, 118, 133, 140). The initial electron temperature T_0 is 5–15 eV, or 50,000–150,000 K (118, 141). For a given metal, N_0 depends on the breakdown voltage; for copper electrodes an increase of V_0 from 20 to 60 kV leads to a variation in N_0 of 1.3×10^{18} cm^{-3} to 6.3×10^{18} cm^{-3} (142).

Values of V_0 and E increase with increasing interelectrode gap width, as appears from expressions (3.2) and (3.6). At the same time the relation between E and the volume V of the material ejected into the gap is expressed by the formula (90)

$$E \propto V^{0.4} \tag{3.9}$$

This means that the specific (per atom) energy E_{sp} spent in the discharge drops when the gap increases:

$$E_{sp} = \frac{E}{n_{at}} \propto \frac{E}{V} \propto E^{-1.5} \tag{3.10}$$

This is a possible reason why, with increasing gap width, the ionization degree of the plasma (90, 114) as well as the relative number of multiply charged ions in the mass spectrum (140) decrease. However, some authors state an increase in the content of multiply charged ions with increasing gap width (57).

Nevertheless, one should not directly relate charge distribution of the ions in the mass spectrum to the initial charge distribution in the plasma. The average initial degree of plasma ionization must exceed 1. This follows from the fact that the average energy (per sample atom) spent during the spark discharge varies from 50 eV (127) to 600 eV (120). However, the charge composition of the plasma changes considerably as a result of strong recombination, accompanying initial moments of plasma dispersal.

3.2.3. Plasma Dispersal

Let us note that the spark discharge imparts energy to the sample atoms only during the first two steps of the mass spectrum formation (Fig. 2.9). Subsequent processes take place in the plasma due to the stored energy. In principle, during the discharge, the plasma can be confined in a channel by the magnetic field of the current flowing through the plasma; however, this can be achieved only if the current is at least 100 A (83). Commercial mass spectrometers do not attain such a high value. This means that exactly at the moment the plasma cloud is being formed its dispersal in vacuum starts.

3.2.3.1. Recombination. When the material leaves the zone of heating, recombination starts to prevail over ionization. Recombination is accomplished mainly according to a three-particle mechanism, when the energy released in the recombination process of ion $A^{(n+1)+}$ with an electron is captured by a second electron (119). Thus, the reverse process begins to prevail in expression (3.8). Recombination accompanied by the emission of light quanta can also take place:

$$A^{(n+1)+} + e \rightarrow A^{n+} + h\nu \qquad (3.11)$$

but its role is significant only at high temperature and low plasma density.

As a result of recombination the charge distribution of ions strongly changes (Fig. 2.14). Computer simulation of this process for a plasma of complex chemical composition has been carried out (143). The possibility of recombination of ions with charge z is high if the characteristic length of the recombination l_z (i.e., the average distance traveled by an ion from the moment of its formation to the moment of its recombination) is less

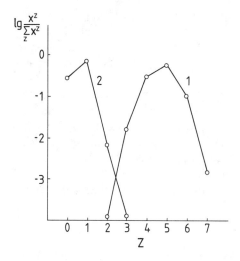

Figure 2.14. The charge distribution of ions: 1, in the plasma formed during the discharge; 2, in the mass spectrum after recombination.

than the original plasma radius R_0, before dispersal; l_z strongly depends on the ion charge,

$$l_z \propto 1/z^3 \tag{3.12}$$

but it is not related to its ionization potential. Plasma dispersal behavior is mainly determined by the matrix atoms. That is why the charge distribution of admixture ions is close to that of matrix ions (121).

The recombination time in the expanding plasma is usually a few tens of nanoseconds (140). During this time interval its density drops rapidly, and at a distance of not more than 1–10 mm from the place of ion formation the recombination ceases. The residual degree of ionization of the plasma does not usually exceed 10^{-2} (114).

3.2.3.2. Chemical Reactions in the Plasma. Neutral atoms formed in the recombination process can interact with ions. Chemical reactions take place in the plasma, giving rise to clusters (144, 145). On the average, the percentage of doubly charged clusters in the mass spectrum is 2 orders of magnitude lower than that of singly charged clusters; triply charged clusters have not been found (60) or only rarely (77). This is understandable, since neutral atoms are formed mainly during the final moments of dispersal, when multiply charged ion concentrations have already fallen off strongly. Besides, the process of cluster formation must proceed more intensely in the low-energy phase of the discharge, when the initial degree of ionization of the plasma is decreasing. From time-resolved measurements it has been concluded that cluster ions are formed later than multiply charged atomic ions (61, 62). The duration of current oscillations after the breakdown is shortened when a resistance is introduced into the discharge circuit; this also results in decreasing the cluster ion concentration (116). The mechanism of cluster formation during plasma dispersal differs from gas-phase reactions taking place at relatively low temperatures (146, 147). This can be explained by the high initial plasma temperature; besides, ions and neutral atoms formed in the process of recombination are more often excited.

3.2.3.3. Ion Acceleration. During dispersal of the plasma, most of its thermal energy is converted into kinetic energy of the particles. Ions are accelerated mainly according to two mechanisms (148):

1. Hydrodynamic acceleration: During free expansion of the plasma into the vacuum, pressure forces accelerate ions and electrons.
2. Acceleration in a "self-consistent" field: Electrons, having a smaller mass, tend to leave the plasma cloud more rapidly than ions. In this way an electric field is generated at the outer surface of the expanding plasma cloud, and as a result, electrons are decelerated and ions are accelerated.

Ion formation cannot be considered as a process imparting a definite amount of energy to a separate particle:

$$E_{total} = E_{atom} + E_{ion} + E_{kin}$$

Energy flux densities q_{atom} and q_{ion} simultaneously influence many particles contained within a given volume of solid, vapor, or plasma. This energy is converted into potential energy (ionization states) as well as into kinetic energy (plasma temperature). During plasma dispersal, part of this energy is converted into ion kinetic energy. However, particles are accelerated as a whole. The kinetic energy that ions gain in mechanism (1) is proportional to the multiplicity of their charge, and that in mechanism (2) is proportional to the square of the multiplicity of their charge.

Van Puymbroeck and Gijbels (149) made a study of the real energy spread of Au^{n+} ($n = 1-7$) ions in the spark by time-resolved measurements. It was shown that the average energy and energy spread increased with increasing multiplicity (Table 2.1), suggesting that the hydrodynamic mechanism dominates. Other dependencies revealed by Franzen (115), Woolston and Honig (150), and others may reflect sideline processes (e.g., atoms of different elements can enter the plasma at different moments).

Table 2.1. Average Energy $\bar{E}$ and Energy Spread ΔE (At Half-Peak) of Ions in the Spark Source, for Gold Electrodes[a]

Ion	$\bar{E}$	ΔE
Au^{7+}	230	310
Au^{6+}	230	315
Au^{5+}	225	325
Au^{4+}	140	195
Au^{3+}	120	175
Au^{2+}	80	115
Au^{+} [b]	90	80
	20	32

From Ref. 149.

[a] Values in electronvolts.

[b] Two maxima were observed for singly charged ions.

It has been shown that, contrary to the assumptions of Franzen (115) and Berthod et al. (151), the electric field between the electrodes does not play a significant role in ion acceleration. A suitable selection of conditions under which the spark discharge influences the sample could in principle allow a reduction in the ion energy spread to <100 eV (half-peak), whereas in previous papers values of 2500 eV (150) or less (152) have been reported.

3.2.4. Ion Beam Formation

Although the plasma expands in all directions, it is mainly concentrated within a solid angle of about 60° around the axis, perpendicular to the sample surface (153). On the other hand, only a relatively narrow particle beam can be directed to the mass analyzer. Besides, in order to obtain a high resolution while detecting ions having a relatively broad energy spread, it is necessary to accelerate the ions up to relatively high energies (in commercial instruments up to 18–30 kV). The function of beam formation and ion acceleration is usually accomplished by a simple system consisting of a plate to which the accelerating potential is applied, followed by a number of slits at ground potential (Fig. 2.2). Due to imperfections in this system, the elemental and charge composition of the beam as well as the ion energy distribution can be noticeably distorted (154–156).

3.2.5. Ion Mass Analysis and Detection

These are well-investigated processes (8) and are briefly discussed in Section 2. However, it is necessary to take into account the possibility of ion mass discrimination, particularly because of the fringe magnetic field influencing the trajectory of the ions during their flight from the source to the magnetic analyzer (157–159).

4. APPLICATIONS

Spark source mass spectrometry has been used for the analysis of a large variety of samples. The problems that have been studied typically include survey and quantitative bulk analysis, as well as microsample, layer, and in-depth analysis. The technique has also been applied in surface contamination studies.

4.1. Bulk Analysis

In comparison with other analytical techniques, the special features of SSMS are its capability of simultaneous detection of nearly all elements, its high detection power, the detection of isotopes, its good accuracy when isotope dilution is used, and the simple sample preparation. SSMS is capable of analyzing practically all solid materials. Because of the electric spark used for ion formation, the sample should, however, be available in the form of conducting electrodes. As a rule, mechanical shaping and cleaning are all that is required for conducting samples. Insulating samples are usually pulverized and either mixed with high-purity powders such as graphite, silver, or gold (14) or used alone (31). Graphite is most commonly used; it can be obtained as a relatively pure powder, compacts into strong electrodes, and is relatively cheap. It suffers, however, from two basic problems: It produces many molecular ions, which can interfere with masses of interest in the spectrum, and it is of low atomic weight, which has the effect of atomically diluting the sample with which it is mixed to a significant extent. Gold has the advantage of having a high mass, so that fogging may occur only in a region that is of less importance in most analyses. Also, its only isotope has an odd mass number so that interferences from matrix ions are rare. A study on the purity of graphite, silver, and gold powders used for mixing showed that graphite is the purest and silver the most contaminated (160). Liquids can also be analyzed after they have been frozen, evaporated or freeze-dried.

Preconcentration procedures are sometimes applied to improve the detection limits of the elements of interest relative to the matrix. The various preconcentration techniques—evaporation, coprecipitation, cocrystallization, ion-exchange chromatography, adsorption, cementation, electrodeposition, and solvent extraction—have been reviewed (16). Special sample preparations are sometimes applied for biological materials (161, 162) and for homogenizing some materials (163). Oxidative acid digestion and dry-ashing procedures are widely used for biological samples.

Typical application fields of SSMS, which have been reviewed by Beske et al. (14) and by Bacon and Ure (16) with a summary of references, are bulk trace analysis in metals, semiconductors, and insulators; multielement analysis of the major, minor, and trace elements in technical alloys and in geochemical and cosmochemical samples; multielement analysis of biological samples and radioactive materials; and environmental analysis.

There is increasing interest in SSMS for the analysis of various high-purity materials, especially from the semiconductor industry. Emission spectroscopy, secondary ion mass spectrometry, and activation analysis are also very useful techniques, but they have some disadvantages for general use compared to SSMS. Emission spectroscopy is useful for the determination of many elements in group III–V compounds, but for germanium and silicon it is less suitable than SSMS, and in general the detection limits are worse. SIMS is not a typical bulk analysis technique; the sensitivities vary over four orders of magnitude, and even semiquantitative results are difficult to obtain; also, detection limits for some elements are worse than with SSMS (164). By activation analysis one can determine elements at concentration levels far below the parts per billion level. Such extreme sensitivities, however, are attained only for specific elements, whereas for other elements the method is not suitable at all or requires tedious radiochemical separations. Therefore, SSMS is of special value for analyzing semiconductor materials, especially since in such applications detection limits are often more important than accuracy.

After preparation and mounting the electrodes in the ion source, optimum experimental conditions should be applied, that is, a selection of spark parameters that allow stable sparking and slit settings that ensure optimum mass resolution and minimum discrimination (165).

The selection of parameters depends greatly on the type of sample and the problem under study. A general rule should be to keep all conditions as constant as possible during sparking in order to improve the precision (see Section 4.1.3). All the information from an SSMS analysis is recorded on a photoplate. This information can be used for qualitative and quantitative analysis. In qualitative analysis it is usually intended to detect one or more impurities semiquantitatively in a short time, and if they are not detected to define upper limits for their concentration. This is done by comparing the plate of interest with a calibration plate. The next step is to determine whether the line is due to an impurity element, which can usually be concluded from the isotope abundances. Concentrations are estimated by comparing the impurity lines to a matrix line or to the line of an impurity element of known concentration. Visual estimation yields uncertainties of up to a factor of 2–3 for the experienced eye (differences in relative sensitivities not taken into account).

For quantitative multielement analysis, on the other hand, accuracy is the major concern. Plates are obtained with successive exposures that differ by factors of $\sim 10^{1/2}$. In this way a dynamic range of 10^7 is obtained with about 15 exposures. After development, the line blackenings are converted to the number of ions striking the plate. A recording microphotometer is essential for quantitative evaluation of the position (mass) and blackening of the lines. Rapid data reduction from plates with a large number of lines can be accomplished only if the microphotometer is computer-interfaced and properly programmed.

The next step is to construct a calibration curve correlating the line transparency (transmission) T with the concentration of the element in the sample. A simple form is known as the transmission curve, where the logarithm of exposure E is plotted versus T. This curve has a limited linear range and is inconvenient to use. However, several linearization methods are described in the literature (16, 166, 167). Among them the empirical function given by Hull, which expresses the entire transmission range as a function of exposure, is often used (168):

$$E = \frac{1 - T}{(T - T_s)^{1/R}} \qquad (4.1)$$

Here T_s is the saturation transmission, and R is a linearization parameter. In the evaluation procedures corrections are also made for abundance, line width, and mass dependence of emulsion blackening and for background transparency (167).

Once the ion intensities have been determined, it is assumed that the obtained value is proportional to the concentration of the element. In bulk analysis it is found that the proportionality factors for the formation of singly charged ions are about the same within a factor of 0.3–3 for most elements in a given matrix. Accurate results can be obtained only if these factors, called relative sensitivity factors, are known.

A way to obtain accurate results without the need to introduce sensitivity factors is to apply the isotope dilution technique. This method was pioneered by Leipziger in 1965 (169), and since then it has been applied for trace analysis in metals and in geological samples. The technique is obviously restricted to elements that have two or more naturally occurring or long-lived isotopes. For each element to be determined, a known amount of spike is mixed with the sample. The unknown concentration of the element in the sample is then calculated from the measured isotopic abundances of the mixture.

A review of the ID-SSMS method can be found in the literature (170).

The measurement of the isotopic ratios is usually carried out by electrical detection in the peak-switching mode, since this gives the best precision.

4.1.1. Relative Sensitivity

It would be desirable in SSMS to make the recorded mass-resolved ion sample represent the solid sample composition as closely as possible, for in this case a standardless quantitative analysis would be possible. Unfortunately, this goal has not yet been achieved. Because of the various processes that occur in the vacuum discharge and because of discrimination effects in the mass spectrometer, the ratio of the number of detected ions corresponding to two elements x and y usually differs from the true concentration ratio of these elements in the sample c_x/c_y.

Consequently, in order to calculate the concentration of impurities from mass spectrometric results, correction factors are introduced that relate these two ratios:

$$\frac{I_x}{I_y} = \frac{c_x}{c_y} \, \text{RSF} \left(\frac{x}{y}\right)_z \tag{4.2}$$

The proportionality factor $\text{RSF}(x/y)_z$ is called the relative sensitivity factor when x is determined relative to y (internal standard) in a matrix z. Its value characterizes the dependence of the sensitivity of the SSMS method on the type of elements and on the sample matrix and includes all the discrimination effects present under the given experimental conditions, that is the relative sensitivities of source, transmission, and detector. If corrections are made for discrimination effects, mostly for variations in line width and detector mass response, then the proportionality factor between I'_x/I'_y and c_x/c_y is called the relative sensitivity coefficient (S_R or RSC).

In practice, sensitivity factors are experimentally determined by analyzing standard samples, and their values in SSMS are usually confined within one order of magnitude (see, e.g., 171, 172). This range is much smaller than in most other physical or chemical methods of analysis. Since it is rather time-consuming to determine differences in relative sensitivity from standard analysis, and because homogeneous standard reference materials are not available for many matrices, attempts have been made to calculate the relative sensitivity coefficients theoretically. Relations between experimental RSCs and physical and chemical properties such as melting point (173), boiling point (71, 174), heat of sublimation (175–177), ionization potential (71, 175, 176), covalent radius (177), ionization cross section (176), vapor pressure (178), and temperature for constant vapor pressure (171) have been studied, and empirical relationships for calculating the RSC have been proposed. Most of these relationships, however, rely upon one matrix, and there are few reports that deal with the influence of matrices (179). It has often been stated that the RSC for an individual element does not change with the matrix (180–182), whereas other studies have reported significant variations with the type of matrix (183–185). The authors support the latter point of view.

Many reports deal with the influence of experimental conditions on the relative sensitivities. These were shown to be influenced by the shape of the sample electrodes (186); spark gap width (187, 188); spark voltage, repetition frequency, and accelerating voltage (189); temperature of the sample electrodes (190); position of the spark relative to the ion optical axis (187); discharge duration (117); current through the discharge gap (112); energy liberated in the gap (120, 191); and other factors. The above considerations led to the conclusions that it is inadequate to express the RSC only as a ratio of some physical property of a measured element to that of an internal standard element and that the equation for calculating the RSC should contain a factor determined by the measurement condition. Vieth (192) has proposed relations between the RSC and the heat of sublimation and ionization potentials, taking experimental conditions into account. With a three-parameter equation, a threefold increase in accuracy for analyzing impurities in metals could be obtained. The above-mentioned influences of the experimental parameters on the RSC also imply that detection limits, absolute sensitivities, accuracy, and precision are affected by them. For optimum analysis, each spark parameter should thus be kept as constant as possible during analysis. In order to arrive at a more reproducible spark, a special capacitor has been included in the circuitry that isolates the first spark stage and excludes the low-voltage arc stage (117). It has often been suggested that the RSC values of different elements should be close to unity because of the high energy release during

Table 2.2. Experimentally Determined Relative Sensitivity Coefficients in Iron, Copper, Aluminum, and Zinc Matrices, Using Standard Reference Materials

	Iron	Copper	Aluminum	Zinc
Al	—	1.3 ± 0.1	1.08 ± 0.06	1.1 ± 0.3
Si	—	0.96 ± 0.09	0.84 ± 0.08	0.8 ± 0.2
Fe	≡1	≡1	≡1	≡1
Cr	1.7 ± 0.1	1.7 ± 0.2	1.6 ± 0.1	1.6 ± 0.6
Ni	0.82 ± 0.06	0.71 ± 0.07	0.74 ± 0.07	0.8 ± 0.2
Cu	1.6 ± 0.1	0.68 ± 0.06	0.67 ± 0.05	0.6 ± 0.1
Sn	4.2 ± 0.7	2.3 ± 0.5	2.2 ± 0.3	2.1 ± 0.5
Pb	—	2.8 ± 0.7	2.8 ± 0.4	2.7 ± 0.8

From Ref. 172.

the spark stage. Other authors, however, have recorded a considerable difference in RSC values under such conditions (193).

Another series of papers have dealt with the surface investigation of sparked alloy electrodes by various techniques such as electron microscopy and Auger electron microscopy (141, 194–197), and it has been shown that the sample surface composition may change with repetitive sparking. In addition it has been demonstrated by optical emission spectroscopy that the vapor composition may differ from that of the solid sample (197, 198). A study of sparked metal electrodes by secondary ion mass spectrometry showed that the ratio of the surface to bulk compositions relative to the same ratio measured for an internal standard element qualitatively agrees with RSCs determined by SSMS (see Fig. 2.12) (132). All these studies suggest that differences in relative sensitivity in SSMS are to some extent due to differences between the sampled material and the bulk. A study on the possible discriminations connected with the main stages in the formation of a mass spectrum led to the conclusion that during the atomization and ionization steps, segregations may occur and cause the ion composition of the plasma to deviate from the atomic composition of the incoming vapor (121). Differences in the composition of the sampled material as compared to the bulk were also reported to be due to selective vaporization (198), segregation (195, 196), oxide layer formation (194, 196, 197) even at low pressures in the sample chamber, fractional condensation (186), selective sputtering, zone refining (57, 132), and other causes.

A summary of references giving experimentally determined RSF or RSC values for various materials can be found in the literature (16, 165). Table 2.2 shows, by way of example, relative sensitivities in steel, copper, and aluminum; it can be seen that the values do not deviate too much from unity. It should be realized that RSFs can seriously differ from one instrument to another, and thus for quantitative analysis RSF values determined with the same instrument should be used.

4.1.2. Detection Limits

One of the most important features of SSMS is its high detection sensitivity. In general, the parts-per-million level is reached with exposures of 1–10 nC. Collection of 10 nC usually takes only 5 min. Important limitations of the sensitivity are the intense halo extending on the high-mass side of the major matrix lines and a general, nearly uniform, blackening of the entire mass spectrum. This latter background, due to nonfocused particles and the faint light emission of the ion beam itself, becomes apparent only at exposures of ~100 nC and is mainly determined by the quality of the vacuum in the analyzers.

Figure 2.15 shows these artifacts for a spectrum of arsenic after a 600-nC collection. The halo results from secondary emission processes: low-energy ions and electrons are sputtered away from the plate emulsion by the high primary ion bombardment.

In addition, soft X ray and fluorescence radiations occur (199). Various procedures have been proposed to reduce or avoid the production of halo, such as cutting plates where the matrix lines appear (200, 201), painting a grounded conducting strip where the matrix lines appear (202), using a thin sheet of magnetic material behind the ion-sensitive plate (203), using gelatin-free photoplates (204), modifying the developer (205), and combining superficial bleaching and internal developer treatments (206, 207). Liu et al. (208) used the cutting procedure for the analysis of arsenic, and with an appropriate developer the halo vanished to a low level even for a very high, 1200-nC, exposure.

From a comparative study on three types of photoplates—Ilford Q2, Kodak SWR, and Ionomet IM—it was concluded that the latter two are more sensitive (209); the Ilford Q2 plate, however, had a much lower background level and better line definition. Chupakhin and Ramendik, on the other hand, reported similar sensitivity for both Ilford and Kodak plates (210).

According to the literature, the smallest number of ions n^+ that are necessary to give a just visible (detectable) line on the Ilford and Kodak plates is ~3000 (210). In order to be able to measure the line photometrically, n^+ should be about 10^5 (210). The volume of sample material

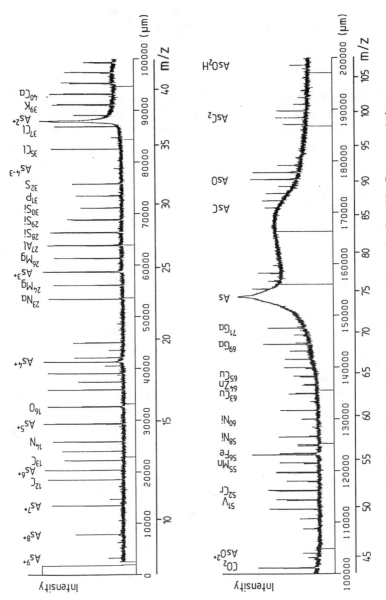

Figure 2.15. Part of the mass spectrum of an arsenic sample after 600-nC exposure (micro-densitometer recording of photoplate).

v (μm^3) consumed in making an exposure of n^+ ions in order to determine an impurity at a concentration c in ppma can be estimated from

$$v = \frac{An^+ \times 10^{20}}{N_A \rho \Delta c I} \qquad (4.3)$$

where A is the matrix atomic weight, I is the abundance (%) of the isotope line measured, N_A is Avogadro's number, and ρ is the density of the matrix (g/cm^3). Δ is the useful yield, the number of atoms that must be atomized to deliver one ion at the detector. The value of Δ depends on the specimen material and on experimental parameters such as breakdown voltage, slit widths, and type of electrostatic sector (57, 165). Usually its value lies in the range 10^{-7}–10^{-8} (57). Equation (4.3) is especially useful for estimating detection limits in the analysis of microsamples or layers, where the amount of sample is limited.

Detection limits have been reported for various materials (e.g., 67, 211–216) and sometimes compared with those of other techniques (164, 215, 216). In general, values are in the range 1–10 ppb, depending on the isotopic abundance, discrimination effects, relative sensitivity coefficients, and detection system. The lowest detection limit that has been reported is 0.3 ppb (217). This sensitivity is not reached for all elements, for the reasons just given, but in general half the elements will have a detection limit below 10 ppb, and cases where this value is higher than 100 ppb are quite rare. Table 2.3 summarizes detection limits for impurities in a gallium matrix (218).

Of course, detection limits may deteriorate because of interferences of spectral lines, depending on the sample; in that case, the analysis should be based on another, interference-free, mass line such as a less abundant isotope of the element of interest, or on the doubly charged ion(s). For the analysis of carbon, nitrogen, and oxygen, the detection limits are influenced by the residual gas pressure in the ion source. As mentioned before, the detection limits can then be lowered by improving the vacuum in the ion source, and detection limits of ~10 ppba have been reported when helium cryopumping was used (67, 219).

When electrical detection in the peak-switching mode is used, the obtainable limits of detection are usually a factor of 10 better than in plate detection. However, since mass resolution in electrical detection is relatively poor, interferences of spectral lines become more important. A quantitative description of detection limits in electrical detection has been given by Hull (220).

When isotope dilution mass spectrometry is applied, detection limits are usually in the range 0.1–1 ppm (170).

Table 2.3. Limits of Detection (ppma) for Elements in a Gallium Matrix, Based on the Intensity of a "Just Visible" Line on an Ilford Q2 Photoplate, at 600 nC Exposure, Using the Pb Concentration as Internal Standard, and Assuming All RSC = 1

Element	DL	Element	DL	Element	DL
H	0.004[a]	Ni	0.0035	Cs	0.004
Li	0.001	Cu	0.0045	Ba	0.04[f]
Be	0.001	Zn	0.006	La	0.045[g]
B	0.0015	Ga	Matrix[c]	Ce	0.005
C	0.0015[a]	Ge	0.0085	Pr	0.0045
N	0.0015[a]	As	0.003	Nd	0.02
O	0.0015[a]	Se	0.006	Sm	0.02
F	0.0015	Br	0.006	Eu	0.009
Na	0.02[b]	Rb	0.015[d]	Hf	0.015
Mg	0.002	Sr	0.01[e]	Ta	—[h]
Al	0.002	Y	0.0035	W	0.015
Si	0.002	Zr	0.007	Re	0.008
P	0.002	Nb	0.004	Os	0.01
S	0.002	Mo	0.015	Ir	0.008
Cl	0.003	Ru	0.01	Pt	0.015
K	0.0025	Rh	0.0035	Au	0.015[i]
Ca	0.0025	Pd	0.015	Hg	0.015
Sc	0.0025	Ag	0.007	Tl	0.007
Ti	0.003	Cd	0.015	Pb	0.01
V	0.003	In	0.0045	Bi	0.015[j]
Cr	0.003	Sn	0.01	Th	0.006
Mn	0.0025	Sb	0.007	U	0.006
Fe	0.0025	Te	0.01		
Co	0.0025	I	0.004		

[a] For the analysis of H, C, N, and O, the detection limits are in practice worse because of the residual gas pressure in the ion source.

[b] Based on Na^{3+} because interference from $^{69}Ga^{3+}$ on $^{23}Na^{+}$.

[c] The photoplate was cut to avoid the halo from the Ga^{+} matrix lines.

[d] Based on Rb^{2+} because of possible interference from GaO^{+} on Rb^{+}.

[e] Based on Sr^{2+} because of possible interference from GaO^{+} and $GaOH^{+}$ on Sr^{+}.

[f] Based on $^{137}Ba^{+}$ because of interference from Ga_2^{+} on $^{138}Ba^{+}$.

[g] Based on La^{2+} because of interference from Ga_2^{+} on La^{+}.

[h] Ta cannot be determined because various critical components of the ion source are made of this element.

[i] Based on Au^{2+} because of possible interference from TaO^{+} on Au^{+}.

[j] Based on Bi^{2+} because of interference from Ga_3^{+} on Bi^{+}.

4.1.3. Precision

Despite the unique merits of spark source mass spectrometry, there is a basic problem that limits the wide use of the technique: the reproducibility and therefore the accuracy of the results of individual measurements are not always satisfactory. The basic sources of error can be attributed to the unstable action of the spark discharge on the sample, that is, to variations in the conditions of ion formation, to changes in the position of the ion beam path, to the nonuniform distribution of impurities in the solid sample, to nonuniformity in the photographic plate emulsions, and to photometric errors (11, 155, 221, 222). Standard deviations larger than 40% usually indicate inhomogeneous element distributions. Special electronic devices incorporated in the ion source allow increases in the amount of material consumed per unit collected charge (20–23). Also, by using rotating electrodes, concentration gradients or inhomogeneities in different areas can be statistically averaged, thus improving the precision (17). Even for an error-free analytical method, the relative standard deviation $R(\%)$ of an analysis will not be zero because of sample heterogeneity. For a given element in a given sample, R is inversely proportional to the square root of the analytical subsample weight w (223):

$$R(\%) = 100\sqrt{K_s/w} \tag{4.4}$$

where K_s is the "sampling constant." This effect is illustrated in Fig. 2.16 for the analysis of Mn in gold at the 40-ppb level by SSMS, using electrical detection in the peak-switching mode (224). Since a typical SSMS analysis with photoplate detection consists of a series of graded exposures, the short exposures are much more subject to this sampling error.

According to Ramendik, the probe technique with monopolar discharge stabilizes the geometry in the interelectrode gap and its position relative to the ion optical axis to a maximum possible degree (11, 28). When, in addition, the sample is scanned, average bulk analysis data can be obtained. Using the probe technique in the analysis of nonconducting objects, relative standard deviations of 5–8% were obtained for geological samples (31). Good precision and accuracy have also been obtained by other authors (70, 71, 225).

In general, a better precision of 3–7% is possible with more elaborate sample preparation (e.g., application of isotope dilution and chemical dissolution of the sample in the case of inhomogeneous materials, see above) and by the application of electrical detection. Van Hoye et al. (15) showed that with electrical detection, autospark, and a beam controller, using homogeneous standard reference materials, a relative standard deviation

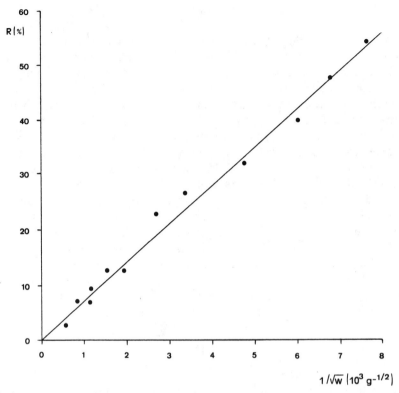

Figure 2.16. Sampling error for Mn determination in gold wire at the 40-ppb level using SSMS in the peak-switching mode (224).

of 5% can routinely be achieved for metals and alloys. In addition it was concluded that changes in repetition frequency, pulse length, and accelerating voltage do not significantly affect the precision of an analysis. Isotope ratios can be measured precisely by electrical detection in the peak-switching mode; in some cases a precision of 2–5% or even better has been obtained (226–229). Therefore the most accurate and precise results are obtained by using ID-SSMS (165, 170, 226–230).

4.1.4. Accuracy

Accuracy implies the elimination of systematic errors, but it is clear that the obtainable accuracy cannot be better than the precision or reproducibility, as discussed in the previous section.

With no standards available, the concentrations must be calculated

Table 2.4. Average Absolute Inaccuracy of Uncorrected Results, RSC Approach
and Three Semiempirical Approaches

Matrix (Number of Samples)	Number of Elements	Average Inaccuracy (%)				
		No Correction	RSC	Calculated		
				Ref. 231	Ref. 232	Ref. 175
Iron and steel (5)	16	220	10	48	48	93
Copper and alloys (5)	21	178	13	56	50	50
Aluminum and alloys (10)	11	178	10	—[a]	36	36

[a] Calculation irrelevant.

from the mass spectrometric data with various correction factors for mass, line width, and so on, using formulas chosen from experience with the particular instrument at hand. In such an approach the analysis is actually semiquantitative, since equal sensitivity is assumed for all elements. As shown in Section 4.1.1, empirical formulas have been proposed for estimating relative sensitivities based on various physical properties of the elements of interest (impurities, internal standard, matrix). The use of such calculated factors improves the average accuracy, but this approach is still inferior to the one based on experimentally determined relative sensitivities (175, 231, 232), as can be seen from Table 2.4.

This illustrates the need for analyzed samples, or rather standard reference materials, with certified minor and trace element concentrations; in addition, these standards should be quite homogeneous because of the low material consumption of SSMS (compare Fig. 2.16). In principle, a single standard of the matrix of interest is sufficient, since the relative sensitivities are independent of the elemental concentrations over at least 5 or 6 orders of magnitude as shown by Clegg et al. (233) and Van Hoye et al. (15).

Still, one would like to have a reference material whose composition is similar to that of the sample, since this would allow direct comparison on the photoplate. Unfortunately, certified standards in the ppm–ppb concentration range are not readily available, and therefore methods suitable for preparing them may be of great value. One of these is ion implantation (234–236).

Sample and standard should be measured under identical conditions: spark parameters, spark position, slit settings, photometric procedure,

Table 2.5. Experimentally Determined Concentrations of Some Lanthanides in USGS Standard Rock BCR-1 by Isotope Dilution–Spark Source Mass Spectrometry

Element	Concentration in μg/g (RSD in %)[a]		Best Estimate (μg/g)
	Photoplate	Electrical Det.	
Eu	2.13 (6.7)	1.97 (2.4)	1.95
Gd	6.17 (5.3)	6.38 (2.7)	6.55
Dy	6.00 (5.3)	6.40 (3.6)	6.39
Er	3.33 (3.6)	3.80 (2.4)	3.70
Yb	3.32 (5.5)	3.49 (2.7)	3.48

[a] RSD = relative standard deviation.
From Ref. 228.

and so on. Differences in ion-sensitive emulsions occur, but it has been shown that accuracies within 5% can be obtained when the spectra of sample and standard are recorded on the same photoplate in alternating exposures (165). In this way few or no corrections are necessary, and results are derived by direct comparison of line densities. Generally speaking, the results of SSMS are significantly influenced by the plate calibration procedure (166, 167); when using the Hull equation [Eq. (4.1)], for instance, the values of $1/R$ and T_s should be carefully determined.

With electrical detection in the peak-switching mode, quite accurate results (within 5%) can be obtained in the concentration range 10^{-5}–1% using reliable relative sensitivity coefficients (15).

When isotope dilution is applied in combination with SSMS, the problem of differences in relative sensitivity is eliminated. The technique is now commonly applied, either through a process in which equilibrium is ensured between a sample solution and an isotopically enriched sample, or else by adding isotopically enriched internal standards to powder samples; isotopic equilibrium is then assumed to occur in the spark plasma.

Isotopically doped graphite is often used as a conducting additive to the sample. In some applications, up to 30 elements have been measured simultaneously. Accuracies of 2–5% can be expected in ultimate performance with photoplate detection and accuracies of 1% when electrical detection is used (170).

In Table 2.5, a few results obtained for some lanthanides in USGS standard rock BCR-1 (228) are compared with the best estimate of the true value (71).

It is obvious that systematic errors may arise as a result of mass spectral interference or when the element of interest has an anomalous isotopic abundance, which is a realistic source of error for elements such as lithium, boron, or uranium, whose isotopic composition is often intentionally modified on a large scale. For example, De Goeij et al. (237) measured $^7Li/^6Li$ ratios ranging from 0.98 to 2.94 in commercial samples, whereas the natural ratio is 12.33.

An important source of systematic error may be the failure to remove surface contaminants from the sample. Chemical etching or mechanical machining are often less effective than in situ cleaning by presparking, in particular for elements such as carbon and oxygen (219). Presparking for a sufficiently long time is often used to reduce memory effects from previous measurements, by coating the source housing and the slits with a thin layer of the material under investigation. Many laboratories use a different set of source housing, electrode holders, and slits for each matrix. In a similar way, other errors can be kept to a minimum by maintaining high standards for laboratory techniques.

4.2. Analysis of Microsamples

Sometimes the amount of sample available for analysis is so small that the conventional methods of analysis cannot be applied, even for the major components. SSMS has obvious potential advantages for this type of problem because of its high sensitivity and the simple sample preparation; complicated and expensive separation procedures required by some other analytical approaches can be avoided. The simultaneous detection capability is an important advantage of SSMS compared to many other analytical techniques where the amount of material necessary for analysis is determined by the number of elements to be analyzed. In bulky samples, where the volume of sample is not restricted, the detection limit depends on the noise level of the detection system, whereas for microsamples the sensitivity is limited by the amount of sample available.

McCrea summarized the various factors that determine the mass of material consumed in mass spectrometric analysis (238). These are (a) the efficiency of conversion of specimen material to ions, (b) the efficiency of the ion optics in gathering the ions to the detector system, and (c) the efficiency of the ion-detection system. The product of (a) and (b) gives the "useful ion yield," Δ [see Eq. (4.3)].

Studies on material consumption in SSMS analyses (57, 90, 142, 210) and on useful ion yield (57, 90, 142, 153, 239) showed that the amount of material transferred into the gap is determined by the melting point of the matrix and that Δ for a given metal increases at lower breakdown voltage and at wider slit settings. Before attempting microsample analysis it is thus desirable to know Δ as a function of experimental parameters in order

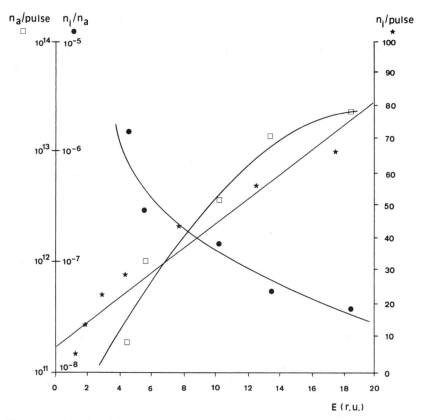

Figure 2.17. Number of atoms atomized per pulse (n_a) and useful ion yield (n_i/n_a) for gold, as a function of the energy released in the spark (E). E is related to the breakdown voltage by Eq. (3.6) (142).

to select appropriate working conditions. Figure 2.17 shows Δ and the corresponding weight loss per nanocoulomb charge collection for a gold matrix as a function of breakdown voltage using two gold electrodes sparked in a plane-to-plane geometry (142). Slit settings were chosen so as to obtain a mass resolution of ~4000 ($M/\Delta M$). These data show that in principle microgram samples of gold can be analyzed to yield 1 nC charge and hence impurity detection limits in the range 1–10 ppma.

Similar studies were also made on other matrices such as aluminum (57), silicon, iron, vanadium, germanium, copper, CuBe alloy, and graphite (142). These data are equally relevant to the analysis of nonconducting samples, which are usually mixed with a conducting powder such as gold, aluminum, or graphite. Bingham and Vossen (240), for instance, determined the abundances of neodymium isotopes in a 1-μg sample of neodymium oxide after mixing it with 30 mg of gold. In practice it is often difficult to do such an analysis; the smaller the sample, the larger the problems. Brown et al. (241) have analyzed microsamples (a 2-mg single radar diode and individual crystals of cadmium sulfide) by inserting them into a split electrode of high-purity graphite. Ahearn (159) used a holder and counterelectrode of high-purity silicon for the analysis of small silicon crystals, and concentrations down to the ppm level could be determined.

Thomas et al. (242) analyzed small graphite samples doped with boron after mounting them in cleft indium rods, whereas Stefani et al. (243) used pure gold wires to hold ruby crystals. Brown et al. (244) described a die for making tipped electrodes that made it possible to spark 0.1-mg samples. The sample to be investigated is pressed on top of one electrode and sparked against a counterelectrode of a pure metal. Swenters (142) used this procedure to analyze doped graphite samples down to the sub-ppm level. A technique for the analysis of very small samples (20 µg) was developed by Wallace and Roboz (245); the sample is folded into a pure platinum foil that serves, after cutting, as both electrodes. Most reports on microsample analysis deal with amounts of 2–5 mg, which allow a series of graded exposures to be obtained on a photoplate. For smaller samples this cannot be done; for such analysis Chastanger (246) has developed a single-exposure method that, using image-broadening data, still allows the estimation of impurity concentrations over a range of 4 orders of magnitude.

Absolute detection limits, when matrix identification is necessary, are very low and are compared in Fig. 2.18 with limits of detection for other analytical techniques (247).

Another type of microanalysis by SSMS, the analysis of thin films, is discussed in the following section.

4.3. Local and In-Depth Analysis

Because of the local effect of the spark on the sample (formation of craters of limited diameter and depth), spark source mass spectrometry can be used to carry out local and in-depth analyses. Analyses of thin solid films and surface contaminants are particular cases of the second type.

4.3.1. Features of the Procedure

Local analysis is accomplished with the help of a pointed counterelectrode. The chosen areas of the sample surface are sequentially positioned

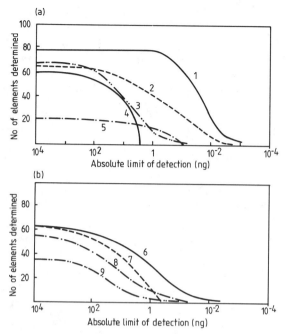

Figure 2.18. Detection limits for various physical methods of analysis (247): 1, SSMS; 2, activation analysis; 3, copper spark emission spectroscopy; 4, dc arc emission spectrometry; 5, graphite spark emission spectrometry; 6, flame spectrometry; 7, absorption spectrometry; 8, atomic absorption; 9, fluorescence.

near the tip of a fixed counterelectrode. In-depth analysis can be carried out in three ways:

1. With the aid of a pointed counterelectrode (91). In this case, in order to obtain a low detection limit and to average information obtained from the maximum available area, one can scan the sample surface by means of rotation (92), rotating-translational motion (so that craters on the sample surface form a spiral groove similar to that on phonograph record) (248), or linear scanning in two mutually perpendicular directions (surface "rastering" as on a TV screen) (249, 250).

2. With the help of a broad counterelectrode having a typical width of 2–4 mm and a thickness of 0.1–1 mm (251). In this case it is possible to scan the surface in just one direction, perpendicular to the broad side of the fixed counterelectrode.

3. With the help of a counterelectrode of the same shape as the sample

(57, 252). Both electrodes are usually machined, as for common bulk analysis, in the shape of rods or parallelepipeds and positioned with their faces parallel to each other. In this case scanning is not applied.

The counterelectrode can be made from the same material as the sample (159) or of different metals including refractory ones (for example, rhenium, tantalum, or rhodium) (250, 251), or it can be made of metals that sputter easily under the influence of the spark discharge (for example, aluminum) (248, 251). The choice of the technique of surface sampling by the spark and the choice of counterelectrode material depend on the analytical task. For example, a maximum depth resolution can be obtained when the sample is scanned in one direction against a wide aluminum counterelectrode. In this way the average depth of sample erosion is significantly reduced (9, 251) and, moreover, maximum stability of the ion current is obtained. Using this method, one can perform in-depth analysis on poorly conducting and insulating samples if their surface is coated with a thin aluminum film (~ 1 μm) prior to analysis (9). In this case, however, the contribution of the counterelectrode material to the ion current may be as high as 50–90% (9, 248), and this results in an increase in the background due to contaminations from the counterelectrode material, as well as in an intensification of the "memory" effects. Material is continuously transferred between the electrodes, and as a result the surface composition of the layers to be analyzed changes. This effect must be taken into account if the impurity concentration gradient is rather large, and a way to achieve this has been proposed (253). The influence of surface roughness induced by the spark on the accuracy of the obtained concentration profile has been investigated by Chupakhin et al. (254). The roughness can be reduced not only by the choice of counterelectrode material but also by decreasing the energy released in the spark discharge. To achieve this, a ballast resistor or a capacitor is introduced into the discharge circuit (84).

Bedrinov and Belousov (255) studied the "edge effect," that is, the influence of the scanning area boundary. A more general discussion on the optimization of in-depth analysis conditions is given by Gerasimov et al. (33).

4.3.2. Local Analysis

Ramendik (120) used spark source mass spectrometry to study the distribution of lithium and boron in a biotite grain of about 400 μm diameter. The analysis was carried out directly on a thin section of a granitic rock

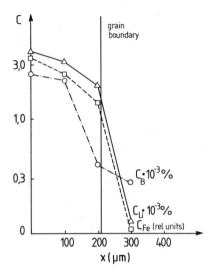

Figure 2.19. The distribution of boron and lithium in a biotite grain: x, distance from grain center; C_B, boron concentration; C_{Li}, lithium concentration; C_{Fe}, iron concentration (120).

(Fig. 2.19). This is the only known example of investigating impurity distribution profiles in natural objects. All other examples in Section 4.3 deal with technical materials.

Local SSMS analysis was used to study the diffusion of various metallic impurities in vanadium (256), thorium (257), aluminum (258), and copper (259). The concentration profile of beryllium extending over several millimeters in a heated diffusion couple is shown in Fig. 2.20 (259). The data have been obtained by means of electrical detection. Good agreement was obtained with the results of in-depth SSMS analysis and lateral ion probe mass analysis of the same samples.

4.3.3. In-Depth Analysis

A number of diffusion profiles were also studied by in-depth SSMS, in particular the diffusion of magnesium in aluminum (57), of beryllium in copper (252) (Fig. 2.20), and of nickel in copper (224). The results agreed well with those obtained by secondary ion mass spectrometry and with literature data.

By investigating the diffusion of antimony from a silicon wafer into a 13-μm thick epitaxial silicon film, the diffusion coefficient at the film-growing temperature of 1250°C was found to be 1.1×10^{-11} cm²/s, which is 15 times higher than for common silicon monocrystals (9).

In-depth analysis was applied by Yudelevich and Shelpakova (260) to study thin films of silicon, germanium, and gallium arsenide. A method

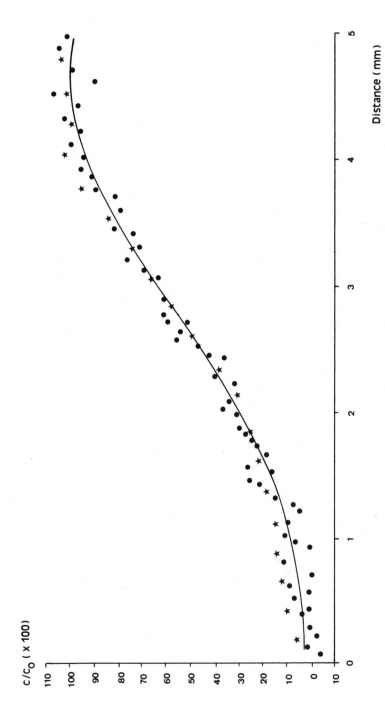

Figure 2.20. Diffusion profile of Be in Cu obtained by lateral (★) and in-depth (•) analysis using SSMS (252).

for the in-depth determination of oxygen, nitrogen, and carbon impurities has been worked out for gallium arsenide (261).

The distribution of sulfur impurity in a synthetic 0.15-μm thick diamond film during growing is a good example of in-depth analysis with a depth resolution of about 0.1 μm. The film was grown on a natural diamond wafer (20).

The analysis of coatings is another type of in-depth analysis. The results of the in-depth analysis of a 10-μm TiC-coated throwaway tip of cemented carbide (WC + Co) have been reported (84). Twenty-five elements were determined, and a peculiar behavior of the most important elements (tungsten and titanium) near the layer–wafer interface was observed. Vasile and Malm (262) investigated thin gold films obtained by the galvanic method, and Liebich and Mai (263) analyzed thin (<0.1 μm) copper films on a glass wafer.

Verlinden et al. (164) used SSMS for contaminant identification in 2.5-μm silicon layers on a pure silicon substrate and in a 3-μm liquid-phase epitaxial layer of GaAs on a pure GaAs substrate. The detection limits of SSMS were compared with those of SIMS, a typical layer analysis technique, and it was shown that SSMS is more sensitive for the determination of some impurities. The same authors also quantitatively determined carbon in 3-μm gold coatings deposited from cyanide solution on a nickel substrate. The results were in good agreement with those obtained by charged-particle activation analysis.

Jansen and Witmer determined the boron distribution in a pyrolytic carbon layer coating a graphite fabric (248). Their paper also gives an example of the determination of surface contamination. The results of surface layer analysis on a polished germanium wafer before and after etching show the possibility of controlling the degree of refining of materials used in microelectronic technology.

Earlier such investigations were carried out by Ramendik et al. on silicon wafers and epitaxial films (see Ref. 9, p. 183).

An interesting utilization of surface contamination analysis, started by Ahearn (91), was proposed by Shelpakova et al. (264). Impurities from water and other liquids were concentrated on the surface of a high-purity silicon wafer after presparking. By means of a counterelectrode, these impurities were determined as surface contaminants, which made it possible to lower the detection limit (recalculated to the original liquid sample) to 10^{-12}–10^{-10} wt%.

4.3.4. Comparison with Other Methods

The minimum diameter of the crater formed on the sample surface during the spark discharge initiation stage is about 10 μm [see expression (3.1)],

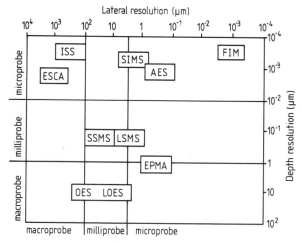

Figure 2.21. Comparison of lateral and depth resolutions of various physical methods of analysis. ISS, ion scattering spectroscopy; FIM, field ion microscopy; SIMS, secondary ion mass spectrometry; ESCA, electron spectrometry for chemical analysis; AES, Auger electron spectrometry; SSMS, spark source mass spectrometry; LSMS, laser source mass spectrometry; EPMA, electron probe microanalysis; OES, optical emission spectrometry; LOES, laser optical emission spectrometry.

and its depth 10^{-2} μm (84). It is difficult to achieve these parameters in practice, a typical value for the lateral resolution being 100 μm (120, 259) and for the depth resolution, about 1 μm (9, 248). As for lateral and in-depth resolutions, spark source mass spectrometry is substantially inferior to typical methods of local or surface analysis (57, 265); this is indicated in Fig. 2.21, where lateral and in-depth resolutions obtained in various surface and in-depth analysis techniques are compared with those of SSMS. These are typical values, and recent developments have made it possible to improve on them with some techniques. For instance, in the case of SIMS, liquid metal ion sources allow a lateral resolution of ~0.1 μm, and in ESCA new commercial instruments make it possible to work with small beams ~150 μm in diameter. Spark source mass spectrometry can be classified among the "milliprobe" techniques (265), whose region is roughly marked in Fig. 2.21.

However, for the analysis of relatively thick layers (2–100 μm) or areas with a diameter of 100–500 μm, spark source mass spectrometry yields more information than microanalytical techniques. A large number of elements can be determined simultaneously, semiquantitative analysis can be done without standards, and detection limits may be 3–5 orders of magnitude lower. For example, in a 1-μm thick layer having an area of 300 mm^2, impurity detection limits reach 10^{-6}–$10^{-5}\%$ atomic (9, 248). In addition, in contrast with other methods, the analytical results are not affected by the sample surface state.

Thus, spark source mass spectrometry is a unique link (265) between microprobe techniques and macroanalytical methods that are characterized by poor lateral resolution (as in optical emission spectrometry) or that have no lateral resolution whatsoever (as in neutron activation analysis).

5. PERSPECTIVES OF FURTHER DEVELOPMENT

At the present time there is a need for characterizing a variety of new materials for the microelectronics industry, based on high-purity gallium, arsenic, selenium, tellurium, indium, and so on, and this need can be expected to persist for some time to come. Although other analytical techniques may be suitable for determining selected impurities at the parts-per-billion level, SSMS is unsurpassed when it comes to obtaining a general view of the entire periodic table. For this type of application, both high sensitivity and high mass resolution are primordial requirements. There are, of course, "pretenders to the throne" (e.g., glow discharge, laser inductively coupled plasma mass spectrometry), but they still have to demonstrate their usefulness at such low concentration levels. For other types of samples, such as rocks, minerals, soils, aerosols, biological samples, and waters, SSMS enjoys vigorous competition from other sensitive optical, electrochemical, and nuclear methods. Unless instrumental and theoretical developments of SSMS become significant enough for a "renaissance" to take place, the technique could well fall into decline. It is the authors' conviction that such developments will become possible in the near future.

In fact, real applications of spark source mass spectrometry started even before fundamental insights were available, a situation that is also characteristic of many other analytical methods (266). In the case of solids mass spectrometry, however, this situation is understandable because, during the production of ions from a solid, a number of complex, interconnected, and not-well-known processes take place. That is why for about 20 years expansion of the field of applications and improvement of the analytical characteristics of the method have been based on a purely empirical approach. Recently the situation has perceptibly changed. On the one hand, the possibilities of the empirical approach are virtually exhausted (11), which leads to the *necessity* to create a theoretical basis. On the other hand, the *possibility* of a new approach is created by the development of concepts in spark discharge physics and on the mechanisms of ion formation in the discharge plasma (see Section 3).

A more profound understanding of these mechanisms gives the scientist a chance to control conditions of ion formation in the vacuum spark discharge (120), which was not thought possible until recently. Such progress should allow improvement in precision and accuracy of results (to 3–5%) in the case of direct analysis (i.e., without employing isotope dilution) using just one standard. If progress is made in the a priori determination of relative sensitivity coefficients and if standardless analysis with an accuracy of no less than 10–20% becomes possible, spark source mass spectrometry could become a respectable standard method of analysis (13).

Reduction of the energy spread as well as an increase in the residual degree of plasma ionization (up to 3 orders of magnitude could be gained here, see Section 3) will contribute to improvement in accuracy and, which is more important, could improve detection limits quite substantially.

The development of position-sensitive electro-optical ion detectors may be the main thrust for double-focusing mass spectrometers (12). Much earlier, one can expect progress in the automation of spectrum evaluation on photoplates. But this does not allow us to solve the main problem, that is, to carry out online analysis.

If these optimistic forecasts can be realized, then spark source mass spectrometry will gain important additional advantages over other analytical techniques. In this case it will be expedient to equip research laboratories with instruments even though their complexity and price will continue to increase because of the introduction of simultaneous electrical detection of many mass spectrum lines.

It seems that only the development of small instruments will open for spark source mass spectrometry the perspectives of wide application in geology, metallurgy, the electronics industry, environment control laboratories, and so on. In this case it will become possible to implement the unique advantages of spark source mass spectrometry and thus to raise the technique to a new stage of development. No effort to develop the theory, technique, and instruments to solve such an important problem could be excessive.

The development of simple, small, relatively cheap, and at the same time highly automated instruments to carry out survey quantitative elemental analysis of natural objects and technical materials is another important direction in instrument design. It requires the construction of spark ion sources with reduced energy spread. The first experiments to combine the spark ion source and a time-of-flight analyzer with ion reflector were promising (267). For some applications, this development could be of interest because it makes possible the detection of the entire mass spectrum from one single spark pulse. Also the quadrupole analyzer has already been combined with the spark source (47); if the energy band width could be reduced, the combination might be viable.

REFERENCES

1. A. J. Dempster, *Proc. Phil. Soc.*, **75**, 755 (1935).

2. A. J. Dempster, *Rev. Sci. Instrum.*, **7**, 46 (1936).

3. A. E. Shaw and W. Rall, *Rev. Sci. Instrum.*, **18**, 278 (1947).

4. J. G. Gorman, E. J. Jones, and J. A. Hipple, *Anal. Chem.*, **23**, 438 (1951).

5. N. B. Hannay, *Rev. Sci. Instrum.*, **25**, 644 (1954).

6. N. B. Hannay and A. J. Ahearn, *Anal. Chem.*, **26**(6), 1056 (1954).

7. R. D. Craig, G. A. Errock, and J. D. Waldron, *Adv. Mass Spectrom.*, **1**, 136 (1959).

8. A. J. Ahearn, Ed., *Trace Analysis by Mass Spectrometry*, Academic Press, New York, 1972.

9. M. S. Chupakhin, O. I. Kryuchkova, and G. I. Ramendik, *Analytical Capabilities of Spark Source Mass Spectrometry*, Atomizdat, Moscow, 1972, p. 224.

10. Working Group Spark Source Mass Spectrometry, *Fresenius Z. Anal. Chem.*, **309**, 257–341 (1981).

11. G. I. Ramendik, in L. Niinisto, Ed., *Euroanalysis IV* (*Reviews on Analytical Chemistry*), Akademiay Kiado, Budapest, 1982, p. 57.

12. F. Adams, *Phil. Trans. Roy. Soc. London*, **305A**, 509 (1982).

13. G. I. Ramendik, *J. Anal. Chem. USSR*, **38**(11), part 2, 1570 (1983); transl. from *Zh. Anal. Khim.*, **38**(11), 2036 (1983).

14. H. E. Beske, R. Gijbels, A. Hurrle, and K. P. Jochum, *Fresenius Z. Anal. Chem.*, **309**, 329 (1981).

15. E. Van Hoye, F. Adams, and R. Gijbels, *Talanta*, **23**, 789 (1976).

16. J. R. Bacon and A. M. Ure, *Analyst*, **109**, 1229 (1984).

17. J. Haemers, *J. Phys. E*, **11**, 897 (1978).

18. J. B. Clegg and E. J. Millett, *Acta Electron.*, **18**, 27 (1975).

19. G. I. Ramendik, Yu. G. Tatsii, and M. S. Chupakhin, *J. Anal. Chem. USSR*, **28**, 653 (1973).

20. J. C. Franklin and A. J. Dean, *Anal. Lett.*, **4**, 857 (1971).

21. M. S. Chupakhin, G. I. Ramendik, V. I. Derzhiev, Yu. G. Tatsii, and M. A. Potapov, *Sov. Phys. Dokl.*, **18**, 430 (1973).

22. G. I. Ramendik, M. S. Chupakhin, and M. A. Potapov, *J. Anal. Chem. USSR*, **30**, 564 (1975).

23. P. F. S. Jackson, J. Whitehead, and P. G. T. Vossen, *Anal. Chem.*, **39**, 1737 (1967).

24. R. A. Bingham, P. Powers, and W. A. Wolstenholme, in K. Ogata and T. Hayakawa, Eds., *Recent Developments in Mass Spectroscopy*, University Park Press, Baltimore, MD, 1970, p. 339.

25. J. Haemers, *J. Phys. E: Sci. Instrum.*, **7**, 294 (1974).

26. C. W. Magee and W. W. Harrison, *Anal. Chem.*, **45**, 220 (1973).

27. V. M. Gladskoi, Yu. N. Kuznetsov, L. M. Babenkov, V. I. Belousov, and I. A. Kuzovlen, *Ind. Lab.*, **40**, 1796 (1975).

28. G. I. Ramendik, M. S. Chupakhin, Yu. G. Tatsii, and V. I. Derzhiev, *J. Anal. Chem. USSR*, **29**, 202 (1974); transl. from *Zh. Anal. Khim.*, **29**, 238 (1974).

29. J. Berthod, A. Cornu, and R. Stefani, *C. R. Acad. Sci. Ser. B*, **285**, 195 (1977).

30. J. Berthod, *Adv. Mass Spectrom.*, **6**, 421 (1974).

31. G. I. Ramendik, *J. Anal. Chem. USSR*, **32**, 1573 (1977); transl. from *Zh. Anal. Khim.*, **32**, 1990 (1977).

32. G. I. Ramendik, V. I. Derzhiev, and Yu. V. Vasyuka, *J. Anal. Chem. USSR*, **34**, 1016 (1979).

33. V. A. Gerasimov, A. I. Saprykin, I. R. Shelpakova, and I. G. Yudelevich, *J. Anal. Chem. USSR*, **33**, 998 (1978); transl. from *Zh. Anal. Khim.*, **33**, 1274 (1978).

34. M. A. Potapov, M. S. Chupakhin, V. I. Shtanov, and V. P. Zlomanov, *J. Anal. Chem. USSR*, **33**, 820 (1978).

35. R. A. Bingham and P. L. Salter, *Anal. Chem.*, **48**, 1735 (1976).

36. H. J. Dietze and S. Becker, *ZFI-Mitt.*, **101**, 5 (1985).

37. H. van Doveren, *Spectrochim. Acta*, **39B**, 1513 (1984).

38. H. J. Dietze, private communication, 1985.

39. E. Watanabe and M. Naito, in K. Ogata and T. Hayakawa, Eds., *Recent Developments in Mass Spectroscopy*, University Park Press, Baltimore, MD, 1970, p. 249.

40. A. E. Banner and B. P. Stimpson, *Vacuum*, **24**, 511 (1974).

41. R. J. Blattner, J. E. Baker, and C. A. Evans, *Anal. Chem.*, **46**, 2171 (1974).

42. J. Franzen, in K. Ogata and T. Hayakawa, Eds., *Recent Developments in Mass Spectroscopy*, University Park Press, Baltimore, MD, 1970, p. 296.

43. W. W. Harrison and C. W. Magee, *Anal. Chem.*, **46**, 461 (1974).

44. J. Mattauch and R. Herzog, *Z. Phys.*, **89**, 786 (1934).

45. L. C. Frees, L. G. Christophorou, H. W. Ellis, and I. Sauers, *Abstr. 28th Annu. Conf. Mass Spectrom. Allied Topics*, New York, 1980, p. 197.

46. S. Auer, Report No. N74-31920, Goddard Space Flight Center, Greenbelt, MD, 1974, p. 12.

47. T. Fujii and K. Fuwa, *Int. J. Mass Spectrom. Ion Phys.*, **33**, 325 (1980).

48. L. S. Dale, I. Liepa, P. S. Rendell, and R. N. Whittem, *Anal. Chem.*, **53**, 2288 (1981).

49. D. L. Donohue, J. A. Carter, and A. Mamantov, *Int. J. Mass Spectrom. Ion Phys.*, **33**, 45 (1980).

50. D. L. Donohue, J. A. Carter, and A. Mamantov, *Int. J. Mass Spectrom. Ion Phys.*, **35**, 243 (1980).

51. G. J. Louter and A. N. Buyserd, *Int. J. Mass Spectrom. Ion Phys.*, **50**, 245 (1983).

52. A. M. Ure, *Trends Anal. Chem.*, **1**, 314 (1982).

53. H. Kishi, *Bull. Chem. Soc. Jap.*, **54**, 703 (1981).

54. H. Kishi, *Bull. Chem. Soc. Jap.*, **54**, 1999 (1981).

55. H. Kishi, *Bull. Chem. Soc. Jap.*, **54**, 2005 (1981).

56. J. F. Jaworski, thesis, Cornell University, Ithaca, NY, 1974, p. 123.

57. J. Verlinden, Ph.D. thesis, Universitaire Instelling Antwerpen, Wilrijk, Belgium, 1984.

58. V. S. Venkatasubramanian, S. Saminathan, and P. T. Rajagopalan, *Int. J. Mass Spectrom. Ion Phys.*, **24**, 207 (1977).

59. J. Burstenbinder, J. Luck, and W. Szacki, *Fresenius Z. Anal. Chem.*, **309**, 325 (1981).

60. S. Becker and H. J. Dietze, *Int. J. Mass Spectrom. Ion Phys.*, **51**, 325 (1983).

61. M. Viczian, I. Cornides, J. Van Puymbroeck, and R. Gijbels, *Int. J. Mass Spectrom. Ion Phys.*, **51**, 77 (1983).

62. J. Van Puymbroeck, R. Gijbels, M. Viczian, and I. Cornides, *Int. J. Mass Spectrom. Ion Phys.*, **56**, 269 (1984).

63. J. Franzen and H. Hintenberger, *Z. Naturforsch.*, **16A**, 535 (1961).

64. H. J. Dietze, S. Becker, I. Opauszky, L. Matus, I. Nyary, and J. Frecska, *ZFI-Mitt.*, **48**, 1 (1981).

65. F. Degreve, *Adv. Mass Spectrom.*, **5**, 529 (1971).

66. J. B. Clegg, I. G. Gale, and E. J. Millett, *Analyst*, **98**, 69 (1973).

67. J. A. J. Jansen and A. W. Witmer, *Fresenius Z. Anal. Chem.*, **309**, 262 (1981).

68. D. L. Dugger and D. W. J. Oblas, *J. Vac. Sci. Technol.*, **17**, 826 (1980).

69. W. W. Harrison and W. A. Mattson, *Anal. Chem.*, **46**, 1979 (1974).

70. A. M. Ure and J. R. Bacon, *Analyst*, **103**, 807 (1978).

71. S. R. Taylor and M. P. Gorton, *Geochim. Cosmochim. Acta*, **41**, 1375 (1977).

72. J. Van Puymbroeck and R. Gijbels, *Bull. Soc. Chim. Belg.*, **87**, 803 (1978).

73. E. Doernenburg, H. Hintenberger, and J. Franzen, *Z. Naturforsch.*, **16A**, 352 (1961).

74. G. Vidal, P. Galmard, and P. Lanuse, *Int. J. Mass Spectrom. Ion Phys.*, **2**, 405 (1969).
75. G. Vidal, *Int. J. Mass Spectrom. Ion Phys.*, **10**, 204 (1974).
76. J. Verlinden, R. Gijbels, H. Silvester, and P. De Bièvre, Commission of the European Communities, Physical Sciences, Report EUR 9653 EN, 1985.
77. L. Morvay and I. Cornides, *Int. J. Mass Spectrom. Ion Processes*, **62**, 263 (1984).
78. V. I. Rakhovsky, *Physical Foundations of the Electric Current Commutation in Vacuum*, Nauka, Moscow, 1970 (in Russian).
79. G. A. Mesyats, *Proc. 10th Int. Conf. Phenomena Ionized Gases*, Oxford, 1971, p. 333.
80. I. N. Slivkov, *Electroisolation and Discharge in Vacuum*, Atomizdat, Moscow, 1972.
81. G. A. Mesyats and D. I. Proskurovsky, *Impulse Electrical Discharge in Vacuum*, Nauka, Novosibirsk, 1984 (in Russian).
82. J. Franzen, in A. J. Ahearn, Ed., *Trace Analysis by Mass Spectrometry*, Academic Press, New York, 1972, p. 11.
83. G. I. Ramendik and V. I. Derzhiev, *J. Anal. Chem. USSR*, **32**(8), 1197 (1977); transl. from *Zh. Anal. Khim.*, **32**(8), 1508 (1977).
84. V. I. Derzhiev, G. I. Ramendik, V. Liebich, and H. Mai, *Int. J. Mass Spectrom. Ion Phys.*, **32**, 345 (1980).
85. H. E. Beske, A. Hurrle, and K. P. Jochum, *Fresenius Z. Anal. Chem.*, **309**, 258 (1981).
86. D. Alpert, D. A. Lee, E. M. Lyman, and H. E. Tomaschke, *J. Vacuum Sci. Technol.*, **1**, 35 (1964).
87. P. A. Chatterton, *Proc. Phys. Soc.*, **88**, 231 (1966).
88. W. P. Dyke and W. W. Dolan, *Adv. Electronics Electron Phys.*, **8**, 180 (1956).
89. J. Verlinden, J. Van Puymbroeck, and R. Gijbels, *Int. J. Mass Spectrom. Ion Phys.*, **47**, 287 (1983).
90. J. Van Puymbroeck, J. Verlinden, K. Swenters, and R. Gijbels, *Talanta*, **31**, 177 (1984).
91. A. J. Ahearn, *J. Appl. Phys.*, **32**, 1197 (1961).
92. T. Kessler, A. G. Sharkey, W. M. Hickam, and G. S. Sweeney, *Appl. Spectrosc.*, **21**, 91 (1967).
93. V. P. Bedrinov, Yu. M. Ukrainskiy, and M. S. Chupakhin, *Zh. Anal. Khim.*, **27**, 1901 (1972).
94. W. P. Dyke, J. K. Trolan, E. E. Martin, and J. P. Barbour, *Phys. Rev.*, **91**, 1043 (1953).
95. I. L. Sokol'skaya and G. N. Fursey, *Radiotech. elektronica*, **7**, 1474 (1962).
96. G. A. Mesyats, *Le Vide*, **143**, 282 (1969).
97. S. P. Bugaev, E. A. Litvinov, G. A. Mesyats, and D. I. Proskurovsky, *Usp. Fiz. Nauk*, **115**, 101 (1975).

98. A. Maitland, *J. Appl. Phys.*, **32**, 2399 (1961).

99. G. Demortier and L. Lamberts, *Int. J. Mass Spectrom. Ion Phys.*, **31**, 135 (1979).

100. K. Suzuki and S. Kobayashi, *J. Phys. D: Appl. Phys.*, **15**, 1227 (1982).

101. W. S. Boyle, P. Kisluk, and L. H. Germer, *J. Appl. Phys.*, **26**, 720 (1955).

102. T. Kesley, *J. Phys. D: Appl. Phys.*, **5**, 569 (1973).

103. R. B. Baksht, N. A. Ratakhin, and B. A. Kablambaev, *Sov. Phys.-Tech. Phys.*, **27**, 1091 (1982); transl. from *Zh. Tekh. Fiz. USSR*, **52**, 1778 (1982).

104. L. Cranberg, *J. Appl. Phys.*, **23**, 518 (1952).

105. D. K. Davies and M. A. Biondi, *J. Appl. Phys.*, **37**, 2969 (1966).

106. D. K. Davies and M. A. Biondi, *J. Appl. Phys.*, **39**, 2979 (1968).

107. D. K. Davies and M. A. Biondi, *J. Appl. Phys.*, **41**, 88 (1970).

108. F. M. Charbonnier, C. J. Bennete, and L. W. Swanson, *J. Appl. Phys.*, **38**, 627 (1967).

109. S. P. Bugaev, R. B. Baksht, E. A. Litvinov, and V. P. Stasiev, *Tech. Vys. Temp.*, **14**, 1145 (1976).

110. R. B. Baksht, N. A. Ratakhin, and M. N. Timofeev, *Pis'ma Zh. Tekh. Fiz.*, **1**, 922 (1975).

111. A. A. Ivanov and L. I. Budakov, in *Voprosy teorii plasmy*, 6th ed., Atomizdat, Moscow, 1972, p. 6 (in Russian).

112. I. Berthod, B. Alexandre, and R. Stefani, *Int. J. Mass Spectrom. Ion Phys.*, **10**, 478, (1972/73).

113. V. I. Derzhiev, V. A. Grechishnikov, G. I. Ramendik, Yu. V. Vasjuta, and V. F. Ivanova, in *Discharges and Electrical Insulation in Vacuum*, Proc. VII Int. Symp., Novosibirsk, USSR, 1976, p. 152.

114. V. A. Galburt, A. E. Zelenin, and G. G. Sikharulidze, *Int. J. Mass Spectrom. Ion Processes*, **55**, 125 (1984).

115. J. Franzen, *Z. Naturforsch.*, **18**, 410 (1963).

116. J. Haemers, N. Van Wassenhove, M. Viczian, I. Cornides, and M. Van Risseghem, unpublished results.

117. M. S. Chupakhin, G. I. Ramendik, V. I. Derzhiev, Yu. G. Tatsii, and M. A. Potapov, *Dokl. Akad. Nauk SSSR*, **210**, 1074 (1973).

118. G. I. Ramendik and V. I. Derzhiev, *J. Anal. Chem. USSR*, **32**, 1204 (1977); transl. from *Zh. Anal. Khim.*, **32**, 1516 (1977).

119. G. I. Ramendik and V. I. Derzhiev, *J. Anal. Chem. USSR*, **34**, 647 (1979); transl. from *Zh. Anal. Khim.*, **34**, 837 (1979).

120. G. I. Ramendik, *Adv. Mass Spectrom.*, **8A**, 408 (1980).

121. G. I. Ramendik, O. I. Kryuchkova, D. A. Tyurin, T. R. Mchedlidze, and M. Sh. Kaviladze, *Int. J. Mass Spectrom. Ion Processes*, **63**, 1 (1985).

122. G. I. Ramendik, *Trace Microprobe Tech.* **2**, 1 (1984).

123. D. I. Proskurovsky and V. P. Rotshtejn, *Izv. Vuzov. Fiz.*, **11**, 142 (1973).

124. A. V. Grechishnikov, V. I. Derzhiev, V. F. Ivanova, G. I. Ramendik, and Yu. A. Surkov, *Prib. Tek. Eksper.*, **5**, 89 (1980) (in Russian).

125. T. F. Ready, *Effects of High-Power Laser Radiation*, Academic, New York, 1971, Ch. 3.

126. S.I. Anisimov, J. A. Imas, G. S. Romanov, and Yu. V. Khodyko, *Influence of High-Power Radiation on Metals*, Nauka, Moscow, 1970, p. 272 (in Russian).

127. G. I. Ramendik, O. I. Krjuchkova, D. A. Tjurin, T. R. Mchedlidze, and M. Sh. Kaviladze, *J. Anal. Chem. USSR*, **38**, 1393 (1983); transl. from *Zh. Anal. Khim.*, **38**, 1749 (1983).

128. V. E. Iljin and S. V. Lebedev, *Zh. Tech. Fiz.*, *USSR*, **32**, 986 (1962) (in Russian).

129. J. Haemers, *Spectrochim. Acta*, **38B**, 1367 (1983).

130. G. I. Ramendik, M. S. Chupakhin, and M. A. Potapov, *Zh. Anal. Khim.*, **30**, 669 (1975) (in Russian).

131. J. Verlinden, K. Swenters, and R. Gijbels, *Spectrochim. Acta*, **39B**, 1573 (1984).

132. J. Verlinden, K. Swenters, and R. Gijbels, *Anal. Chem.*, **57**, 131 (1985).

133. K. Swenters, J. Verlinden, and R. Gijbels, *Adv. Mass Spectrom.*, **1985B**, 1007.

134. O. I. Kryuchkova, V. I. Derzhiev, G. I. Ramendik, N. S. Stroganova, and E. B. Strel'nikova, *Dokl. Akad. Nauk SSSR*, **245**, 1166 (1979) (in Russian).

135. V. I. Derzhiev, A. Yu. Zakharov, and G. I. Ramendik, On the influence of the chemical composition on the kinetics of ionization and recombination in spark and nuclear plasma, Institute of Applied Mathematics, Academy of Sciences, Preprint No 23, 1980, p. 22 (in Russian).

136. R. B. Baksht, A. P. Kudinov, and E. A. Litvinov, *Sov. Phys.-Tech. Phys.*, **18**, 94 (1973); transl. from *Zh. Tekh. Fiz. USSR*, **43**, 146 (1973).

137. L. P. Mix, J. G. Kelly, G. W. Kuswa, *J. Vac. Sci. Technol.*, **10**, 951 (1973).

138. V. I. Derzhiev, V. Liebich, G. I. Ramendik, Yu. V. Vasjuta, A. V. Grechishnikov, and V. F. Ivanova, *Sov. Phys.-Tech. Phys.*, **26**, 428 (1981); transl. from *Zh. Tekh. Fiz.*, *USSR*, **51**, 719 (1981).

139. G. I. Ramendik and A. G. Blokin, *Zh. Tekh. Fiz. USSR*, in press.

140. V. I. Derzhiev and G. I. Ramendik, *Sov. Phys.-Tech. Phys.*, **23**, 291 (1978); transl. from *Zh. Tekh. Fiz.*, *USSR*, **48**, 312 (1978).

141. J. Haemers, *Spectrochim. Acta*, **38B**, 859 (1983).

142. K. Swenters, Ph.D. thesis, Universitaire Instelling Antwerpen-Wilrijk, Belgium, 1986.

143. V. I. Derzhiev, A. Yu. Zakharov, and G. I. Ramendik, *Sov. Phys. Tech. Phys.*, **23**, 1068 (1978); transl. from *Zh. Tekh. Fiz.*, *USSR*, **48**, 1877 (1978).

144. I. Cornides and T. Gal, *High Temp. Sci.*, **10**, 171 (1978).

145. G. I. Ramendik, O. I. Kryuchkova, V. I. Derzhiev, N. S. Stroganova, and E. B. Strel'nikova, *Dokl. Akad. Nauk SSSR*, **245**, 1166 (1979).

146. W. L. Grady and M. M. Bursey, *Int. J. Mass Spectrom. Ion Processes*, **56**, 161 (1984).

147. E. S. Ackerman, W. L. Grady, and M. M. Bursey, *Int. J. Mass Spectrom. Ion Processes*, **55**, 275 (1984).

148. G. I. Ramendik, V. I. Derzhiev, Yu. A. Surkov, V. E. Ivanova, A. V. Grechishnikov, A. A. Sysoev, V. A. Oleinikov, and V. A. Aleksandrov, *Int. J. Mass Spectrom. Ion Phys.*, **37**, 331 (1981).

149. J. Van Puymbroeck and R. Gijbels, *Arbeitstreffen Funkenmassenspektrometrie*, Wilrijk, Belgium, March 1984.

150. J. R. Woolston and R. E. Honig, *Rev. Sci Instr.*, **35**, 69 (1964).

151. J. Berthod, A. M. Adreani, and R. Stefani, *Int. J. Mass Spectrom. Ion Phys.*, **27**, 305 (1978).

152. V. M. Gladskoi, V. N. Nevolin, L. P. Spak, and V. I. Belousov, *Sov. Phys.-Tech. Phys.*, **23**, 786 (1978); transl. from *Zh. Tekh. Fiz.*, *USSR*, **48**, 1394 (1978).

153. V. A. Gerasimov, I. R. Shelpakova, and N. S. Rudaya, *J. Anal. Chem. USSR*, **34**, 13 (1979); transl. from *Zh. Anal. Khim.*, **34**, 20 (1979).

154. V. A. Gerasimov, M. S. Chupakhin, and I. R. Shelpakova, *J. Anal. Chem. USSR*, **35**, 138 (1980); transl. from *Zh. Anal. Khim.*, **35**, 224 (1980).

155. I. R. Shelpakova, V. A. Gerasimov, and I. G. Yudelevich, *J. Anal. Chem. USSR*, **36**, 1655 (1981); transl. from *Zh. Anal. Khim.*, **36**, 2299 (1981).

156. L. Vos and R. Van Grieken, *Int. J. Mass Spectrom. Ion Phys.*, **59**, 221 (1984).

157. G. Mathieu, J. Lamberts, M. Thomas, and G. Demortier, *Int. J. Mass Spectrom. Ion Phys.*, **38**, 35 (1981).

158. V. I. Belousov, *Zh. Anal. Khim.*, in press.

159. A. J. Ahearn, *Proc. Xth Coll. Spectrosc. Int.*, Washington, DC, 1963, p. 769.

160. E. Michiels, thesis, Universitaire Instelling Antwerpen, Wilrijk, 1981 (in Dutch).

161. L. Vos, Ph.D. thesis, Universitaire Instelling Antwerpen, Wilrijk, 1984 (in Dutch).

162. A. Verbueken, E. Michiels, and R. Van Grieken, *Fresenius Z. Anal. Chem.*, **309**, 300 (1981).

163. G. H. Morrison and A. M. Rothenberg, *Anal. Chem.*, **44**, 515 (1972).

164. J. Verlinden, R. Vlaeminck, F. Adams, and R. Gijbels, in W. Katz and P. Williams Eds., *Applied Materials Characterization*, Materials Research Society, Pittsburgh, PA, 1985, p. 331.

165. H. Farrar IV, in A. J. Ahearn, Ed., *Trace Analysis by Mass Spectrometry*, Academic, New York, 1972, p. 239.

166. K. D. Schuy and J. Franzen, *Fresenius Z. Anal. Chem.*, **225**, 260 (1967).

167. B. P. Datta, V. A. Raman, V. L. Sant, P. A. Ramasubramanian, S. K. Agarwal, and H. C. Jain, *Int. J. Mass Spectrom. Ion Processes*, **64**, 139 (1985).

168. C. W. Hull, *Abstr. Tenth Mass Spectrom. Conf., New Orleans*, 1966, p. 404.

169. F. D. Leipziger, *Anal. Chem.*, **37**, 171 (1965).

170. I. Cornides, in L. Niinisto, Ed., *Reviews on Analytical Chemistry*, Akademiay Kiado, Budapest, 1982, p. 105.

171. J. M. McCrea, *Int. J. Mass Spectrom. Ion Phys.*, **5**, 83 (1970).

172. E. Van Hoye, R. Gijbels, and F. Adams, *Adv. Mass Spectrom.*, **8A**, 357 (1980).

173. M. Ito and K. Yanagihara, *Bunseki Kagaku*, **22**, 10 (1979).

174. N. H. W. Addink, *Fresenius Z. Anal. Chem.*, **206**, 81 (1964).

175. J. Kai and M. Miki, *Mass Spectrosc. (Tokyo)*, **12**, 81 (1964).

176. G. Vidal, P. Galmard, and P. Lanusse, *Methodes Phys. Anal. GAMS*, **4**, 404 (1968).

177. R. H. Honig, *Adv. Mass Spectrom., Inst. Pet. London*, **3**, 101 (1966).

178. W. D. Bratton and C. H. Wood, *Appl. Spectrosc.* **24**, 509 (1970).

179. M. Ito, S. Sato, and K. Yanagihara, *Anal. Chim. Acta*, **120**, 217 (1980).

180. J. F. Jaworski and G. H. Morrison *Anal. Chem.*, **46**, 2080 (1974).

181. D. A. Griffith, R. J. Conzemius, and H. J. Svec, *Talanta*, **18**, 665 (1971).

182. G. Ehrlich, U. Stahlberg, and H. Scholze, *Spectrochim. Acta*, **37B**, 45 (1982).

183. E. I. Hamilton and M. J. Minski, *Int. J. Mass Spectrom. Ion Phys.*, **10**, 77 (1972).

184. J. C. Franklin, thesis, University of Tennessee, Knoxville, 1971, p. 121.

185. J. Kai, Y. Ogata, K. Ohi, and M. Watanabe, in K. Ogata and T. Hayakawa, Eds., *Recent Developments in Mass Spectroscopy*, University Park Press, Baltimore, MD, 1970, p. 314.

186. J. Franzen and K. D. Schuy, *Adv. Mass Spectrom.*, **4**, 499 (1968).

187. C. W. Magee and W. W. Harrison, *Anal. Chem.*, **45**, 852 (1973).

188. K. Yanagihara, S. Sako, S. Oda, and H. Kamado, *Anal. Chim. Acta*, **98**, 307 (1978).

189. N. Yamaguchi, R. Suzuki, and O. Kammori, *Bunseki Kagaku*, **18**, 3 (1969).

190. E. Van Hoye, F. Adams, and R. Gijbels, *Int. J. Mass Spectrom. Ion Phys.*, **30**, 75 (1979).

191. I. R. Shelpakova, A. I. Saprykin, V. A. Gerasimov, and I. G. Yudelevich, *J. Anal. Chem. USSR*, **35**, 413 (1980); transl. from *Zh. Anal. Khim.*, **35**, 629 (1980).

192. W. Vieth, *Int. J. Mass. Spectrom. Ion Processes*, in press.

193. R. J. Conzemius and J. M. Capellen, *Int. J. Mass Spectrom. Ion Phys.*, **34**, 197 (1980).

194. G. Herberg, P. Holler, and A. Koster-Pflugmacher, *Spectrochim. Acta*, **23B**, 363 (1968).

195. S. Brewer and J. P. Walters, *Anal. Chem.*, **41**, 1980 (1969).

196. A. Van Oostrom and L. Augustus, *Vacuum*, **32**, 127 (1982).

197. P. Holler, *Spectrochim. Acta*, **23B**, 1 (1967).

198. L. S. Palatnik and A. A. Levchenko, *Sov. Phys.-Tech. Phys.*, **10**, 680 (1965).

199. A. Cornu, *Adv. Mass Spectrom.*, **4**, 401 (1968).

200. A. J. Ahearn and D. L. Malm, *Appl. Spectrosc.*, **20**, 411 (1966).

201. H. Mai, *J. Sci. Instrum.*, **42**, 339 (1965).

202. N. H. W. Addink, *Nature*, **211**, 1168 (1966).

203. A. J. Ahearn and P. J. Paulsen, *Anal. Chem.*, **40**, 75A (1968).

204. R. E. Honig, J. R. Woolston, and D. A. Kramer, *Rev. Sci. Instrum.*, **38**, 1703 (1967).

205. J. Franzen, K. H. Maurer, and K. D. Schuy, *Z. Naturforsch.*, **A21**, 37 (1966).

206. P. Kennicott, *Anal. Chem.*, **38**, 633 (1966).

207. A. Cavard, *Adv. Mass Spectrom.*, **4**, 419 (1968).

208. X. D. Liu, J. Verlinden, F. Adams, and E. Adriaenssens, *Anal. Chim. Acta*, **180**, 341 (1986).

209. J. I. Masters, Report No. NSF/RA-780507, Ionomet Co., Waban, MA, 1978, p. 36.

210. M. S. Chupakhin and G. I. Ramendik, *J. Anal. Chem. USSR*, **26**, 1241 (1972); transl. from *Zh. Anal. Khim.*, **26**, 1390 (1971).

211. M. Gauneau, *Spectra 2000*, **7**(53), 52 (1979).

212. A. Hurrle and J. Dietl, *Fresenius Z. Anal. Chem.*, **309**, 277 (1981).

213. A. M. Andreani, J. C. Brun, J. P. Mermoud, and R. Stefani, *Methodes Phys. Anal. (GAMS)*, **7**, 269 (1971).

214. L. N. Shabanova, I. R. Shelpakova, and I. G. Yudelevich, *Izv. Sib. Otdel. Akad. Nauk SSSR, Ser. Khim. Nauk*, **6**, 126 (1979).

215. Yu. A. Karpov and I. P. Alimarin, *J. Anal. Chem. USSR*, **34**, 1085 (1979); transl. from *Zh. Anal. Khim.*, **34**, 1402 (1979).

216. P. F. Kane, *Chem. Tech.*, **1971**, 532.

217. R. Brown, R. D. Craig, and R. M. Elliott, *Adv. Mass Spectrom.*, **2**, 141 (1963).

218. K. Swenters, unpublished results, 1986.

219. J. B. Clegg and E. J. Millett, *Philips Tech. Rev.*, **34**, 344 (1974).

220. C. W. Hull, *Int. J. Mass Spectrom. Ion Phys.*, **3**, 293 (1969).

221. J. Franzen and K. D. Schuy, *Fresenius Z. Anal. Chem.*, **225**, 295 (1967).

222. A. Pilate and F. Adams, *Fresenius Z. Anal. Chem.*, **309**, 295 (1981).

223. C. O. Ingamells and P. Switzer, *Talanta*, **20**, 547 (1973).

224. R. Gijbels, J. Verlinden, and K. Swenters, *JEOL News*, **20A**, 5 (1984).

225. E. Van Hoye, R. Gijbels, and F. Adams, *Talanta*, **23**, 369 (1976).

226. K. P. Jochum and M. Seufert, *Geol. Rundschau*, **69**, 997 (1980).

227. K. P. Jochum, M. Seufert, and S. Best, *Fresenius Z. Anal. Chem.*, **309**, 308 (1981).

228. J. Van Puymbroeck and R. Gijbels, *Fresenius Z. Anal. Chem.*, **309**, 312 (1981).

229. K. P. Jochum, A. W. Hofmann, E. Ito, H. M. Seufert, and W. M. White, *Nature*, **306**, 431 (1983).

230. J. Van Puymbroeck, J. Verlinden, and R. Gijbels, *J. Microsc. Spectrosc. Electron.*, **7**, 303 (1982).

231. R. K. Willardson and A. J. Socha, Aerospace Res. Lab. Report No. ARL-65-130, 1965.

232. B. B. Goshgarian and A. V. Jensen, *Abstr. 12th Annual Conf. Mass Spectrom. and Allied Topics*, Montreal, 1964, p. 350.

233. J. B. Clegg, F. Grainger, and I. G. Gale, *J. Mater. Sci.*, **15**, 747 (1980).

234. M. Gauneau, A. Rupert, M. Minier, O. Regreny, and R. Coquille, *Anal. Chim. Acta*, **135**, 193 (1982).

235. W. H. Christie, J. A. Carter, R. E. Eby, and L. Landau, *Anal. Lett.*, **12**, 1123 (1979).

236. L. Matus, I. Opauszky, I. Nyary, and E. Pasztor, *Fresenius Z. Anal. Chem.*, **309**, 316 (1981).

237. J. J. M. De Goeij, J. P. W. Houtman, and J. B. W. Kanij, *Radiochim. Acta*, **5**, 117 (1966).

238. J. M. McCrea, in P. F. Kane and G. B. Larrabee, Eds., *Characterization of Solid Surfaces*, Plenum, New York, 1976, p. 577.

239. F. Adams, J. Verlinden, and R. Gijbels, *Proceedings II, 6th Intern. Symp. High-Purity Materials in Science and Technology*, Dresden, 1985, p. 1.

240. R. A. Bingham and P. G. T. Vossen, *Int. J. Mass Spectrom. Ion Phys.*, **14**, 259 (1974).

241. R. Brown, R. D. Craig, J. A. James, and C. M. Wilson, in M. S. Brooks and J. F. Kennedy, Eds., *Conference on Ultrapurification of Semiconductor Materials*, MacMillan, New York 1962, p. 279.

242. J. M. Thomas, C. Roscoe, G. G. Cookson, T. H. Owen, and R. Tushingham, *Carbon*, **4**, 457 (1966).

243. R. Stefani, A. Cornu, R. Bourguillot, A. M. Robin, and j. C. Brun, *Chim. Anal.*, **48**(5), 253 (1966).

244. R. Brown, W. J. Richardson, and H. W. Somerford, ASTM E-14, *Ann. Conf. Mass Spectrom. Allied Topics, 15th*, Paper No. 49, p. 157.

245. J. Roboz, *Mass Spectrometry*, Interscience, New York, 1968, p. 398.

246. P. Chastagner, *Anal. Chem.*, **41**(6), 796 (1969).

247. G. H. Morrison and R. K. Skogerboe, in G. H. Morrison, Ed., *Trace Analysis, Physical Methods*, Interscience, New York, 1965, p. 16.

248. J. A. J. Jansen and A. W. Witmer, *Fresenius Z. Anal. Chem.*, **309**, 305 (1981).

249. F. Berkey and G. H. Morrison, Mater. Sci. Center Rep. 1129, Cornell Univ., Ithaca, NY, 1969.

250. J. Clegg, E. J. Millett, and J. A. Roberts, *Anal. Chem.*, **42**(7), 713 (1970).

251. M. S. Chupakhin, G. I. Ramendik, and O. I. Kryuchkova, *Zh. Anal. Khim. USSR*, **24**, 965 (1969) (in Russian).

252. K. Swenters, J. Verlinden, and R. Gijbels, *Spectrochim. Acta*, **40B**, 769 (1985).

253. M. S. Chupakhin, G. I. Ramendik, and E. V. Venitsianov, *Zh. Anal. Khim. USSR*, **25**, 855 (1970) (in Russian).

254. M. S. Chupakhin, G. I. Ramendik, and A. N. Javrjan, *Zh. Anal. Khim. USSR*, **25**, 1301 (1970) (in Russian).

255. V. P. Bedrinov and V. I. Belousov, *J. Anal. Chem. USSR*, **31**, 1709 (1976); transl. from *Zh. Anal. Khim.*, **31**, 2315 (1976).

256. F. A. Schmidt, R. J. Conzemius, D. N. Carlson, and H. J. Svec, *Anal. Chem.*, **46**, 810 (1974).

257. W. N. Weins and O. N. Carlson, *J. Less-Common Met.*, **66**, 99 (1979).

258. J. Verlinden and R. Gijbels, *Adv. Mass Spectrom.*, **8A**, 485 (1980).

259. K. Swenters, J. Verlinden, and R. Gijbels, *Spectrochim. Acta*, **39B**, 1577 (1984).

260. I. G. Yudelevich and I. R. Shelpakova, *Mikrochim. Acta* I 547 (1978).

261. A. N. Dorokhov, I. R. Shelpakova, I. G. Yudelevich, and S. T. Khristianova, *Izv. Sib. Otdel. Akad. Nauk USSR, Ser. Khim. Nauk*, **2**, 61 (1977) (in Russian).

262. M. J. Vasile and D. L. Malm, *Anal. Chem.*, **44**(4), 650 (1972).

263. V. Liebich and H. Mai, *Adv. Mass Spectrom.*, **6**, 655 (1974).

264. I. R. Shelpakova, A. I. Saprykin, T. A. Chanysheva, and I. G. Yudelevich, *J. Anal. Chem. USSR*, **38**, 439 (1983); transl. from *Zh. Anal. Khim.*, **38**, 581 (1983).

265. V. Liebich, in *High Purity Materials in Science and Technology*, 5th Int. Symp., Poster Abstracts, Akademie der Wissenschaften der DDR, Zentralinstitut für Festkörperphysik und Werkstofforschung Dresden, 1980, p. 194.

266. H. A. Laitinen, *Anal. Chem.*, **44**(14), 2305 (1973).

267. G. I. Ramendik, A. G. Blokin, V. G. Samoilovich, V. L. Volkov, A. Yu. Khromov, and V. S. Fainberg, *Zh. Tekh. Fiz. USSR*, **55**, 1985 (in Russian).

CHAPTER

3

GLOW DISCHARGE MASS SPECTROMETRY

W. W. HARRISON

University of Virginia, Charlottesville, Virginia

1. INTRODUCTION

1.1. Development of the Glow Discharge Ion Source

The glow discharge has been known for well over 50 years as an ion source in mass spectrometry. Gas discharges were a natural choice for early mass spectrometric studies, being simple in operation and yet yielding intense ion currents. They proved to be particularly valuable during the early

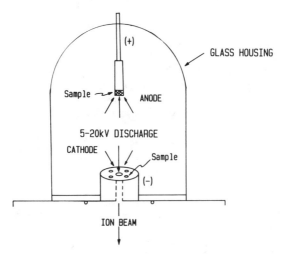

Figure 3.1. Early glow discharge ion source showing two alternative sample introduction schemes.

elemental isotopic investigations (1, 2) in the 1920s and 1930s. In fact, this type of source was the standard in the field until the development of the vacuum spark (3). The early popularity of the glow discharge was followed by a rapid and extended decline into relative obscurity as a source for mass spectrometry, aside from certain fundamental studies (e.g., ionization and collision processes in gases), but it has found renewed interest in recent years. Even as the development of the spark source led analysts away from the glow discharge, the spark's eventual limitations caused a reexamination of the gas discharge, a turn of events that has now led to the commercialization of the glow discharge as an ion source for elemental analysis.

This review will emphasize the present status and potential of glow discharge mass spectrometry, but it is worthwhile to recognize that the principles involved in the first glow discharge ion sources form the basis of today's modified and updated successful sources. A typical example of the former is shown in Fig. 3.1. This was essentially a high-voltage glow discharge designed to produce isotopes of all the known elements (2). The 5–20 kV neon discharge produced "cathode rays" to bombard an anode that was either composed of the desired element or covered with a graphite paste containing a compound of the element. The high-energy electron beam vaporized a sufficient quantity of the sample to permit ionization and isotopic determinations. Alternatively, "accelerated anode rays" could be used to sputter a metal cathode or material deposited into the cathode. In either approach, a fraction of the vaporized atoms was then ionized in the discharge and extracted through the cathode orifice. Anode vaporization appeared to be the favored approach, even to the extent of diluting the discharge gas with CO_2 to reduce cathodic sputtering, which could melt the cathode under these intense conditions. Some remarkably good mass spectra were obtained. Nevertheless, the practitioners did not seem overly fond of the technique. Aston noted (1):

After over thirty-five year's experience . . . , the writer knows no method of obtaining, still less of reproducing, first-class results with any certainty. . . . the extreme complexity of the mechanism of this form of discharge renders its use much more of an art rather than a science. Both its phenomena and its limitations abound in the curious and the unexpected.

While at certain times today's glow discharge practitioners might nod in agreement with Aston, for the most part this source has overcome the earlier design problems.

Mass spectroscopists temporarily lost interest in discharges, but the optical emission literature shows sustained use and their development both as line sources and as direct analytical emission sources (4–7). The hollow cathode mode of the glow discharge was of particular interest. Work in the Soviet Union (8–10) has continued to demonstrate that high-sensitivity analyses are possible, and the development of the Grimm glow discharge (11–13) has led to many applications. These successes in optical spectroscopy showed that stable, reproducible glow discharge devices were attainable. The presence of large populations of excited sample atoms also suggested the likely existence of ions for mass spectrometric sampling. Beyond this, strong ion emission lines were detected in the optical sources.

Coburn and coworkers (14–16) brought the glow discharge source to the attention of mass spectroscopists as an analytical source for the analysis of solids using both dc and rf discharge modes. Other forms of the discharge were investigated by Oechsner and Gerhard (17, 18), who worked at low pressures (10^{-4}–10^{-5} torr) with a high-frequency source for elemental analysis of solids, while Harrison and coworkers (19–21) developed the hollow cathode source for both bulk solids and solution residue analysis. Evans and coworkers (22, 23) have also studied the analytical capabilities of the hollow cathode discharge. Hecq et al. used a conventional glow discharge to investigate reactive sputtering processes (24) and ionization phenomena (25). Over recent years various ion source modifications and applications (26–30) have demonstrated the versatility and analytical potential of the technique. This has been reinforced by the appearance of a commercial glow discharge mass spectrometer (31), a final step in the "arrival" of any new analytical method.

This chapter is intended as an overview of the present status of glow discharge mass spectrometry, particularly its analytical role, and is not by any means an exhaustive review of glow discharges in general. Two recent reviews (32, 33) treat hollow cathode devices in considerable depth, more oriented toward emission studies where the greater number of applications have been found. Other reviews have appeared (26, 34), and a number of useful books and monographs (35–37) have been devoted to the fundamental aspects of the glow discharge.

The major commercial application of glow discharges is in process control and the monitoring of sputter etching and deposition techniques (38) in the thin-films industry. A review of the literature finds these reports dominating the citations. Increasingly, references to glow discharges in conjunction with gas lasers (39) are also encountered. In addition, many studies have explored various fundamental aspects of glow discharge ionization (40–42). This review will not include these areas, treating instead only the use of glow discharge mass spectrometry for elemental analysis or related applications.

1.2.   Role of Glow Discharge Ion Source in Elemental Analysis

Glow discharge mass spectrometry has its greatest potential application in the elemental analysis of solids. This type of sample has been traditionally analyzed by arc or spark optical emission (43) or spark source mass spectrometry (44). More recently, the inductively coupled plasma has been sufficiently advantageous for elemental analysis to justify the necessary sample-dissolution steps, but the more convenient direct analysis of the solid would be preferable if comparable analytical results were obtainable. The arc and spark emission techniques still find considerable use, especially in the metals industries, where long-established methodology is drawn upon for routine analysis. Spark source mass spectrometry won a certain deserved popularity beginning in the 1960s due to such advantages as parts-per-billion sensitivity, generally uniform sensitivity, simple spectra, and its applicability to both metals and nonmetals. The cost, experimental complexity, and quantitative uncertainty of the spark source prevented its widespread application and created a limited market that eventually could not sustain itself. Only one company continues to make spark source mass spectrometers, and many of those still in operation are becoming aged, but the technique still finds many significant applications and will likely continue in importance for some time. Spark source masss spectrometry has enough inherent advantages over optical emission to encourage the development of a suitable successor, one that would need to exhibit the advantages of the spark source without being constrained by its significant limitations. The renewed interest in the glow discharge has resulted from such considerations. It is a simple source that produces a stable and intense ion beam of the analytical sample while maintaining the described spark source advantages.

Solids analysis by mass spectrometry persists as an important application. Excellent analytical techniques are available for the elemental analysis of solution samples, but direct solids analysis has not kept pace. The interest in this subject is seen in the numerous attempts to adapt solution techniques, such as the ICP, to solids analysis (45). Glow discharge mass spectrometry offers promise as filling the role of a general-purpose multielement method for solids analysis.

2. FUNDAMENTALS

2.1. The Glow Discharge

The glow discharge is one form of a plasma, a partially ionized gas consisting of approximately equal concentrations of positive and negative charges plus large numbers of neutral species. In its simplest form, the glow discharge consists of a cathode and anode immersed in a low-pressure (~0.1–10 torr) gas medium. Application of an electric field across the electrodes causes breakdown of the gas (usually one of the rare gases) and the acceleration of electrons and positive ions toward the oppositely biased electrodes. Reproduced in many monographs (35) is an illustration showing up to nine distinct zones produced in the resultant discharge plasma. Ion sources used in analytical spectroscopy are better represented by the simpler model shown in Fig. 3.2 where only two plasma regions are of significance. A thin layer adjacent to the cathode exhibits a high positive-ion density and shows relatively little optical emission. This region is known as the cathode dark space (CDS) and accounts for nearly all of the voltage drop (cathode fall) across the discharge due to the high positive space charge. In the large negative glow (NG) region, which fills much of the volume in small analytical sources, the greater part of the current is carried by electrons, many of which have sufficient energy to create sustaining ionization and to excite discharge gas atoms to emit the characteristic glow. A Faraday dark space may be seen near the anode, depending upon cathode–anode separation. Discharge gas ions formed at the CDS/NG interface are accelerated to the cathode surface, where upon collision an assortment of secondary particles is released, including elec-

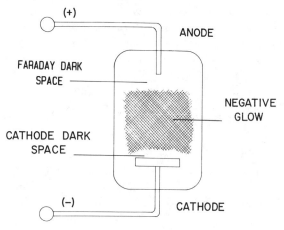

Figure 3.2. Basic depiction of a glow discharge, including the cathode dark space and negative glow.

trons to help sustain the discharge. Except in the CDS, the discharge is essentially field-free with ambipolar diffusion of ions and electrons to confining surfaces.

The glow discharge can operate in the normal or abnormal mode. In a normal discharge, the voltage does not increase with current because the glow area does not completely cover the cathode surface. Until all the surface is covered the current density remains constant, so although the current rises the driving voltage remains constant. Once the cathode area is completely covered, any further current increase requires a higher voltage, since the current density now must rise. Small surface area analytical electrodes usually operate in the abnormal mode. The voltages encountered in glow discharges may range from a few hundred volts to thousands of volts, depending mainly on discharge pressure; voltage needs increase sharply as pressure decreases. Argon and neon are probably used more than any other gases as the discharge medium, although the other rare gases and, indeed, many other gases will suffice. Reactive glow discharges may employ air, oxygen, or other, more specialized gases (38) designed to achieve an optimum environment for the application at hand.

The hollow cathode discharge (HCD) is a specialized form of glow discharge and one that is important in analytical spectroscopy because of its unique properties. The descriptive name of this discharge mode arises from a cathode that has been drilled out to form a hollow, cylindrical cavity closed at one end as shown in Fig. 3.3. The hollow cathode effect

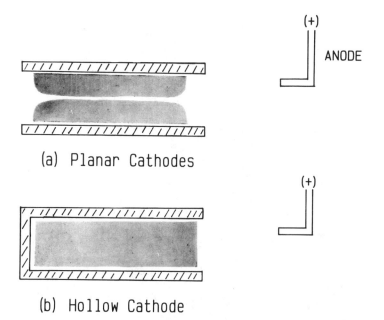

(a) Planar Cathodes

(b) Hollow Cathode

Figure 3.3. Hollow cathode discharge phenomenon. (*a*) Two negative glows from planar cathodes are brought together to achieve a coalescence as shown in (*b*), a typical hollow cathode discharge.

can be visualized in Fig. 3.3*a* as a glow discharge with two parallel cathode plates being brought sufficiently close to each other that the two cathode glow regions overlap and begin to coalesce. An increase of several orders of magnitude in current density is observed compared to a conventional glow discharge at the same conditions. Of significance to the analyst, a large increase in excitation and ionization is observed within the hollow cathode cavity.

2.2. Glow Discharge Sputtering

Sputtering is the phenomenon that makes a glow discharge so useful in analytical spectrometry, providing the means of obtaining directly from a solid sample an atomic population for subsequent excitation and ionization. As shown in Fig. 3.4*a*, sputtering involves directing an energetic particle onto a surface where upon collision it transfers its kinetic energy in a series of lattice collisions, likened sometimes to three-dimensional billiards. Atoms near the surface can receive sufficient energy to overcome the lattice binding and be ejected, normally as neutral atoms (46)

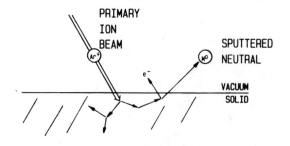

a) Ion Beam Sputtering

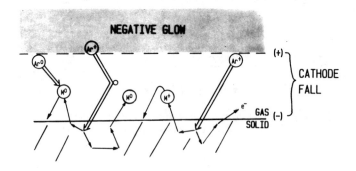

b) Glow Discharge Sputtering

Figure 3.4. Comparison of sputtering processes. (*a*) A directed ion beam in a vacuum environment; (*b*) glow discharge sputtering in a gaseous environment.

with average energies in the 5–15 eV range. The bombarding projectiles are normally ions, since they can be easily accelerated by electric fields, although fast neutral particles (47) are also effective. The sputter yield, the number of ejected atoms per bombarding ion, depends upon several factors, most critically the mass and energy of the incoming ion. The lattice structure of the sputter target and the electronic configuration of the target atoms will also affect yield. Noble gas ions are routinely used as sputter gases, with argon and neon most commonly encountered. The sputter process has some threshold value in the tens of electronvolts and shows an increase in sputter yield with ion energy up to the low thousands of electronvolts (keV) (48). A significant population of atoms can be obtained in the 100–500 eV range. Ion guns allow control of ion flux, energy, and angle of incidence for maximum control of physical sputtering.

Sputtering in a glow discharge relies upon the same fundamental processes just described, but the operational differences are many—simpler in some aspects, more complex in others. As shown in Fig. 3.4b, the glow discharge serves both to form the required bombarding ions in the negative glow and to accelerate them across the cathode fall onto the sample surface. A simple two-electrode discharge powered by an inexpensive low-voltage supply suffices to yield generous atom populations. However, the simplicity is somewhat deceptive in terms of the concomitant complications the glow discharge can bring. In contrast to the high-vacuum controlled sputtering mode shown in Fig. 3.4a, the glow discharge setup depicted in Fig. 3.4b operates at relatively high pressure (e.g., 0.1–10 torr) in a very "busy" collisional environment. Gas ions near the edge of the negative glow are accelerated across the cathode fall and collide with the cathode, causing sputter release of surface atoms, ions, and electrons. The energies of the bombarding ions are far from uniform, however, because of collisions (charge-exchange reactions) in the dark space. Given the short mean free path at 1 torr, few ions negotiate the cathode fall without collisions. The average energy of the ions striking the cathode is reported to be much smaller than the discharge voltage (49), perhaps roughly 100–200 eV for a 500–600 V cathode fall potential, but this is sufficient to well exceed the sputter threshold. In addition to the noble gas ions, the cathode is bombarded by fast neutrals arising from charge exchange in the cathode fall. Some self-sputtering also occurs as secondary ions or ions formed in the cathode fall are returned to the cathode.

The glow discharge sputter process releases the same type of secondary particles as those observed in ion beam sputtering, but they are subjected to a very different environment upon release. Instead of a high-vacuum, field-free region, glow discharge sputtered particles are perturbed by short mean free paths and an intense electric field. Secondary electrons and sputtered negative ions are accelerated away from the cathode; secondary positive ions are attracted back to the cathode by the cathode fall potential (Fig. 3.4b). The neutral sputtered atoms ricochet their way through the dark space into the negative glow, although significant collisional redeposition occurs. Thus, a true sputter yield is difficult to determine, but a net sputter yield may be calculated in which the net cathode weight loss is compared to an integrated discharge current. Impurities in the discharge gas can severely alter sputter rates. Water vapor will dissociate to form H^+, which then carries a significant fraction of the ion current but causes no effective sputter release due to the small mass of the ion.

2.3. Ionization in the Glow Discharge

We have seen that glow discharge sputtering can introduce into the plasma a representative population of cathode (sample) atoms. To permit use of

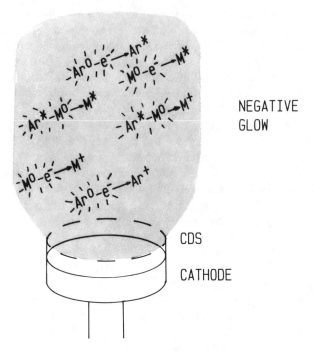

NEGATIVE
GLOW

CDS

CATHODE

Figure 3.5. Typical ionization reactions in the negative glow region. CDS = cathode dark space.

this device as an analytical source for mass spectrometry, the discharge must then ionize a fraction of the sputtered atoms for elemental analysis. Figure 3.5 shows a representation of the glow discharge and the most significant excitation and ionization processes that occur. The discharge must, of course, sustain itself by producing enough charge carriers to make up for the discharge loss processes such as recombination and diffusion to container walls. While it is true that the bulk of the ionization processes in a glow discharge involve the noble gas species, our interest will focus more on the much smaller population of sputtered atoms and their ionization mechanisms.

At the pressures used in glow discharges, collisional processes determine the observed excited and ionized states. Secondary electrons are released by noble gas ions striking the cathode. While some ionization undoubtedly occurs in the cathode dark space (CDS, Fig. 3.5), the electrons are rapidly accelerated and, owing to their small cross section at these energies, have only a small probability of ionizing either free gas atoms or sputtered atoms in the thin sheath. These fast electrons may

Table 3.1. Ionization Processes in the Glow Discharge

I. Primary Ionization Processes[a]
  A. Electron impact
     $M^0 + e^- \to M^+ + 2e^-$
  B. Penning ionization
     $M^0 + Ar^{m*} \to M^+ + Ar^0 + e^-$
II. Secondary Ionization Processes
  A. Charge transfer
     1. Nonsymmetric
        $Ar^+ + M^0 \to M^+ + Ar^0$
     2. Symmetric (resonance)
        $A^+ \text{ (fast)} + A^0 \text{ (slow)} \to A^+ \text{ (slow)} + A^0 \text{ (fast)}$
     3. Dissociative
        $Ar^+ + MO \to M^+ + O + Ar^0$
  B. Associative ionization
     $Ar^{m*} + M \to ArM^+ + e^-$
  C. Photoionization
     $M^* + h\nu \to M^+ + e^-$
  D. Cumulative ionization
     $M^0 + e^- \to M^* + e^- \to M^+ + 2e^-$

[a] M^0 = sputtered neutral; Ar^{m*} = metastable Ar.

even traverse the negative glow and strike the anode without an intervening collision. However, many electrons do undergo collisions in the negative glow, exciting and ionizing both noble gas atoms and sputtered atoms, creating electrons of lower energy for further interaction. The negative glow will contain electron densities of up to 10^{14} cm^{-3} (50) with energies ranging from fast to thermal. Electron impact is likely the major ionization mode in the glow discharge. Metastable atoms, with their long lifetimes, may also play a significant role in the ionization (51) of sputtered neutrals. Argon, for example, has a metastable energy of 11.55 eV, a value exceeding nearly all of the elemental first ionization potentials. Metastable densities of 10^{11} cm^{-3} have been reported in glow discharges, and metastable atoms were identified as the dominant ionization species in some configurations (15), although others have calculated (37) a more modest contribution.

Ionization of sputtered cathode atoms thus occurs mainly in the negative glow by electron impact and Penning ionization steps. Table 3.1 shows other possible ionization contributors, but these are considered to be secondary factors. Ionization of sputtered atoms by collision with ions is not considered important (37) in the essentially field-free negative glow but could be more significant in the cathode dark space, where the ions

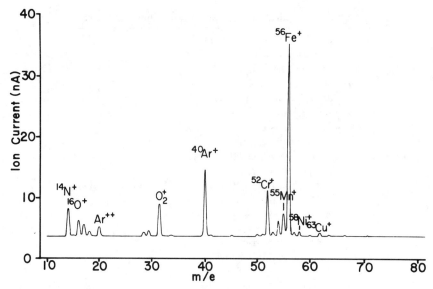

Figure 3.6. Representative mass spectrum from a stainless steel cathode in an argon glow discharge.

are accelerated by the fall potential. All the ionization processes combined still produce only a low net ionization in the discharge, estimated at 1% or less (52). Attempts have been made to utilize the large ground-state atomic population by introducing supplementary ionization means, including an electron-emitting filament (53), secondary electrode discharges (54), and tunable dye lasers (55). In these cases, the glow discharge serves more as an atom source than an ionization source.

A typical mass spectrum taken with a glow discharge source is shown in Fig. 3.6. Dominating the ion spectrum are isotopes of the sputtered cathode atoms and the noble gas, usually argon. Ar^+, Ar^{2+}, and Ar_2^+ might typically appear in a 100:20:1 proportion. Almost no doubly ionized metal ions are observed. Residual gases such as oxygen, nitrogen, and water vapor always contribute. Glow discharge sputtering produces an atomic vapor representative of the cathode constituents, and the discharge ionization processes are also relatively nonselective. This results in mass spectra in which relative sensitivity factors (44) are quite similar, usually within a factor of 3–5. As one example, the stainless steel spectrum shown in Fig. 3.6 has peak heights that are closely representative of their bulk concentrations. Spectra taken under more sensitive conditions (higher detector voltage) than Fig. 3.6 will show trace elemental constituents in the low and sub-ppm range. Certain diatomic ions (e.g., M_2^+, MAr^+) are formed but are normally observed only for the major constituents. Spectral interferences can result if alternative isotopes are not available for the element to be analyzed. The spectra can be strongly affected by the ion energy window selected for transmission. Ions of the sputtered atoms exhibit a higher energy than the background gas ions, permitting energy discrimination to favor a spectrum richer in the ions of analytical interest.

3. INSTRUMENTATION

Interfacing a glow discharge to a mass spectrometer requires proper cognizance of the critical operating conditions of both the discharge and the spectrometer. Unlike the case of optical spectroscopy wherein the discharge can be sampled as a closed, sealed-off system, mass spectrometry requires the physical transport of particles (ions) from the plasma to the detector, so that the discharge and spectrometer become a common flow system. The analytical sources considered here operate in the 0.01–1 torr range, relatively high pressure for mass spectrometers, which need vacuums of $<10^{-5}$ to avoid disrupting collisions in the ion flight path. Differential pumping is required to permit this coupling.

Important experimental parameters that define or affect instrumental systems include:

Type of discharge—dc, rf, pulsed.

Discharge pressure—low pressures require high sustaining voltages.

Discharge current and power supply requirements.

Electrode configuration—conventional glow discharge vs. hollow cathode.

Mass resolution—magnetic vs. quadrupole.

No "standard" glow discharge ion source exists. The technique has been used by practitioners in many areas of interest who have applied a wide range of operating modes. This section will review some of the more representative analytical systems that have been used in recent years.

3.1. Glow Discharge Ion Sources

Three main discharge modes have been used in glow discharge ion sources: (a) capacitively coupled rf, (b) dc, and (c) pulsed dc. Many different electrode configurations are successfully employed in a variety of

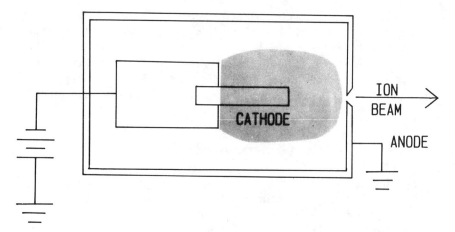

Figure 3.7. A pin cathode configuration for an analytical glow discharge ion source.

ion sources; each has certain real or perceived advantages in the given application, but no obviously superior design has emerged.

Coburn and Kay (15) have used the planar diode configuration common in plasma etching methods. Figure 3.7 shows the basic components of this source, which is powered by a 13.56-MHz rf voltage capacitively coupled to the target cathode. The source may also be used in a dc voltage mode (14) with comparable results. In each case, a glow discharge is established between the cathode and the grounded ion-sampling aperture through which ions are extracted from the negative glow region. Operating pressures are in the 0.01–0.1 torr range. Lower pressures are possible in the rf mode because, unlike the dc case where electrons can gain ionizing energy only in the cathode fall, the varying rf potentials can create ionization throughout the discharge plasma. These reduced pressures also produce less redeposition of sputtered atoms, permitting depth profile studies (56).

Harrison and coworkers (57–59) have experimented with a wide variety of glow discharge sources with a number of electrode configurations. In the most functional of these, illustrated in Fig. 3.7, a rod or pin serves as a negatively biased cathode coupled with a coaxial grounded anode. Ions are sampled from the negative glow region just in front of the ion exit orifice. Argon at 0.4–1.0 torr is the normal discharge gas. A simple dc power supply suffices at voltages of 500–1000 V (depending upon pressure) and 1–5-mA currents. This source lends itself to diverse cathode shapes, including rods, wires, disks, and pins. Nonconducting materials are compacted with graphite into a disk or pin electrode for analysis (60).

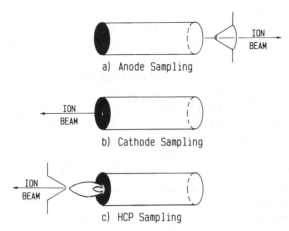

a) Anode Sampling

b) Cathode Sampling

c) HCP Sampling

Figure 3.8. Three configurations for ion sampling from a hollow cathode discharge. See text.

Solution residues are analyzed by drying a solution sample onto a graphite electrode (60). A Grimm glow discharge configuration (11) can also be adapted to mass spectrometric analysis for similar applications.

A significant advantage may be gained by pulsing the discharge (61), which permits higher peak voltages and currents than dc operation at the same average power. The result is higher sputtering yields and larger ion currents. Pulsed operation also permits time-resolved data acquisition and spectral discrimination against background ion contribution.

3.2. Hollow Cathode Ion Sources

As described in Section 2.2, hollow cathode discharges are a special type of glow discharge that provides more effective ionization in the negative glow. In addition, the sputtering is concentrated into the hollow cathode cavity, creating high atom densities in the confined volume. On the other hand, preparation of a hollow cathode from a bulk metal is more tedious than the preparation of the disk, pin, or rod used in the conventional glow discharge ion source. Also, some types of samples, such as wires or sheet metal, simply do not lend themselves to hollow cathode formation.

Ion sampling from a hollow cathode can be accomplished in several ways, as shown in Fig. 3.8. Initial hollow cathode ion sources (19, 22) used the scheme in Fig. 3.8a. The negative glow, which contains sputtered atoms from the cathode cavity, extends beyond the mouth of the hollow cathode and is sampled by the adjacent ion aperture. These sources operate at low voltages in the 1-torr regime. Discharge currents up to 100–

200 mA have been employed. Figure 3.8*b* shows a second means of sampling ions from the hollow cathode. The density of electrons and sputtered atoms is greatest at the base of the cathode cavity. With a small orifice drilled in the cathode base, ions may be sampled directly from the inner cavity. This mode easily handles solution samples by evaporation to a residue in the base of the hollow cathode. A recent development (62) is shown in Fig. 3.8*c*, which illustrates the hollow cathode plume (HCP). By appropriate differential pressures, a flamelike plume is ejected from a hollow cathode orifice. This extension has the effect of creating intense sputter erosion in the microchannel orifice, carrying a large atom population into the base of the plume. Subsequent excitation and ionization yield a sensitive source for atomic emission and mass spectrometry. Hollow cathode discharge plume conditions are about 2 torr pressure, 200–300 V, and 50–200 mA. A thin disk at the cathode base may serve as the sample in this configuration.

3.3. Instrumentation Requirements

The mass spectrometric sampling of any gas discharge requires the incorporation of a differential pumping system. While this may add some small additional cost to the system, differentially pumped chambers pose no particular experimental difficulty. Other factors requiring consideration include the selection of ion energies and type of mass analysis equipment. Experimental systems described in the literature have often arisen less from optimized design than from judicious application of already available apparatus. In principle, a relatively simple ion source like the glow discharge would seem to call for its coupling with the simplest mass analysis unit consistent with experimental needs. For the most part, this has meant quadrupole spectrometers, although magnetic instruments have also found application.

3.3.1. Ion Energy Analysis

Atoms sputtered into the negative glow are rapidly thermalized by atomic collisions. A small fraction of these atoms are ionized by energetic electrons or metastable atoms. Plasma potentials can impart variable amounts of energy to the ions before extraction, leading to an ion beam having a significant range of ion energies. Also, ions arising from primarily different ionization processes may exhibit different energy maxima. The ability to select an ion transmission window of a particular energy and variable width can greatly aid in attaining spectral resolution and selectivity.

Two general types of energy analyzers have been employed: the radial

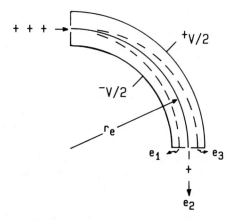

a) Electrostatic Analyser

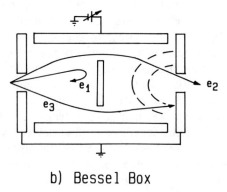

b) Bessel Box

Figure 3.9. Two common ion energy analyzers used to provide an energy-window bandpass in mass spectrometry. e_1 = lower-energy ions; e_2 = ions within energy bandpass; e_3 = higher-energy ions.

electrostatic analyzer (63) and the Bessel box (64). Figure 3.9 shows a schematic representation of each. In Fig. 3.9a, positive ions enter a radial electrostatic field established between two coaxial spherical plates and are deflected toward the negative plate, describing a radius r_e,

$$r_e = \frac{2V}{E}$$

where V is the acceleration potential applied to the ions and E is the field strength between the plates. At the exit orifice, only ions of a given kinetic energy (shown as e_2 in Fig. 3.9*a*) are permitted to leave the analyzer. Note that no mass term appears; the unit serves as an energy filter.

Figure 3.9*b* shows the components of a Bessel box energy analyzer. As opposed to the 90° deflection required in Fig. 3.9*a*, the Bessel box permits an in-line axial configuration. Energy analysis results from a retarding potential applied to the center element, which slows down entering ions. Ions with energy e_1 below a selected window potential are repelled, ions at energy e_3 above the retarding potential will be only partially deflected and strike the analyzer body, while ions of energy e_2 very close to the retarding potential are slowed greatly and deflected at the proper angle for transmission. A center stop precludes the straight-line passage of ions, fast neutrals, and photons.

3.3.2. Mass Analysis

A comparative analysis of the benefits of using magnetic sector and those of quadrupole mass analysis centers on the well-known advantages and limitations of the two systems described elsewhere in this monograph. Most significantly, magnetic instruments permit high-resolution spectra and accurate mass identification, while quadrupole systems are more compact and less expensive, allow rapid mass scanning, and feature high transmission rates at the lower mass ranges. Coburn has used a simple residual gas analyzer (RGA) quadrupole instrument that appeared to have approximately unit resolution. Clearly, these devices will not resolve two ions with the same nominal mass, such as $^{14}N_2^+$ and $^{28}Si^+$, but such interferences can usually be avoided by use of an alternative isotopic line. Another advantage of the quadrupole lies in its ability to function with very small sampling voltages. The glow discharge and associated utilities (power supplies, gas feed line, gauges) are not required to float at the high potential used with magnetic sector instruments. Except for some early (19, 22) and recent (29) work with double-focusing mass spectrometers, most of the literature reports on glow discharge mass spectrometry have featured quadrupole instrumentation. For the low-mass range involved in elemental analysis (1–250 amu), the quadrupole offers high ion throughput, modest cost, and configurational flexibility, which account for its popularity in glow discharge monitoring.

The higher resolution available with magnetic sector instruments does offer a significant advantage for the analysis of complex matrices where the possibility of isobaric interferences increases. A double-focusing instrument of the reverse Nier–Johnson geometry has been introduced as a commercial glow discharge mass spectrometer: the VG9000 (VG Isotopes, Cheshire, England), shown schematically in Fig. 3.10. This unit

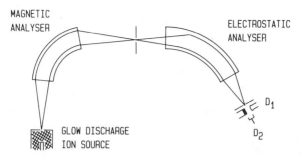

Figure 3.10. Representation of VG9000 glow discharge mass spectrometer. D_1, Daly detector; D_2, Faraday detector.

achieves resolution up to 6000 (10% valley definition). In its low-resolution mode, fast scanning can be obtained by stepping the spectrometer along each mass unit with a short, selectable dwell time at each mass. For quantitative analysis, longer integration times are employed. An integration time of 1 s per peak is reported to detect down to 10 ppb. A complete VG9000 system with data packages costs about $500,000.

4. APPLICATIONS

The increasing use of the glow discharge ion source arises from the need for fast, reliable bulk metal analyses. These applications were formerly covered by spark source mass spectrometry (and still are in many laboratories), but in many cases this aging instrumentation is not being replaced because of its cost and the complexity of the technique. The recent appearance of a commercial glow discharge mass spectrometer (29) offers analysts the opportunity to perform the same rapid qualitative elemental surveys available from the spark source with the added benefit of the glow discharge's simplicity, lower cost, and greater precision. While bulk metals analysis would appear to be the major need to be met, other applications in which the glow discharge has found use will also be described in this section. These include the examination of thin films, solution residues, and compacted powders. Glow discharge ion sources have also been used in conjunction with lasers, utilizing the particular advantage of both methodologies for selective, sensitive analysis. This review of applications offers illustrative examples of each area as a demonstration of the potential of glow discharge mass spectrometry. A previous review (26) described several applications in more depth.

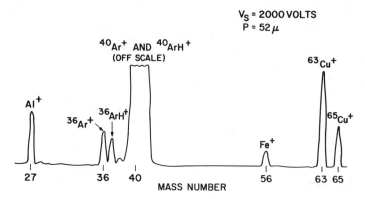

Figure 3.11. Glow discharge mass spectrum of an aluminum–bronze sample in a dc argon discharge (14).

4.1. Bulk Metals

To be effective in bulk metal analyses, a technique must consume or probe sufficient sample to negate localized elemental inhomogeneities. A surface-oriented method such as the glow discharge must sputter its way through a sufficient cathode depth to meet these requirements. Sputter rates depend on many factors (see Section 2.2), but a typical ablation rate is 0.1–10 μg/s, yielding adequate sample consumption while still permitting small sample size. Sputtering can serve also as an initial self-cleaning step to strip away surface contamination before initiating the true bulk analysis. Averaging of repetitive scans during the analysis will also aid in smoothing out sample inhomogeneities.

Coburn and coworkers reported a series of applications (14–16) in the early to mid-1970s, which returned attention to the glow discharge as an analytical ion source. They showed that metal cathodes could be characterized in a glow discharge from a variety of sputtered species. Copper and aluminum–bronze cathodes were used in a dc argon discharge at voltages from 1000 to 2000 V depending on pressure. Figure 3.11 shows a portion of the mass spectrum from the aluminum–bronze sample in which the three major constituents (Cu, Al, and Fe) are evident along with large peaks from the argon species. These investigators were not primarily interested in trace element components in the sample but were able to show the potential for such analyses. Poorly conducting samples, such as EuO and even quartz, were successfully run in an rf discharge. The intensity of the Si^+ and SiO^+ ions is shown in Fig. 3.12, which includes spectra taken in argon and neon. The major difference between

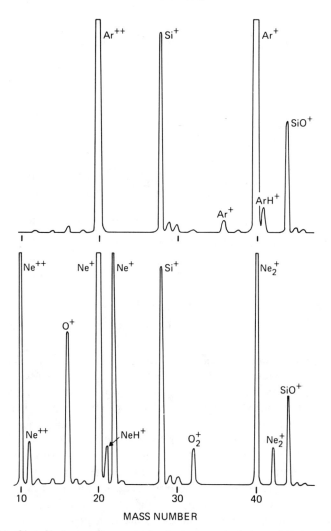

Figure 3.12. Glow discharge mass spectrum of a quartz cathode in an rf discharge (15) using argon (top) and neon (bottom).

the two spectra, other than the discharge gas used, lies in the presence of the O^+ and O_2^+ peaks in the neon discharge spectrum, which were attributed to Penning ionization. Neon has sufficient energy in its metastable forms (16.62 and 16.71 eV) to cause ionization of O^+ (IP = 13.55 eV), while argon metastable values (11.55 and 11.72 eV) are too low.

Over the past decade, Harrison and coworkers have developed the glow discharge ion source in many different versions, most of them falling

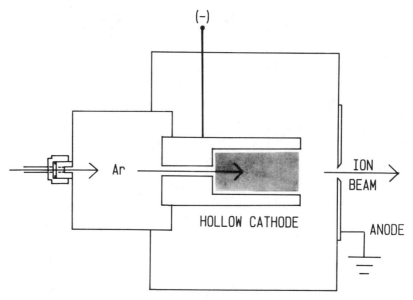

Figure 3.13. Hollow cathode ion source with a flow-through mode for argon discharge gas.

under the general categories of hollow cathode (27) or coaxial (57) sources. The first analytical use of a hollow cathode ion source in 1973 (19) used the argon discharge gas flowing directly through the cathode cavity (Fig. 3.13) to assist the diffusion of sputtered atoms to the ion exit orifice. Using a double-focusing Mattauch–Herzog geometry unit, survey analyses of a stainless steel sample showed sensitivities to the parts-per-billion level with few molecular ions formed. Ar^+ and ArH^+ ions were an order of magnitude higher than the Fe^+ ions of the sputtered matrix. An rf power supply was applied to the same source configuration to analyze zinc, tin, and Pyrex glass cathodes (21). Evans and coworkers (22, 23) also used a hollow cathode ion source for the elemental analysis of solid conducting materials, including molybdenum, copper–beryllium, stainless steel, and NBS cartridge brass. Interferences were found to be less severe than with SSMS or SIMS. Twenty elements were examined at concentrations from the percent level to low parts per million. A percent RSD of 26% was observed for 15 repetitive spectra. That laboratory also studied the effect of hollow cathode bore depth and diameter on ion signal. Ions of sputtered material were reported to increase sharply as bore diameter increased and depth decreased.

The hollow cathode ion source has also been coupled to a quadrupole mass analyzer to produce a simple solids mass spectrometer (65) as shown

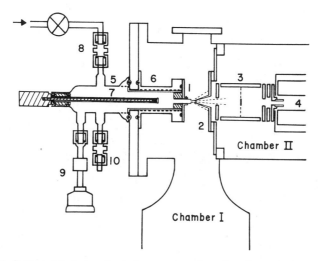

Figure 3.14. Reversed hollow cathode ion source and quadrupole measurement system. 1, Hollow cathode; 2, skimmer cone; 3, Bessel box energy analyzer; 4, quadrupole; 5, glass housing; 6, stainless steel chamber; 7, anode; 8, gas inlet; 9, vacuum gauge; 10, auxiliary port.

in Fig. 3.14. Cathodes were made from NBS steel and brass reference materials. Using these samples, the effects of current and pressure on sputter rate and ionization were determined. Reproducibilities of selected isotopes ranged from 1.7 to 10% for 10 consecutive scans, while sensitivities were in the low to sub-ppm range. Quantitative analyses were performed using two standardization approaches: (a) Relative sensitivity factors (RSFs) were computed from an NBS steel and used in the "analysis" of a second standard of similar composition; (b) ion intensities from one NBS reference material were used as external standards for direct computation of the elemental concentrations in a second NBS reference used as the unknown. Both methods were satisfactory, with the RSF technique yielding greater accuracy.

Many materials do not readily lend themselves to a hollow cathode configuration; a more conventional glow discharge permits the utilization of rods, pins, wires, disks, fragments, and so on as the active cathode so that elemental analysis may be performed. A compact, simple arrangement is illustrated in Fig. 3.15, which shows the cathode in a coaxial mode with the surrounding anode. This arrangement provides more versatility than the hollow cathode source when the analyst is confronted with the need to analyze many different types of materials. By adding side ports to the source, atomic absorbance (66) or fluorescence of solids samples can be accommodated.

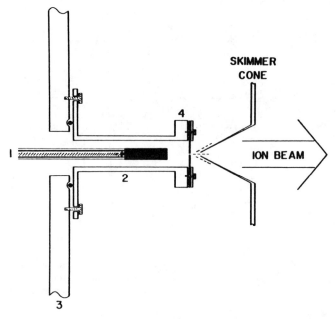

Figure 3.15. Coaxial cathode glow discharge ion source. 1, Cathode; 2, anode cylinder; 3, flange; 4, ion exit plate.

Cantle and associates (29, 31) have suggested many potential applications for glow discharge mass spectrometry (GDMS). Using a commercially available magnetic instrument (see Section 3.3.2), they have analyzed for impurities in several metals, including copper, gold, indium, and gallium. Figure 3.16 shows a portion of a glow discharge mass spectrum for high-purity copper. Evident are the lead and bismuth impurities at 10 ppm and 150 ppb, respectively. As an example of assays on precious metals, a gold analysis revealed up to 200 ppm of the trace metals copper, titanium, ruthenium, rhodium, zirconium, and iridium. Of particular note also is the ability of GDMS to measure chlorine and carbon impurities,

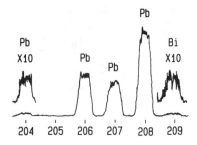

Figure 3.16. Portion of a glow discharge mass spectrum showing lead (10 ppm) and bismuth (150 ppb) impurities in copper (31).

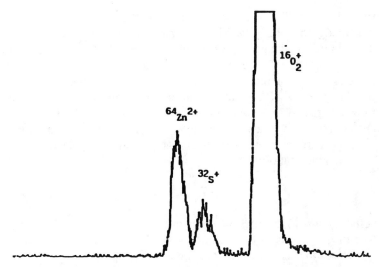

Figure 3.17. Portion of a glow discharge mass spectrum showing the separation of sulfur and doubly ionized zinc in a double-focusing magnetic instrument (31).

elements usually not assayed by other elemental techniques due to their difficulty. A technique was also reported by which liquid gallium could be analyzed for copper, zinc, indium, mercury, lead, and bismuth with up to 10 ppb sensitivity.

The VG application notes also describe the use of GDMS for the analysis of semiconductor materials such as gallium arsenide in which iron, phosphorus, cobalt, and nickel were determined at sub-ppm concentrations. Taking advantage of the high resolution of the double-focusing instrument, $^{54}Fe^+$ and $^{40}Ar^{14}N^+$ are shown to be easily resolved, eliminating the potential interference. The analysis of sulfur in zinc also requires resolution ($\sim$6000) sufficient to separate the doubly charged zinc isotopes from the desired sulfur isotope. Figure 3.17 shows that it is possible to resolve $^{32}S^+$ from $^{64}Zn^{2+}$ and $^{16}O_2^+$. Another troublesome analysis is the determination of carbon, nitrogen, and oxygen in metals, made difficult by the usual background contribution seen for each of these elements in an imperfect vacuum system. By thorough cleanup of the discharge gas and careful sample preparation, background levels for the three elements can be reduced to the sub-ppm levels, permitting analysis of C, N, and O at the parts-per-million concentrations in copper, according to VG application notes.

Other GDMS applications to bulk solids should be included here even though the primary thrust of the laboratories is not analytical in the sense of striving for a detailed analysis of trace elements in the cathode material. These investigators study the glow discharge as a tool in plasma etching processes in which an analysis of the plasma, including sputtered cathode constituents, is very useful in plasma diagnostics and process control. These plasmas may be inert (noble gases) or, as is often the case, reactive (O_2, methane, etc.). Hecq and coworkers have published a series of interesting papers (24, 25, 67) that describe the mass spectrometric sampling of plasma sputter systems. They have shown (25) that the intensity of M^{2+} ions, which increases with discharge pressure, is possibly due to a "hybrid Penning reaction":

$$Ar^{2+} + M \rightarrow Ar^+ + M^{2+} + e^-$$

In another report (67), they have shown that the ratio of ionic species in reactive discharges is not the same as the ratio of neutral gas species, another effect that is pressure-dependent.

Aita (30) has studied the reactive glow discharge as a sputter deposition device for various types of thin films. Her interest has been to use GDMS to study the relationship of discharge chemistry to film growth, including the determination of mass and energy of those positive ions incident on a substrate during sputter deposition. Important discharge parameters, such as the effect of rf power on sputter deposits (68), have also been examined. While such studies are not of direct analytical relevance, a better grasp of the fundamental processes involved in the glow discharge is vital to a greater appreciation of the analytical potential and limitations of GDMS.

4.2. Thin-Film Analysis

An application of particular commercial interest is the elemental analysis of thin surface layers of solids. Techniques such as the ion microprobe (69) and nuclear backscattering (70) are useful in obtaining composition profiles but suffer from matrix effects and high equipment costs. In principle, glow discharge mass spectrometry should serve as a competitor for those methods based on its highly sensitive detection of surface-sputtered particles. A concern would be the attainable resolution, in light of the plasma collisions that create redeposition of sputtered atoms. This effect would tend to cause overlap of ion signals from two adjacent thin films that may be less than 100 Å thick. Clearly, low-pressure rf glow discharges in the submillitorr regime would have advantages over the 1-torr conventional dc discharges. Coburn (71) has shown that an average surface removal rate of 0.1 Å/s can be attained with a minimum detectable ion

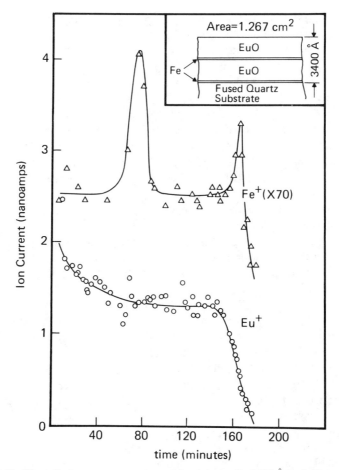

Figure 3.18. Glow discharge mass spectrometric analysis of a 3400-Å EuO film containing two thin layers of iron (71). Rf discharge in argon.

current corresponding to about 10^{11} atoms/s, a removal rate of 10^{-3} monolayer/s from a 1-cm^2 area. On this basis, 100 ppm impurities could be detected. The results of one thin-film analysis are shown in Fig. 3.18, which depicts the detection of two 100-Å iron films separating EuO layers. The removal rate in this study was about 0.35 Å/s. It should be recognized, however, that much higher penetration rates can result from other glow discharge systems. Hecq and coworkers (72) used a 2000-Hz oxygen glow discharge in the millitorr range to examine a target consisting of thin metallic films of tantalum, silver, gold, and cobalt in 5000–20,000-Å thicknesses. Over a 150-min sputter period, they were able to resolve the four layers, although significant ion signal overlap occurred. Even with high-pressure (1 torr) dc discharges, some depth profiling is possible. The removal of a surface coating of V_2O_5 from a vanadium sample was reported (73) to proceed at 5 μm/min (833 Å/s) in argon discharge. Obviously, this approach is not useful for thin films, but the substitution of helium for argon may reduce the sputter rate by an order of magnitude.

Coburn et al. (74) summarize the possible advantages of GDMS for elemental profiling, including (a) lack of matrix effects, (b) generally uniform sensitivity for all elements, (c) ability to vary the sputter-etch rate by changing discharge gas and applied power, (d) little surface damage from the low-energy sputter ions, (e) sensitivity to sub-ppm, and (f) stable analytical ion signal of low energy, so that double-focusing mass spectrometers are not required. Disadvantages include lack of spatial resolution in the sample plane and possible interferences from the plasma gases. As glow discharge mass spectrometry continues to evolve, particularly with the development of commercially available equipment, compositional depth profiling may become an important application.

4.3. Solution Residues

Solution samples are not easily amenable to mass spectrometric analysis because of the incompatible nature of solutions and the high vacuum necessary for the spectrometer. It is necessary to condense the sample in some differential manner, concentrating the analytical constituents at the expense of the solution matrix. For elemental analysis of solutions, mass spectrometry may not even be the method of choice. There exist in analytical chemistry many good ways to analyze solutions for trace elements, most notably ICP atomic emission (75). Recent developments also suggest an increased importance for ICP-MS (76) in solution analysis. Glow discharge mass spectrometry cannot tolerate the introduction of large, variable gas volumes that result from solution injection, so samples must be dried to a residue upon a suitable cathode for subseqeuent sputter-atomization into the plasma. While this extra desolvation step introduces additional time per analysis, a significant analytical advantage results. The residue from the sample lies concentrated into a thin film, which is quickly sputtered (often within seconds) and ionized. In this manner—somewhat analogous to furnace atomic absorption (77)—a sample that would yield a weak steady-state signal is caused to produce a large, short-lived ion pulse that can be evaluated for peak height or area. Very sensitive measurements at the parts-per-billion level have been observed. Thus, while GDMS serves primarily as a solids analysis technique, its application for solution analysis has not been neglected.

The best cathode material for GDMS solution analysis is a dense conducting medium. High-purity metals work quite well if there is no interference from the sputtered matrix and no chemical interaction with the sample. Perhaps the best cathode substrate found to date is pyrolytic graphite (78) or glassy carbon (79). These materials are impervious to the solution so no residue is lost by absorption into the cathode lattice, and they are essentially inert to solutions applied to their surfaces. Furthermore, carbon sputters very poorly (80) and thus contributes little background to the mass spectrum.

Harrison and Magee have studied barium and strontium solutions (19) in a stainless steel cathode. An interesting feature of such spectra is the relatively minor contribution of the cathode matrix during the short measurement cycle. The deposited inorganic residue, bound very loosely to the cathode surface, quickly sputters away before significant substrate removal is observed. Daughtrey and Harrison (20), Donohue and Harrison (21), and Mattson et al. (57) extended these studies to extensive multielement applications, determining relative sensitivity factors, sensitivities, and measurement precision. Keefe (60) studied various discharge parameters and matrix effects in the analysis of lanthanide solutions. Foss and associates (81) reported a unique application of GDMS for solutions, in a cryogenic ion source. Aqueous samples (20–50 μL) were frozen into a hollow cathode and held at liquid nitrogen temperatures during analysis. The discharge was initiated in an oxygen support gas at 1 torr but quickly became a water vapor discharge upon discharge stabilization. Nine standard solutions were examined; they contained a total of 70 elements, all but 7 of which could be determined. Sensitivities ranged from ppb to ppm values, with precision estimated at ±20% for the photographic detection. The method was applied to an EPA water sample for 12 selected elements.

4.4. Compacted Powders

The glow discharge ion source serves quite well for the analysis of solid conducting metal electrodes and even permits the use of solid semiconductor samples (82). However, many important types of samples do not present themselves in such convenient forms. For example, geological samples are often in a granular or powder state and are nonconducting. Although little in this regard has been reported in GDMS, methodology exists (44) for the incorporation of nonconducting powders into a conducting solid electrode matrix. High-purity graphite or metal powder is mixed thoroughly with the sample in a fixed ratio, and the mixture is compacted in a molding die by application of high pressure using a hydraulic press. Pin electrodes (1 × 7 mm) can be readily formed with

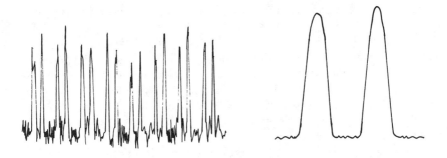

a) Repetitive Single Scans   **b) 200 Accumulated Scans**

Figure 3.19. Repetitive single scans versus multiple scan averaging for 0.01% Eu as Eu_2O_3 in graphite, showing increase in signal-to-background ratio.

sufficient structural integrity to permit mounting in the cathode holder and sputter-atomization in the glow discharge.

A metal matrix has produced sensitivities better than those obtained with graphite. High-purity powdered silver and gold (60) yield strong electrodes, high conductivity, and excellent sputter characteristics, but they suffer from two major limitations: (a) The particle size of the metal is larger than desirable for homogeneous electrode preparation, and (b) the metals produce significant spectral interference (M^+, M_2^+, MO^+, MAr^+, etc.). Graphite is available in very high purity, extremely small particles and produces little mass spectral background. Loving (28) showed that a 9:1 graphite–arsenic mixture gave excellent discharge characteristics and long-term stability. The arsenic signal remained stable over a 30-min sputter period, indicating no differential sputter effects. Keefe (60) has examined compacted electrodes in much greater detail, concentrating on analysis of the lanthanides. He has found that 0.01% Eu (as Eu_2O_3) in graphite yields the spectra shown in Fig. 3.19. The europium isotopes are easily seen in the repetitive single scans in Fig. 3.19a and the signal-to-noise enhancement of multiple scan averaging is clearly evident in Fig. 3.19b Signal intensities have been monitored for extended periods using electrodes containing many different lanthanides. The ion signal stability indicates both homogeneous electrode preparation and a lack of electrode surface enrichment or depletion from the plasma sputter process. The extension of GDMS to compacted samples has the potential of greatly extending the scope of its applications.

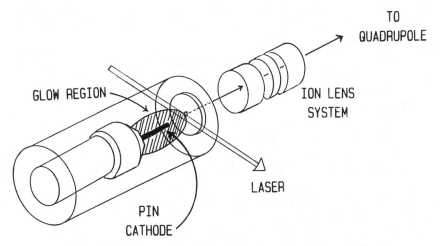

Figure 3.20. Laser-enhanced ionization in a glow discharge (55).

4.5. Laser-Assisted Methods

A laser has been used recently in conjunction with a glow discharge to yield two new approaches to mass spectrometry: (a) The sputter atomization properties of the glow discharge make it an advantageous atom reservoir for selective ionization by a tuned laser, and (b) a laser can serve as an atomization device to ablate solids directly into a glow discharge plasma for subsequent ionization.

4.5.1. Resonance Ionization Mass Spectrometry (RIMS)

RIMS has found increasing application in recent years using a variety of atomization sources, including a filament (83, 84), graphite furnace (85), flame (86, 87), and ion gun (88). The glow discharge has particular advantages for the RIMS analysis of solids in that it is a simple and stable source of atoms, the atom density can be controlled by adjustment of current and pressure, and the sputter yields produce generally similar sensitivities. The glow discharge is well known as an atomization source for thin-film production and has been used as an atomization source for solids in atomic absorption (89). The ease of adapting a glow discharge for RIMS (55) is evident in Fig. 3.20. Two quartz windows allow positioning of the laser beam in the negative glow region directly in front of the exit orifice. High ionization efficiency is possible in the small effective extraction volume, and by use of a tunable dye laser the ionization can be made very selective, tuning the laser to a specific optical transition of the analysis element. In this way, isobaric interferences (84) and other spectral interferences may be minimized or even eliminated. The smaller intrinsic ion signal from the glow discharge can be negated by electronic subtraction. RIMS spectra comprise merely the ions of the selected element, requiring only a simple, low-resolution mass spectrometer for analysis.

4.5.2. Laser Ablation

It is well known that lasers can provide sufficient power to atomize and ionize solid materials thermally (90). A laser mass spectrometer is commercially available (91) and has found many applications (92, 93). Multiple laser approaches have also been used for solids analysis: one laser to ablate material, a second laser fired microseconds later to ionize a selected element from the ablated atoms. A simpler approach utilizing the glow discharge employs a single laser as a nonselective thermal ablation source for a solid sample. The pulse of ablated material enters directly into the plasma of a glow discharge, as shown in Fig. 3.21a. In this way a greatly enhanced atom production is introduced into the negative glow for ionization and mass spectral analysis. The solid sample, if a conductor, may serve as the glow discharge cathode, so that both sputter atomization and laser ablation contribute to the net atom yield. If the sample is a nonconductor, laser ablation offers a particular advantage in that it would be impossible to run this sample by GDMS. However, by placing the nonconductor in or adjacent to the negative glow (Fig. 3.21b), laser ablation injects material into the glow discharge plasma, where it is subjected to the same ionization steps as if it had arisen from a sputtered cathode. Use of a low-background cathode, such as pyrolytic graphite or glassy carbon, yields a discharge that can generate a large ion signal for the ablated material and a relatively small contribution from the cathode. This methodology would lend itself to such difficult samples as geologicals and other nonconducting powders that could be compacted into a disk or pellet for laser ablation into a glow discharge.

4.6. Analytical Summary

It is hazardous to attempt to summarize aspects of any rapidly developing technique. Studies under way and others not yet reported may—one might even say should—cause such a review to be somewhat out of date before its publication. With this caveat in place, brief comments follow on some general areas in which analytical techniques are most often evaluated.

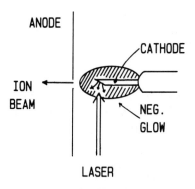

(a) DIRECT LASER ABLATION GDMS

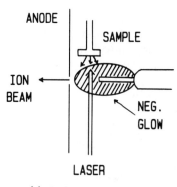

(b) INDIRECT LASER ABLATION GDMS

Figure 3.21. Laser ablation of cathode material into a glow discharge for mass spectrometric analysis.

Applicability. Glow discharge mass spectrometry offers broad, general application to metals and alloys; there is no question but that this is the major anticipated use. Potentially important extension to compacted sample materials awaits development. Meaningful applications to surface studies, thin films, and solutions are more problematical at this time.

Sensitivity. Parts-per-billion sensitivities have been demonstrated in several studies involving metal analysis. For high-sputter materials such as metals and alloys there is little doubt that such sensitivities are realistic, particularly with state-of-the-art commercial instrumentation. Metal lattices tend to yield up their atoms rather efficiently by the sputter process. Less sensitive detection might be anticipated for more troublesome samples like graphite-compacted geological materials, where the glow discharge may have some difficulty in totally breaking up certain refractory compounds. Still, there exist applications that do not require trace capabilities and for which the ability to perform fast solids analysis may be attractive.

Precision. Stability of the glow discharge itself is on the order of 1–2%. Trace elements in the parts-per-million range often show 5% RSD over 10–20 repetitive scans. Signal integration could further improve this.

Relative Sensitivity. Few studies of relative sensitivity factors (RSFs) have appeared, but evidence suggests that the elemental sensitivities range over a maximum of about $10\times$ in metal–alloy matrices, with a 3–$5\times$ average. In a nonmetallic matrix, where two elements may exist in quite different chemical combinations, larger differences in relative sensitivity will be expected.

Interferences. Background gaseous contributions do not generally cause serious interference; their spectral locations are critical for only a few elements. Problems may arise from combinations involving sputtered atoms and the discharge gas, which can appear at a nominal elemental m/e (e.g., $^{63}Cu^{40}Ar^+$ on $^{103}Rh^+$). Except for monoisotopic elements, alternative isotopes are usually available for analysis.

Matrix Effects. No real matrix effect data have been reported. A trace element might be expected to differ in sensitivity between two rather different matrices (e.g., steel vs. brass), but a more relevant study would be a comparison of trace element sensitivities as a function of composition changes within a given matrix (e.g., stainless steel vs. low-alloy steel). Standards would normally, of course, be selected to be as close in composition as possible to the analytical sample.

Qualitative Analysis. This is a very strong feature of glow discharge mass spectrometry. Survey elemental scans can be obtained within minutes showing both all metals and all nonmetals in the sample. By running successive scans at increasing sensitivity, major through parts-per-billion trace constituents can be resolved. Elemental identification is by the m/e location and the correct isotopic distribution. Multiple scan accumulation may be necessary to develop fully the isotopic patterns of components present in very low concentrations.

Quantitative Analysis. No truly detailed studies have taken place to explore the many facets of quantitative analysis by glow discharge mass spectrometry. Piecing together the scattered results, it appears that $+5\%$ quantitative results are possible, assuming suitable standards are available for direct comparison of ion currents. Sources of error that may contribute to quantitative uncertainty include sample inhomogeneity, spectral interferences, matrix differences, and changes in discharge conditions.

5. RETROSPECTIVE

Most new instrumental techniques must pass through a significant gestation period before they become accepted by the analytical community and put into routine practice. Such caution is wise, because advocates may claim (and, indeed, believe) the technique to be capable of more than it may ultimately deliver. Only through continued research, application to real-world problems, and dissemination of these results to the prospective practitioners can a technique hope to gain acceptance. Furthermore, without commercially available equipment at an affordable cost, the technique is destined to remain less than fully developed, since relatively few laboratories would be willing to design, build, and evaluate an experimental system for a yet unproven technique. Glow discharge mass spectrometry has now received enough attention in various laboratories to demonstrate reasonably well its potential for the elemental analysis of solids. Commercial instrumentation, albeit rather expensive, now exists and has yielded excellent results. Very important also is that a need exists for an improved technique for broad multielement analysis of materials directly in the solid state. Glow discharge mass spectrometry permits a quick qualitative scan of the periodic table with generally similar sensitivity for all elements and quantitative analyses with proper standards. The next few years will likely determine whether these seemingly persuasive advantages will result in its more widespread application.

ACKNOWLEDGMENT

Appreciation is expressed to the Department of Energy, Division of Chemical Sciences, for financial support of our research in glow discharge mass spectrometry.

REFERENCES

1. F. W. Aston, *Mass Spectra and Isotopes*, 2nd ed., Longmans, New York, 1942.

2. K. T. Bainbridge and E. B. Jordon, *Phys. Rev.*, **50**, 282 (1936).

3. A. J. Dempster, *Proc. Am. Phil. Soc.*, **75**, 755 (1935).

4. F. T. Birks, *Spectrochim. Acta*, **5**, 322 (1952).

5. W. W. Harrison and N. J. Prakash, *Anal. Chim. Acta*, **49**,151 (1970).

6. P. W. J. M. Boumans, *Anal. Chem.*, **44**, 1219 (1972).

7. J. E. Greene, F. Sequeda-Osorio, and B. R. Natarajan, *J. Vac. Sci. Technol.*, **12**, 336 (1975).

8. I. A. Berezin, *Zavod. Lab.*, **27**, 859 (1961).

9. G. A. Pevtsov, V. Z. Krasilshchik, and A. F. Yabovlera, *J. Anal. Chem. USSR*, **23**, 1569 (1968).

10. D. E. Maksimov and N. K. Rudnevskii, *Zh. Prikl. Spektrosk.*, **39**, 5 (1983).

11. W. Grimm, *Spectrochim. Acta*, **23B**, 443 (1968).

12. R. A. Kruger, R. M. Bombelka, and K. Laqua, *Spectrochim. Acta*, **35B**, 589 (1980).

13. K. H. Koch, M. Kretschmer, and D. Grunenberg, *Mikrochim. Acta*, **1983II**, 225.

14. J. W. Coburn, *Rev. Sci. Instrum.*, **41**, 1219 (1970).

15. J. W. Coburn and E. Kay, *Appl. Phys. Lett.*, **18**, 435 (1971).

16. J. W. Coburn and E. Kay, *J. Appl. Phys.*, **43**, 4965 (1972).

17. H. Oechsner and W. Gerhard, *Phys. Lett.*, **40A**, 211 (1972).

18. H. Oechsner and W. Gerhard, *Surface Sci.*, **44**, 480 (1974).

19. W. W. Harrison and C. W. Magee, *Anal. Chem.*, **46**, 461 (1974).

20. E. H. Daughtrey, Jr., and W. W. Harrison, *Anal. Chem.*, **47**, 1024 (1975).

21. D. L. Donohue and W. W. Harrison, *Anal. Chem.*, **47**, 1528 (1975).

22. B. N. Colby and C. A. Evans, Jr., *Anal. Chem.*, **46**, 1236 (1974).

23. J. R. Wallace, D. F. Natusch, B. N. Colby, and C. A. Evans, Jr., *Anal. Chem.*, **48**, 118 (1976).

24. M. Hecq, A. Hecq, and M. Liemans, *J. Appl. Phys.*, **49**, 6176 (1978).

25. M. Hecq and A. Hecq, *J. Appl. Phys.*, **56**, 672 (1984).

26. J. W. Coburn and W. W. Harrison, *Appl. Spectrosc. Rev.*, **17**, 95 (1981).

27. C. G. Bruhn, B. L. Bentz, and W. W. Harrison, *Anal. Chem.*, **51**, 673 (1979).

28. T. J. Loving and W. W. Harrison, *Anal. Chem.*, **55**, 1526 (1983).

29. J. E. Cantle, E. F. Hall, C. J. Shaw, and P. J. Turner, *Int. J. Mass Spectrom. Ion Phys.*, **46**, 11 (1983).

30. C. R. Aita, T. A. Myers, and W. J. LaRocca, *J. Vac. Sci. Technol.*, **18**, 324 (1981).

31. *VG 9000 Glow Discharge Mass Spectrometer*, VG Isotopes, Cheshire, England, descriptive brochure.

32. S. Caroli, *Prog. Anal. Atom. Spectrosc.*, **6**, 253 (1983).

33. M. E. Pillow, *Spectrochim. Acta*, **36B**, 821 (1981).

34. P. J. Slevin and W. W. Harrison, *Appl. Spectrosc. Rev.*, **10**, 201 (1975).

35. A. M. Howatson, *An Introduction to Gas Discharges*, Pergamon, Elmsford, NY, 1976.

36. J. D. Cobine, *Gaseous Conductors: Theory and Engineering Applications*, Dover, New York, 1941.

37. B. Chapman, *Glow Discharge Processes*, Wiley, New York, 1980.

38. J. W. Coburn, *Plasma Etching and Reactive Ion Etching*, AVS Monograph Series, American Vacuum Society, New York, 1982.

39. Z. Xu-hui and L. Jian-bang, *Appl. Phys.*, **29B**, 291 (1982).

40. M. Pahl, W. Lindinger, and F. Howorka, *Z. Naturforsch.*, **27a**, 678 (1972).

41. T. D. Märk, *Z. Naturforsch.*, **28a**, 1397 (1973).

42. A. Stamatovic, K. Stephan, and T. D. Märk, *Int. J. Mass Spectrom. Ion Processes*, **63**, 37 (1985).

43. M. Slavin, *Emission Spectrochemical Analysis*, Wiley-Interscience, New York, 1971.

44. A. J. Ahearn, Ed., *Trace Analysis by Mass Spectrometry*, Academic, New York, 1972.

45. J. W. Carr and G. Horlick, *Spectrochim. Acta*, **37B**, 1 (1982).

46. R. V. Stuart and G. K. Wehner, *J. Appl. Phys.*, **35**, 1819 (1964).

47. M. Barber, R. S. Bordoli, G. J. Elliott, R. D. Sedjwick, and A. N. Tyler, *Anal. Chem.*, **54**, 645A (1982).

48. G. Carter and J. S. Colligon, *Ion Bombardment of Solids*, Elsevier, New York, 1968.

49. W. D. Davis and T. A. Vanderslice, *Phys. Rev.*, **131**, 219 (1963).

50. F. M. Penning, *Naturwiss.*, **15**, 818 (1927).

51. J. M. Brackett, J. C. Mitchell, and T. J. Vickers, *Appl. Spectrosc.*, **38**, 136 (1984).

52. W. D. Westwood, *Prog. Surf. Sci.*, **7**, 71 (1976).

53. B. L. Bentz, Ph.D. dissertation, University of Virginia, Charlottesville, 1980.

54. W. W. Harrison and B. L. Bentz, *Anal. Chem.*, **51**, 1853 (1979).

55. P. J. Savickas, K. R. Hess, R. K. Marcus, and W. W. Harrison, *Anal. Chem.*, **56**, 817 (1984).

56. J. W. Coburn, E. W. Eckstein, and E. Kay, *J. Appl. Phys.*, **46**, 2828 (1975).

57. W. A. Mattson, B. L. Bentz, and W. W. Harrison, *Anal. Chem.*, **48**, 489 (1976).

58. W. W. Harrison and B. L. Bentz, "Trace Element Analysis by Glow Discharge Mass Spectrometry" in P. Bratter and P. Schramel, Eds., *Trace Element Analytical Chemistry in Medicine and Biology*, Walter de Gruyter, New York, 1980.

59. B. L. Bentz and W. W. Harrison, *Anal. Chem.*, **54**, 1644 (1982).

60. R. B. Keefe, Ph.D. dissertation, University of Virginia, Charlottesville, 1983.

61. P. J. Savickas, Ph.D. dissertation, University of Virginia, Charlottesville, 1984.
62. R. K. Marcus and W. W. Harrison, *Spectrochim. Acta B*, **40**, 933 (1985).
63. F. A. White, *Mass Spectrometry in Science and Technology*, Wiley, New York, 1968, p. 21.
64. Extranuclear Laboratories, Inc., Pittsburgh, PA, Technical Manual for Bessel Box, 1976.
65. C. G. Bruhn, B. L. Bentz, and W. W. Harrison, *Anal. Chem.*, **50**, 16 (1978).
66. T. J. Loving and W. W. Harrison, *Anal. Chem.*, **55**, 1523 (1983).
67. M. Hecq, A. Hecq, and M. Fontignies, *Thin Solid Films*, **115**, L45 (1984).
68. C. R. Aita, R. L. Lad, and T. C. Tisone, *J. Appl. Phys.*, **51**, 6405 (1980).
69. H. Liebl, *Adv. Mass Spectrom.*, **7A**, 807 (1978).
70. M. A. Nicolet, in J. Zemel, Ed., *Nondestructive Evaluation of Semiconductor Materials and Devices*, Plenum, New York, 1979, pp. 581–627.
71. J. W. Coburn and E. Kay, *Appl. Phys. Lett.*, **19**, 350 (1971).
72. M. Hecq, A. Hecq, and M. Fontignies, *Anal. Chim. Acta*, **155**, 191 (1983).
73. VG Isotopes, Ltd., Application Note No. 5, "Glow Discharge Mass Spectrometry," Cheshire, England.
74. J. W. Coburn, E. Taglauer, and E. Kay, *J. Appl. Phys.*, **45**, 1779 (1974).
75. M. Thompson and J. N. Walsh, *A Handbook of Inductively Coupled Plasma Spectrometry*, Methuen, New York, 1983.
76. A. R. Date and A. L. Gray, *Spectrochim. Acta*, **38B**, 29 (1983).
77. S. R. Koirtyohann and M. L. Kaiser, *Anal. Chem.*, **54**, 1515A (1982).
78. C. L. Mantell, *Carbon and Graphite Handbook*, Interscience, New York, 1968.
79. J. Heller, *Thin Solid Films*, **17**, 163 (1973).
80. N. Laegreid and G. Wehner, *J. Appl. Phys.*, **32**, 365 (1961).
81. G. O. Foss, H. J. Scev, and R. J. Conzemius, *Anal. Chim. Acta*, **147**, 151 (1983).
82. VG Isotopes, Ltd., Application Note GD005, "Glow Discharge Mass Spectrometry," Cheshire, England.
83. J. P. Young, G. S. Hurst, and S. D. Kramer, *Anal. Chem.*, **51**, 1050A (1979).
84. D. L. Donohue and J. P. Young, *Anal. Chem.*, **55**, 378 (1983).
85. A. S. Gonchakov, N. B. Zorov, Y. Y. Kuyzyakov, and I. O. Matveev, *J. Anal. Chem. USSR*, **34**, 1792 (1980).
86. G. C. Turk, J. R. Devoe, and J. C. Travis, *Anal. Chem.*, **54**, 643 (1982).
87. J. C. Travis, G. C. Turk, and R. B. Green, *Anal. Chem.*, **54**, 1006A (1982).
88. N. Winograd, J. P. Baxter, and K. M. Kimock, *Chem. Phys. Lett.*, **88**, 581 (1982).
89. A. Walsh, *Spectrochim. Acta*, **35B**, 639 (1980).

90. R. J. Conzemius and J. M. Capellan, *Int. J. Mass Spectrom. Ion Phys.*, **34**, 197 (1980).

91. R. Wechsung, F. Hillenkamp, R. Kaufmann, R. Nitsche, and H. Vogt, *Microsc. Acta Suppl.*, **2**, 611 (1978).

92. E. Denoyer, R. Van Grieken, F. Adams, and D. F. S. Natusch, *Anal. Chem.*, **54**, 26A (1982).

93. D. M. Hercules, R. J. Day, T. A. Dang, and C. P. Li, *Anal. Chem.*, **54**, 280A (1982).

CHAPTER

4

SECONDARY ION MASS SPECTROMETRY

ALEXANDER LODDING

Chalmers University of Technology, Gothenburg, Sweden

1. INTRODUCTION

The first mention of sputtered secondary ions in the literature was made in 1910 by Sir J. J. Thomson (1). The first regular secondary ion mass spectrometer was based on a patent by Herzog in 1942 (2, 3), and the first successful studies of surface compositions using mass-analyzed sputtered ions were made by several teams in the early 1950s (4, 5). An accelerated development of the field was stimulated by new efficient designs of narrow-beam primary ion columns (6) and of ion optics for "direct" imaging (7). The late 1960s saw the emergence of the first commercial instrumentation (8–10) and the coining of the SIMS acronym (10). Today SIMS (secondary ion mass spectrometry) has an acknowledged place

125

among the major techniques of surface analysis and microstructural characterization of solids.

SIMS is particularly noted for its outstanding sensitivity of chemical and isotopic detection. Quantitative or semiquantitative analysis can be performed for small concentrations of most elements in the periodic table, including the lightest. However, the high versatility of SIMS is mainly due to the combination of high sensitivity with good topographic resolution both in depth and (for imaging SIMS) laterally.

These assets (see Table 4.1), together with recent advances in quantitation, have led to a rapid expansion of SIMS in industry as well as in interdisciplinary science. Of the commercially built "three-dimensional" SIMS equipment, about 200 instruments since 1968, some 75% have been supplied since 1980. The global boost in SIMS applications has been especially connected with fast developments in semiconductor engineering. Presently more than 50% of existing SIMS instrumentation is employed in the field of microelectronics and computer materials, ~25% in metallurgy, ceramics, and other construction materials; ~10% in earth and space sciences, solid-state sciences, and biology; and the rest in interdisciplinary and nondedicated research.

The topic of SIMS has been repeatedly presented in review articles (5, 11–13). Moreover, the "state of the art" receives regular and thorough

Table 4.1 Typical Features of Major Surface-Analytical Techniques[a]

Method[b]	Element Range (Atomic No.)	Lateral Resolution (μm)	Information Depth (Å)	Detection Limit (atom ppm)
EMP, AEM	≥ 5	2	10^4	10^3
AES, SAM	≥ 3	10^{-1}	10	10^2
ESCA, XPS	> 2	10^3	20	10^4
LEIS	> 2	10^2	5	10^3
HEIS, RBS	≥ 5	2	10^2	10^2
PIXE	≥ 10	5	10^3	1
AINR	≤ 10	5	10^3	10
FIM-AP	All	10^{-2}	10	10^3
SIMS, IMMA	All	$1-10^{-1}$	5	$10^{-4}-10$

[a] Order-of-magnitude figures, applying to routine analysis of elements with average ease of detection.

[b] EMP, electron microprobe; AEM, analytic electron microscopy; AES, Auger electron spectroscopy; SAM, scanning Auger microprobe; ESCA, electron spectroscopy for chemical analysis; XPS, X-ray photoelectron spectrometry; LEIS, HEIS, low- and high-energy ion scattering spectrometry; RBS, Rutherford backscattering; PIXE, proton-induced X-ray emission; AINR, accelerator-induced nuclear reactions; FIM-AP, field ion microscopy–atom probe; SIMS, secondary ion mass spectrometry; IMMA, ion microprobe mass analysis.

debate in the proceedings of dedicated conferences (14–18). The object of this chapter is an up-to-date orientation, on a nonspecialist level, of the main principles, assets, and intricacies of SIMS from the viewpoint of practical applications to inorganic materials. Illustrations are drawn mainly from experience with the second-generation ion microscope type of equipment, roughly representing the development of commercially available SIMS instrumentation up to 1985.

2. FUNCTIONING PRINCIPLES OF SIMS

Secondary ion mass spectrometry is based on the following functions:

(a) Bombardment of the sample surface by focused primary ions, with sputtering of the outermost atomic layers.

(b) Mass spectrometric separation of the ionized secondary species (sputtered atoms, molecules, clusters) according to their mass-to-charge ratios.

(c) Collection of the separated secondary ions as quantifiable mass spectra, as in-depth or along-surface profiles, or as distribution images of the sputtered surface.

The primary ions are normally produced by a duoplasmatron type (11, 19) of gas source (O_2^+, O^-, N_2^+, Ar^+), by surface ionization (20) (Cs^+, Rb^+), or by liquid-metal field ion emission (21) (Ga^+, In^+). The ions are accelerated and focused to a selected impact area on the specimen. The collision cascade following the incidence of a primary ion results in the implantation of the primary particle, reshuffling of some 50–500 matrix atoms, and emission of secondary particles, neutral or ionized. Secondary ions from the specimen are extracted into the mass spectrometer, which is based either on electric/magnetic deflection fields or on the quadrupole/time-of-flight principle. Secondary ions with given mass-to-charge ratio and within a certain interval (window) of kinetic energy are collected for pulse or current measurement, ion-optic imaging, and data processing. Three main classes of SIMS instrumentation may be distinguished:

1. *Nonimaging ion probes* are used for depth profiling on laterally homogeneous specimens (often accessory to other surface-analytical systems, such as AES, ESCA, or electron microscopy) or for specialized studies of outermost surface layers ("static" SIMS, here only to be mentioned in passing; see Ref. 22).

2. *Imaging ion microprobes* (8, 23–26) utilize a narrow beam ($\lesssim$10 μm) of primary ions. Lateral resolution is given essentially by the beam size. Imaging and microscopy can be effected by a TV-type raster of the beam over the surface of the sample. The production of ion microprobes has received new impetus with the development of efficient narrow-beam liquid-metal ion sources.

3. *"Direct-imaging" microscopes–microanalyzers* (27, 28) are based on relatively wide ($\sim$5–300 μm) primary beams. A point-to-point ion microscope function is achieved by a set of ion optic lenses. The lateral resolution is given either by gating the imaged area or by image processing. Recently produced ion microscope instrumentation (28) has proved efficient in numerous applications and has been acquired by more than 100 laboratories since 1978.

Some elements of SIMS instrumentation are illustrated in Fig. 4.1, which shows the schematic design of a commercially produced direct-imaging secondary ion microscope–microanalyzer. The mass spectrometer is double-focusing (an electrostatic sector for energy focusing, a magnetic sector for focusing in mass), with high mass resolution. The two primary ion sources (one for plasma ionization of gases, one for Cs^+) are interswitchable, and a mass filter safeguards the purity of the primary species. The accessories include an electron gun for neutralization of the positive charge buildup on insulating specimens. The transfer optics system provides an ion-optical "zoom," allowing the choice of very small areas ($\simeq$0.5 μm diameter) for separate imaging or analysis without essential loss of transmission (29). Because of the energy discrimination facility, low-energy secondary ions can be prevented from entering the mass analyzer, which suppresses the presence of polyatomic species in the mass spectrum (see Section 3.4).

For the imaging mode, the ion lens system of the mass spectrometer is designed so as to yield, on the multichannel plate, a point-to-point representation of the bombarded surface (27), with simultaneous information from all parts. The channel plate, consisting of "bundled" electron-multiplying capillaries, converts each arriving ion event to a laterally localized avalanche of electrons visible as a flash on a screen beyond the plate. The mass-spectrometrically resolved image on the screen can be viewed by binoculars or video or collected photographically.

In the "quantitative" mode of analysis, the trajectory of the separated ions is deflected by another electrostatic sector toward an electron multiplier or a Faraday cage. An online computer stores and processes the registered ion currents. This may entail, for example, mass spectra of selected mass numbers, intensities as functions of time (in-depth profiles,

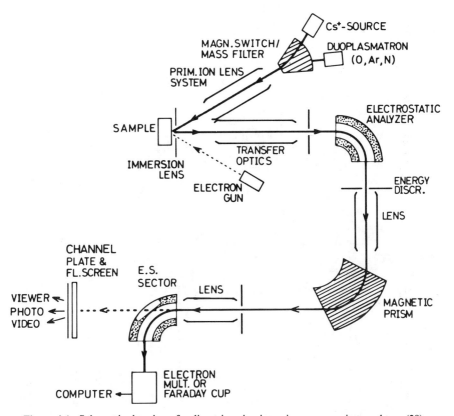

Figure 4.1. Schematic drawing of a direct-imaging ion microscope–microanalyzer (28).

or along-surface line or step profiles), or energy distributions of the secondary ions.

As a complementary example, Fig. 4.2 shows the principles of an advanced instrument of the ion microprobe type (26), which is efficient in the microscope mode due to its excellent lateral resolution. A very narrow and concentrated primary beam (Ga^+) is extracted from a liquid-metal source. The ion currents are here considerably lower than in the previous example; however, an impact area that may approach the theoretical possibilities of optimal lateral resolution (30) receives a comparable or greater current density. Digital rastering of the primary beam is provided for imaging.

The secondary ions pass an electrostatic prism for energy analysis and are focused, by means of a system of transport optics, to the entrance of a quadrupole mass filter. (Quadrupole or time-of-flight separation systems

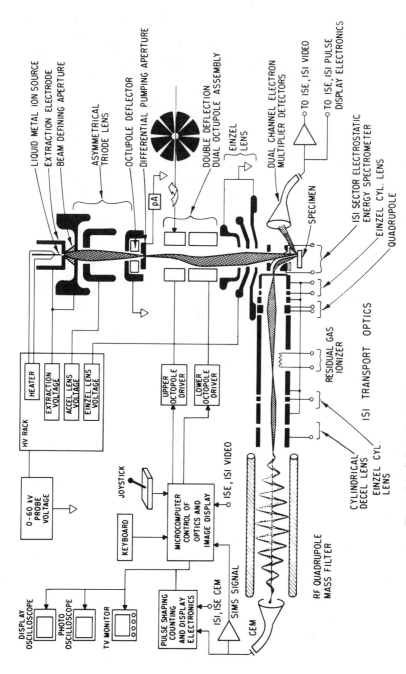

Figure 4.2. Schematic drawing of a scanning ion microscope–microprobe (26).

130

are generally cheaper to produce than electrostatic and magnetic sector analyzers, but at the expense of mass resolution and transmission.) The separated ions are led to an electron multiplier, the signal from which is processed for pulse counting or pulse-mode image display (mapping). As a complement to the SIMS system, a secondary electron counting and imaging facility is also incorporated.

The chief characteristics and specifications of the foregoing examples

Table 4.2. Characteristic Renderings of Two Designs of "Three-Dimensional" SIMS Instrumentation

	Instrument 1[a]	Instrument 2[b]
Primary ions	O_2^+, O^-, Ar^+, N_2^+, Cs^+	Ga^+, In^+
Primary ion impact energy	2–18 keV	40–60 keV
Primary ion current	Max ~10 μA	1.6 pA (at highest resolution) to ~20 pA
Probe diameter d_p	2–250 μm	min. ~20 nm
Max. current density ($\mu A/mm^2$)	~10^2 (O^-) to 10^3 (O_2^+)	5×10^3
Optimal size, imaged area (μm^2)	20×20 to 200×200	$10^3 \times d_p$; min. 20×20
Best lateral and image resolution	~0.5 μm	~20 nm
Sputtering rate in useful profiling (μm/hr)	~10^{-2}–10^2	~$10^{-2} - 5$
Optimal depth resolution in profiling (nm)	~0.5	~1
Mass resolution $M/\Delta M$	~400–10^4	~200–500
Secondary ion energy window (eV)	0.5–120	0.1–15
Secondary ion energy offset (eV)	0–130	0–10
Maximal instrument transmission η	~10^{-2}	~5×10^{-2}
Typical element detection limits (in absence of spectral contamination, at γ ≅ 0.01, mole ppm)	0.03 (insulator) to 0.003 (semiconductor)	~10–100

[a] Direct image microanalyzer (see Fig. 4.1).
[b] Scanning microprobe (see Fig. 4.2).

of modern SIMS instrumentation are summarized in Table 4.2, where instrument 1 is the Cameca IMS-3F stigmatically imaging secondary ion mass spectrometer (28) and instrument 2 is the UC-HRL scanning ion microprobe developed at the University of Chicago in collaboration with Hughes Research Laboratories (26).

3. ASPECTS OF SENSITIVITY AND QUANTITATION

3.1. Element Detection Limits

A SIMS spectrum shows mass peaks that are characteristic of the sputtered solid but conditioned by experimental factors such as the type, intensity, energy, and incidence angle of the primary ions; the acceptance and selectivity of the analyzer and collector; and the ambient vacuum. From the point of view of elemental analysis, one may distinguish between the *intrinsic* spectrum, given by the sputtered matrix, and the superposed *impurity* spectrum. Figure 4.3 shows an example of a positive secondary ion spectrum obtained from a germanium-coated silicon wafer. The peaks due to the germanium intensity are partially affected by contributions from the intrinsic silicon spectrum (such as Si_2O^+, SiO_3^+ ions), a background that has to be accounted for if the impurity concentration is to be measured. The spectrum is seen to contain, in addition to the elemental Si^+, Ge^+, O^+, H^+ contributions (the last two mainly from implantation by primary beam and from residual atmosphere), a multitude of dimers, trimers, and other molecular peaks of the form $Si_kGe_lO_mH_n$ (the subscripts varying, in the detected spectrum of Fig. 4.3, from zero to ~8). One may also see "half-mass" peaks due to the doubly ionized species (Si^{2+}, Ge^{2+}, Si_2O^{2+}).

Figure 4.4 shows the positive spectra from a metalic glass standard (31) at two acceptance ranges of secondary ion energies.

The task of analytical SIMS is to quantify or quantitate the secondary ion currents, that is, to convert the intensity of one or several peaks characteristic of the element L to the corresponding molar concentration c_L. Let a primary ion beam with a current density i_p strike the sample. Collision cascades are initiated, resulting in, among other things, the emission of secondary ions, which are partially detected with an instrument transmission η as a mass spectrum of ions from an analyzed area A. The detected positive or negative current of an ionic species M at the mass number M will be

$$I_M^{+-} = I_p SP_M \eta_M^{+-} \gamma_M^{+-} b_M c_L \qquad (1)$$

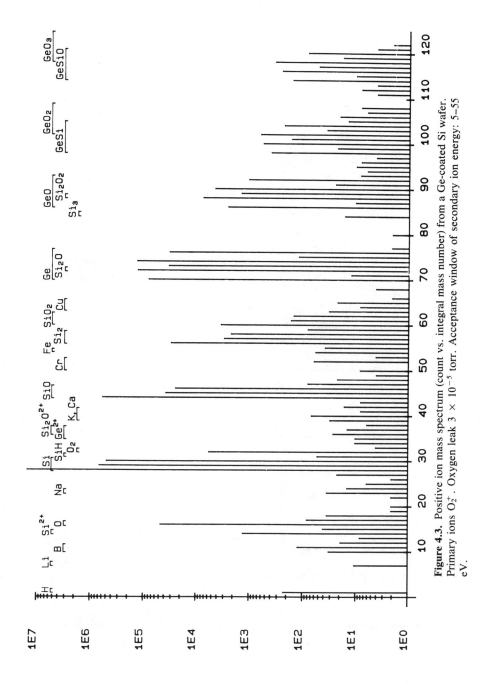

Figure 4.3. Positive ion mass spectrum (count vs. integral mass number) from a Ge-coated Si wafer. Primary ions O_2^+. Oxygen leak 3×10^{-5} torr. Acceptance window of secondary ion energy: 5–55 eV.

133

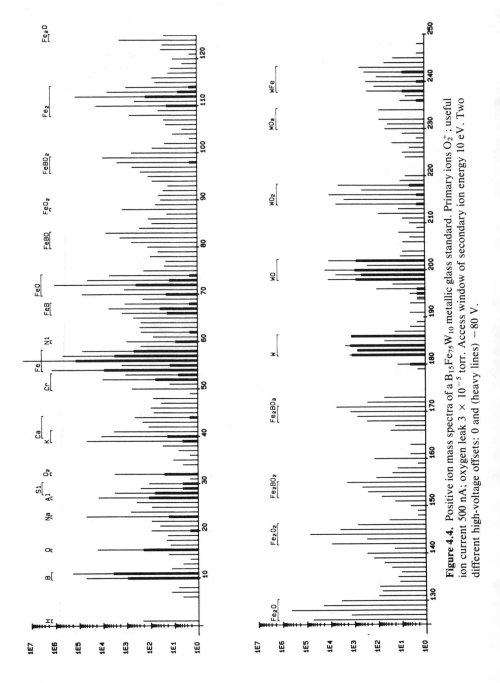

Figure 4.4. Positive ion mass spectra of a $B_{15}Fe_{75}W_{10}$ metallic glass standard. Primary ions O_2^+; useful ion current 500 nA; oxygen leak 3×10^{-5} torr. Access window of secondary ion energy 10 eV. Two different high-voltage offsets: 0 and (heavy lines) -80 V.

where $I_p = i_p A$, P_M is the probability that the particle (atomic or molecular) will emerge as the last step of the sputtering and recombination cascade, S is the sputtering yield (secondary particles per primary ion), γ_M^{+-} is the positive or negative ionizability of M (ions per atom or molecule), and b_M is the isotopic abundance of M in the element L.

In principle, quantitation can be effected from dimers or larger ions as well as from atomic peaks. For example, the $(^{12}C^{14}N)^-$ dimer can be used to detect nitrogen in a matrix where the carbon content is known, as the ionizabilities to N^+ as well as N^- are poor. In Eq. (1), one may then substitute $P_{26}c_L = c_{12}c_{14}r_{CN}$, where $c_{12,14}$ are the moles of the respective isotopes of carbon and nitrogen, and r_{CN} is a measure of the binding probability of carbon and nitrogen in the emerging cascade. The negative ionizability, γ_{CN}, is considerable, and the mass 26 peak may be quite prominent in the spectrum.

Molecular peaks are of particular interest in surface structure and adsorption–desorption energetics ("static SIMS," see Ref. 22) and have been discussed by several authors (11, 12, 22, 32, 33). Normally, however, the monatomic peaks of the elements to be quantified are dominant in the spectra and are preferred for element analysis. The arguments to follow, although extendable to dimers and larger species, will be limited to monatomic ions. It will also be assumed that the sputter rate of the impurity atoms is about equal to that of matrix atoms (equilibrium conditions).

For an isotope M of mass M_L of an element L, then, $P_M = s_L$, where $1 - s_L$ is the (usually slight) probability that atoms of L may be emitted, for example, in abundant molecular combination. From Eq. (1) (dropping the polarity superscripts from here on),

$$I_M = K_{iM}^{-1} b_M c_L I_p \qquad (2a)$$

where

$$K_{iM} = (S s_M \eta_M \gamma_M)^{-1} \qquad (2b)$$

The sensitivity, or the reciprocal of the *lower limit of detection*, for an element L may be arbitrarily defined (13, 34) as the concentration c_{min} corresponding to a multiple f, say, of the background (electronic noise or spectral contamination) J_M at mass number M. Then, from Eq. (2a),

$$(c_{min})_L = f b_M^{-1} K_{iM} J_M (i_p A)^{-1} \qquad (3a)$$

The primary ion current density i_p determines the *erosion rate* of sputtering, v_{sp}, via

$$v_{sp} = i_p \frac{Sm}{z q_e N_0 \rho} \qquad (4)$$

where zq_e is the charge of the primary ion, N_0 Avogadro's number, m the mean atomic mass of the matrix, and ρ the matrix density. Hence Eq. (3) may be rewritten as

$$(c_{min})_L = fb_M^{-1}K_{vM}J_M(v_{sp}A)^{-1} \tag{3b}$$

where

$$K_{vM} \simeq 10^{-5}m \times (z\rho s_M\eta_M\gamma_M)^{-1} \tag{3c}$$

If intrinsic or other spectral background on a mass number M is very low, then J_M is given by the electronic noise. The sensitivity, $(c_{min})_L^{-1}$, is then directly proportional to I_p, that is, to the total "useful" primary ion current, or to $v_{sp}A$, the *rate of material consumption by sputtering from the analyzed area*. The consequence is to be noted: Very high sensitivity of element detection over a *small area* requires a *high sputtering rate*; and when *slow sputtering* is desired (as in "static SIMS" or for shallow profiles) the same sensitivity requires a correspondingly *extended area* of analysis. This competitive quality of sensitivity on the one hand and spatial resolution on the other is illustrated in Fig. 4.5, where the following values have been adopted for the parameters of Eqs. (2) and (3): $f = 3$; $b_M \simeq 1$; $J_M \simeq 3 \times 10^{-19}$; $S \simeq 1$; $s_M \simeq 1$; $\eta_M \simeq 3 \times 10^{-3}$; $z \simeq 1$; $m \simeq 30$; $\rho \simeq 3$; and $10^{44} \le \gamma_M \le 10^{-1}$.

Hence $K_i \simeq 10^{-5}$, and $K_v \simeq 10$ cm^3/A-s. In practical analysis, K_i and K_v are normally found within a couple of powers of 10 from these values, most of the variations being given by the γ and η parameters.

If the background on mass number M from the intrinsic or contamination spectrum is much above the electronic noise level, J_M is likely to be proportional to I_p. Under such conditions it may be useful (34) to relate the impurity ion current to a major matrix peak at mass R, say, whose intensity is measured as I_R. Applying Eq. (2) to this peak and combining with Eq. (3), one finds (with s and η assumed equal for R and M).

$$c_{min} = f \frac{J_M\gamma_R}{b_MI_R\gamma_M} c_R \tag{5}$$

In a realistic analysis, most background peaks can be suppressed (see Fig. 4.4 and Section 3.4) so as to give $J_M/I_R \simeq 10^{-4}–10^{-7}$. Under such premises, in measurements with moderate mass resolution, c_{min} may be

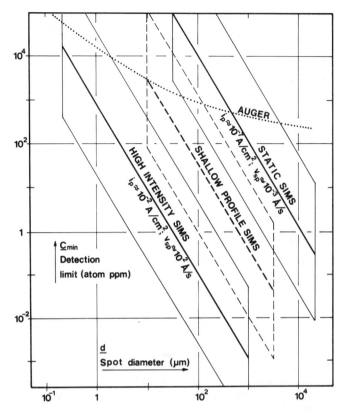

Figure 4.5. Isotope detection limits of SIMS, showing dependence on analyzed area and on sputtering intensity.

expected to lie between $\sim 10^{-2}$ and 10^4 mole ppm, depending mainly on the ionizabilities of the components in the M peak.

3.2. Sensitivity Factors

Quantitation in SIMS, by external standards or otherwise (35, 36) involves the concept of sensitivity factors, which may be defined for isotope M_L of L at mass M from Eq. (2) as

$$SF_M = \frac{dI_M}{dc_L} b_M^{-1} = I_p \left(K_{iM}^{-1} + \frac{d(K_{iM}^{-1})}{d \ln c_L} \right)_{c_L} \qquad (6)$$

Under scrupulously reproducible conditions of analysis, and using external standards with compositions and microstructures not too different from the analyzed samples, the factors S, s, η, and γ in K_{iM} may be considered constant, and useful calibration factors may be obtained empirically, using only the first term in the parenthesis. However, long-term instabilities in analysis (instrumental drift, changes in primary beam conditions, vacuum effects, crystalline effects) make the use of absolute sensitivity factors, for other than very well defined systems, hazardous. It is generally found to be both very feasible and more reliable to utilize the simultaneously measured ion current, I_R, of a matrix reference element R. In the expression for the *relative sensitivity factor* (RSF),

$$RSF_M = SF_M/SF_R \qquad (7a)$$

the primary ion current cancels out. In most practical cases the derivative term in Eq. (6) is also near zero.

It has been found that in many systems the RSFs remain practially constant within quite wide ranges of concentrations, that is, that the different parameters in

$$RSF_M \simeq (s\eta\gamma)_M/(s\eta\gamma)_R \qquad (7b)$$

are only weakly dependent on concentration. Excellent quantitation with RSFs has been reported, for example, for steels (35, 37), binary alloys (38), apatites (13, 39), glasses (40, 41, 31), and semiconductors (42, 43).

In general, however, and particularly in concentrated element mixtures, one must expect RSFs to be concentration-dependent, which may entail tedious routines of external standard preparation and/or careful, theoretically based iterative routines (44).

The dependence of s_M/s_R on concentration is usually slight but may to some extent be influenced by experimental conditions such as ambient vacuum and crystallographic orientations of the samples. The transmission ratio η_M/η_R depends on the energy distribution of the M and R ions and on the acceptance passband of secondary ion energies (energy window and offset, see Section 3.4). Reproducibility of quantification requires that these parameters be controlled (40). Other factors likely to influence η_M/η_R are concentration-dependent effects such as counting deadtime and collector selectivity.

The dominant sources of variation and irreproducibility in absolute and relative sensitivity factors are connected with the ionizabilities γ. The following sections will deal with various parameters relevant to elemental secondary ion yields.

3.3. Ionic Yields of Elements in Sputtering

The differences between elements with respect to ionizability can be very considerable. It is common experience that the yields of positive ions, γ^+, are abundant for the alkalis, alkaline earth elements, and trivalent metals, for example, but very low for the halogens, chalcogens, and group IB and IIB elements. In the negative spectrum, on the other hand, γ^- is generally found to be high for the halogens, chalcogens, and noble metals but low for most other elements. A reasonable semiquantitative prediction of the observed γ behavior is usually rendered by the so-called LTE (local thermal equilibrium) formalism (45), based on the Saha–Eggert equation, drawing an arbitrary analogy between the sputter cascade and the "plasma" in a flame and implying

$$\gamma_{\mathrm{L}}^+ \sim n_{\mathrm{e}}^{-1} \left(\frac{B^+}{B_0}\right)_{\mathrm{L}} \exp\left(\frac{-E_{\mathrm{iL}}}{k_{\mathrm{B}}T_{\mathrm{i}}}\right) \qquad (8\mathrm{a})$$

and

$$\gamma_{\mathrm{L}}^- \sim n_{\mathrm{e}} \left(\frac{g^-}{g_0}\right)_{\mathrm{L}} \exp\left(\frac{E_{\mathrm{aL}}}{k_{\mathrm{B}}T_{\mathrm{i}}}\right) \qquad (8\mathrm{b})$$

Here n_{e} is the electron density in the cascade; B and g denote, respectively, the thermodynamic partition functions and the statistical weights in the first ionized and ground states; E_{iL} is the first ionization potential of element L; E_{aL} its electron affinity; and T_{i}, the "ionization temperature," is an entity of the order of 10^3–10^4 K, characteristic of the matrix.

The thermal equilibrium postulate is difficult to accept as physically rigorous for as fast a process as a sputter cascade. Moreover, T_{i}, as obtained by fitting Eqs. (8) to experimental data, is found to depend not only on the matrix but also on the polarity and charge of the secondary ions (13, 46), crystallographic effects (13, 47), and the energy passband of the secondary ions (40, 41). Nevertheless, the dependence of the elemental yields on E_{i} and E_{a} is so well supported by practical experience that the formalism of the LTE model can be used as a rule-of-thumb orientation for the orders of magnitude of γ to be expected for different elements in a given matrix. In particular, for predicting relative sensitivity factors [neglecting s and η in Eq. (7b)], it has been found useful to extend Eq. (8a) to

$$\mathrm{RSF}_{\mathrm{L,R}} \simeq \left(\frac{M_{\mathrm{L}}}{M_{\mathrm{R}}}\right)^{-\alpha} \frac{(B^+/B_0)_{\mathrm{L}}}{(B^+/B_0)_{\mathrm{R}}} \exp\left(\frac{E_{\mathrm{iR}} - E_{\mathrm{iL}}}{k_{\mathrm{B}}T_{\mathrm{i}}}\right) \qquad (9)$$

[and analogously for (8b)], M_R and M_L being the atomic weights, and the exponent α empirically found (see Section 3.5) to lie between ~0.5 and ~2. If the formalism is obeyed, then a logarithmic plot of experimentally obtained RSFs (divided by the first two terms on the right) versus the first ionization potential should yield a straight line, from the slope of which T_i can be calculated. This is illustrated in Fig. 4.6 for various elements in borosilicate glass (41). For similar empirical systematics regarding negative ion yields (γ^- vs. E_a), see, for example, Refs. 13, 38, and 48.

The predictions of the LTE model, such as those suggested in Fig. 4.6, may be particularly crude in cases where high proportions of L are emitted as molecular rather than monatomic species (non-negligible s_L, see Section 3.1; as often seen for lanthanides, actinides, and other elements, emerging dominantly as oxides) or if they are abundantly ionized in the opposite polarity. In the latter case, Eq. (9) or its negative ion counterpart should be modified by taking into account the *effective* ionizabilities $\gamma_L(\text{eff}) = r_L\gamma_L$, with $1 - r_L$ being the probability of L monatomic ions emerging with another polarity than the observed spectrum.

A caution against the approximate predictions via Eq. (9) is also warranted for the eventuality that the monatomic ion may be a dissociation product from a "directly emitted" molecular species (45, 47), as appears to be the case, for example, for F^+ from calcite or fluorapatite, where much higher ion yields are obtained than would be expected from the LTE model (34).

The parameters of the LTE model may also be affected by crystallographic orientations (13, 44). Further, there is no obvious correlation between the respective fitting parameters T_i (ionization temperature) for positive and negative ions from a given matrix (13).

It should be pointed out here that the measurements in Fig. 4.6 were made under a particular choice of energy passband for the secondary ions, an aspect to be discussed in Section 3.4.

3.4. Energy Distribution and Discrimination of Secondary Ions

The kinetic energies of the ions emerging from a sputtered surface exhibit a more or less sharp maximum, corresponding to the mean information depth in the specimen. The effective range of energies of an ionic species sputtered by primary ions with energies in the order of 2–20 keV may be typically between some 10 and 10^3 eV, narrow distributions characterizing mainly polyatomic clusters. The position of the maximum may also differ somewhat depending on the nature of the species. Thus, molecular ions formed by combination in relatively late stages of the cascade tend to

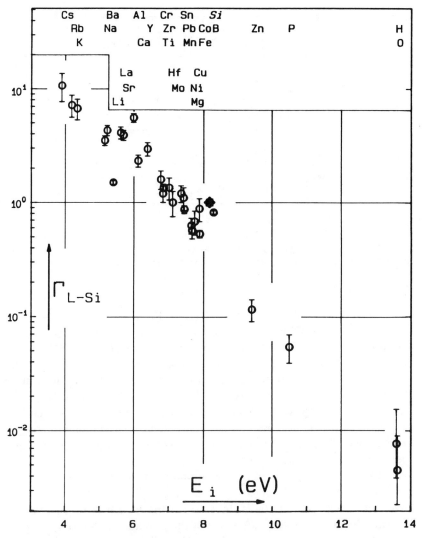

Figure 4.6. Specific ion yields (normalized to Si^+) from borosilicate glasses (41), plotted against first ionization potential. Γ from experimental RSF, last term in Eq. (9), assuming $\alpha = 1$.

form the peak at lower energies than do atomic ions (50, 51). Moreover, the energy distributions of molecular ions are generally more strongly peaked than those of monatomic ions, as illustrated by Fig. 4.7a. This is of considerable significance in practical quantitation routines. By introducing an "offset" in specimen potential, so that only ions with energies

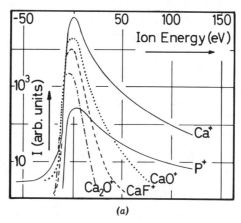

(a)

Figure 4.7. Examples of secondary ion energy distributions. (a) Fluorapatite; primary ions O^- (46). (b) A dental amalgam. Primary ions O_2^+. Note high-intensity emission at low-energy tail for Hg^+. (c) Alkali borosilicate glass; primary ions O^- (41). Axis adjusted for all profiles to coincide at the maximum.

above a certain limit are allowed to reach the analyzer, the molecular contributions to a doublet, triplet, etc., mass peak may be effectively discriminated away. Although this also decreases the total intensity of the monatomic component [i.e., reduces the transmission factor η in Eqs. (2) and (3)], it may often be more essential for sensitivity that the relative intensity of spectral background [i.e. J_M/I_R in Eq. (5)] be lowered. Often this technique of background suppression proves more efficient than the use of extremely high mass resolutions to separate the multiple components at one given mass number. Figure 4.4 is an example of a compound spectrum at two different "offsets."

At the low-energy side of the maximum the drop is usually steep, but the distribution also exhibits a negative energy tail. The ions with energies lower than the product of their charge times the accelerating potential are probably formed at a very late stage of the cascade or "plasma" expansion, or outside the specimen surface. The tail may be particularly dominant for ions from fast-sputtered and volatile systems (see Fig. 4.7b). As the pickup of impurities from residual atmosphere may be considerable in these energy ranges, the usefulness of the "offset" is again evident. However, the low-energy tail has recently been exploited (52) for quantitative analysis essentially on the basis of the idea of postionization of sputtered neutral atoms; if the ionizability in the sputter process itself is kept low, the emitted species can instead be ionized as vapor outside the surface, with far less elemental selectivity than in the sputtering cascade.

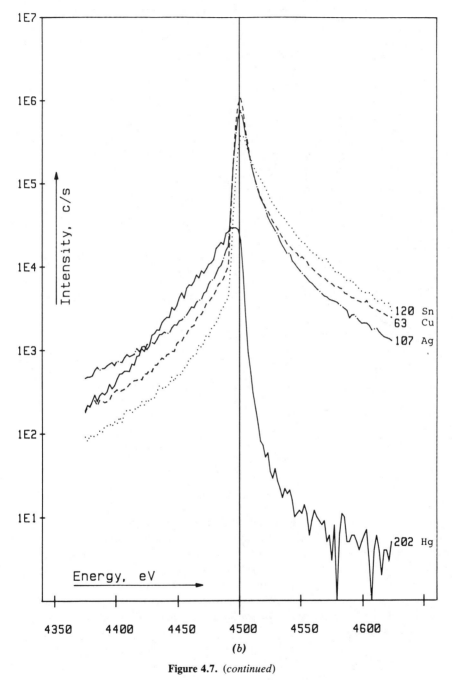

Figure 4.7. (*continued*)

143

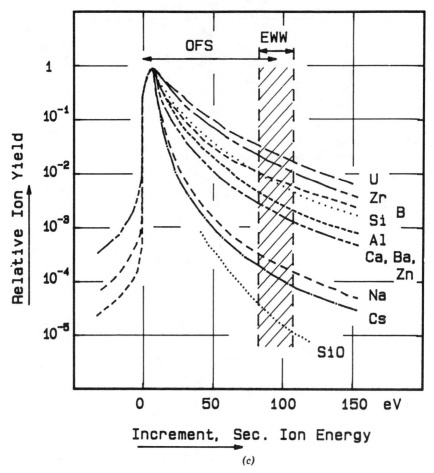

(c)

Figure 4.7. *(continued)*

The post-sputter-ionization ion currents achieved in this way are reported to be only an order of magnitude lower than those from dedicated SNMS [secondary neutral atom mass spectrometry (53, 54)], where postionization is achieved by plasma or laser stimulation.

Figure 4.7c shows that even among monoatomic ions from a given matrix the energy distributions of elements may differ considerably. An often observed tendency is that elements of high valency (e.g., P, Zr, U, Si) exhibit broader distributions than mono- and divalent elements. It is therefore essential in quantitation that the position and width of the pass-band (OFS and EWW in the figure) be well defined. For example, the experimental points in Fig. 4.6 apply at EWW = 50 eV, OFS = 100 eV. A consequence of the elemental differences in kinetic energy distributions is that the fitting parameter T_i in the LTE formalism not only differs from one matrix to another but in a given matrix is dependent on the choice of OFS and EWW. For example, in alkali borosilicate glasses (40) the "ionization temperature" was found to be ~7500 K at very low offset (near the distribution maximum), but of the order of 15,000 K at OFS ≅ 100 eV.

A major problem may arise when insulating materials are studied. A surface charge, due to impinging primary ions and emitted secondary electrons, builds up gradually, changing the specimen potential and shifting the EWW past the distributions of energies. This considerably influences the effective RSF values. Reliable analysis therefore requires automatic compensation for shifts in offset voltage (40, 49).

Some effort has been made in recent years toward understanding the origin of different energy distributions of sputtered ions, but the theory is still in a pilot stage (44, 51, 55–58).

3.5. Mass Effects in Secondary Ion Yields

Secondary ion intensity is "classically" expected to depend on the ion mass M by way of the sputtering yield S_M, viewed as a knock-on process (58). The value of M may also affect the transmission factor η_M by way of, for example, the solid angle of emergence or the discrimination of the collector. Furthermore, the ionization process itself may entail kinetic effects involving the particle mass. The exponent in Eq. (9) therefore normally expresses the superposition of several effects, depending on matrix, mode of bombardment, and mode of detection. The mass fractionation of SIMS, conveniently expressed by

$$I_M = M^{-\alpha} \tag{10}$$

thus may present a non-negligible complication in practical quantitation and is also of interest in any theoretical study of the ionization mechanism. The factor α has therefore received increased attention, particularly in studies of isotope systems (44, 56, 58–60). In practically all cases α was found to lie between ~0.1 and 3 [although theory (58b, 59b, 60) does not altogether exclude the possibility of negative α]. The fractionation is found to be dependent on, among other things, the kinetic energy of the ions, approaching a plateau at high E_{kin}, as illustrated in Fig. 4.8 (59a). An interdependence has been noted (44) among the mass fractionization α, the ionic yield γ, and the half-width $E_{1/2}$ of the kinetic energy distribution: elements with *low* γ appear to display *high* $E_{1/2}$ and *high* α. The evidence

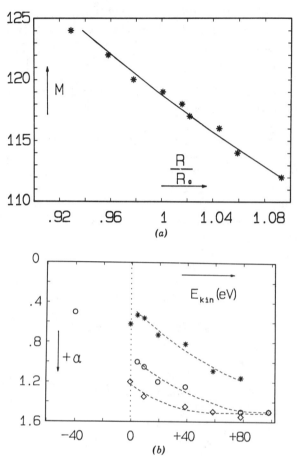

Figure 4.8. (*a*) Specific yields of Sn$^+$ isotopes from tin metal, emerging with energies >70 eV. R/R$_0$, ratio of SIMS-observed to real isotope abundance. From slope of line, fractionization factor α is obtained (59). (*b*) Isotope fractionization factor versus secondary ion kinetic energy for elemental Sn (O), Ge (*), and Cu (◇) (59).

gives some quantitative support to a "bond-breaking" model of ionization (44, 61). More recent work (59b) suggests further systematics and several mechanisms contributing to α.

3.6. Chemical Enhancement of Ionic Yields; Matrix Effects

High and constant yield of secondary ions, as well as reasonable agreement with systematics such as suggested in Eqs. (8) or Fig. 4.6, is found only in ionic matrices or when the specimen surface is purposely contaminated by the introduction of a reactive species. Thus, in metals, the presence of oxygen, either as implant from the primary beam or from an O_2 gas leak at the surface, greatly stimulates the *positive* ion yield. Reproducibility of quantitation can be maintained under such conditions only when the surface layers are saturated with oxygen. This "chemical enhancement" effect is of dominant importance in practical SIMS analysis. The ion yield from a metal bombarded by oxygen is often higher by about 4 orders of magnitude than if it were instead bombarded, for example, by an Ar^+ beam of the same intensity. Another advantage of saturation by oxygen is a certain amorphization of the specimen that prevents undesirable crystallographic effects such as channeling (62).

The physical background of the yield enhancement by oxygen is still imperfectly understood, but the effect appears qualitatively compatible with a bond-breaking model of ionization (44, 63).

The essential role of the chemical enhancement effect naturally underlines the potential influence of various kinds of surface contaminants on quantitation. Again, saturating the specimen surface with purposely introduced reactive species may reduce local effects of impurities.

For *negative* secondary ions, a similar yield enhancement may be achieved if the specimen surface is implanted or flushed with an alkali metal (38, 64), a fact that has been exploited by the development of Cs^+ primary ion guns expressly for the detection of electronegative dopants in semiconductors. The physical basis of this effect is generally believed to be a lowering of the electron exit work function—that is, an increase in electron concentration in the region of the sputter cascade (38).

If the reactive species is introduced only by implantation from the primary beam, then the yield enhancement is effective only at depths corresponding to the projected range of the implant—under normal SIMS conditions, $\sim$50–500 Åu. Closer to the surface the secondary yields may be low and/or erratic. The use of a gas or vapor leak may therefore be preferable if an analysis of the outermost layers is intended. The use of an oxygen leak, up to $\sim$10^{-4} torr at the specimen surface, is a standard procedure in quantitative SIMS analysis of metals.

Introducing the enhancement effect only by means of the primary beam has a further limitation: the efficiency of implantation may be low, especially in fast-sputtered matrices (65, 13). This may be seen from the relation (13) for the steady-state concentration of the implanted species:

$$c_{impl} \simeq 0.6 \times 10^{24} \, \rho\beta/SM \qquad (11)$$

where ρ, S, and M are defined as in Eqs. (1)–(3), and β, the implantation factor, is of the order of 0.5 and only slightly decreasing with increasing i_p. Ion yield studies (65) on impurities in different intermetallic or elemental semiconductor matrices bombarded with O^- or Cs^+ ions have suggested that the specific elemental yields γ^{+-} are proportional to c_{impl}^{-y}, where y is of the order of 3 or 4. Also evident from Eq. (11), if the reactive species is introduced by primary beam only, is that the effective detection sensitivity of a given element is given mainly by the sputtering yield S and only slightly influenced by other differences in matrix properties.

On the other hand, the deposition of the reactive species by means of gas or vapor backfilling may be inefficient at surfaces with a low "sticking coefficient" and is also inhibited if the primary ion intensity is high. The latter fact can be connected with the ratio (13) linking the atom arrival frequency at the analyzed surface from the residual atmosphere or from a gas leak, f_r, with that of the impinging primary ions, f_i:

$$f_r/f_i \cong 10^2 p_r/i_p \qquad (12)$$

where p_r is the effective pressure in torr at the bombarded surface and i_p is expressed in amperes per square centimeter.

4. ASPECTS OF IN-DEPTH PROFILING

Several analytical methods use sputtering for determining variations of concentrations with the depth in solids. However, while other techniques (such as AES and XPS) stepwise sputter away layers of (nonanalyzed) material and analyze the surface of what is left, SIMS continuously studies all material that is being eroded. This specimen economy, together with information depths of the order of one atomic layer and very high sensitivity of elemental or isotopic detection, contributes to making SIMS a major technique of depth profiling, and profiling a major application mode of SIMS.

As mentioned above, sensitivity on the one hand and resolution on the other are two mutually competitive assets. However, in suitable impurity–matrix systems and with good experimental care, an excellent combination of the two can be achieved. Figure 4.9 illustrates this for the profiling of boron in silicon, where the B concentrations are recorded quantitatively with a sensitivity of the order of 1 ppb and in a dynamic range of more than 7 powers of 10, while the depth is reliably indicated with a resolution better than ~5 nm.

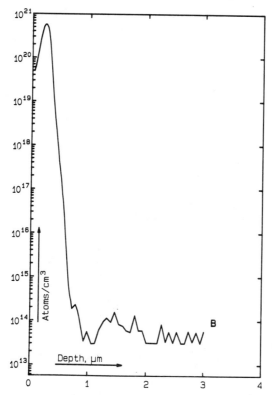

Figure 4.9 SIMS in-depth profile of boron implant in silicon. Positive spectrum. Primary ions O_2^+. Beam raster adjusted at ~0.5 μm depth to increase dynamic range.

Figures 4.10–4.12 show some other typical examples of profiling applications on semiconductors.

The numerous factors affecting the quality of depth profiling by SIMS may be said to entail two main groups of topics: (i) sensitivity, dynamic range, and quantitation; and (ii) depth resolution and depth assessment. In any discussion, some overlap of aspects from i and ii is unavoidable. Thorough reviews of the subject have been published in recent years (66–70). The relatively brief presentation to follow is intended as an application-oriented summary.

4.1. Dynamic Range: Sensitivity Limits and Detector Saturation

At the low-concentration end of a concentration profile, the limit of the dynamic range is given by the signal-to-background ratio. If contamination

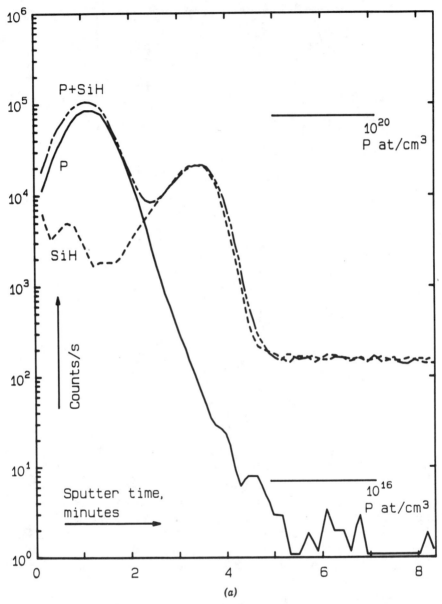

Figure 4.10. SIMS in-depth profiles. (*a*) Phosphorus, implanted together with hydrogen in silicon. High mass resolution ($M/\Delta M \cong 5000$) was employed to separate SiH from P on mass number 31. Negative spectrum. Primary ion O_2^+. (*b*) Chromium implant in silicon. The low detection limits have been achieved by Ta- and Pt-plated ion optics adjacent to the specimen (88).

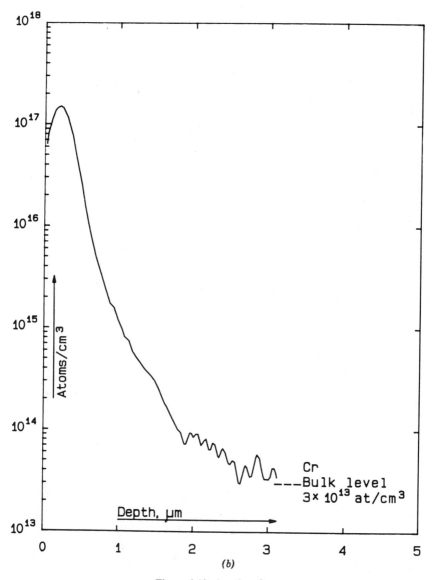

Figure 4.10. (*continued*)

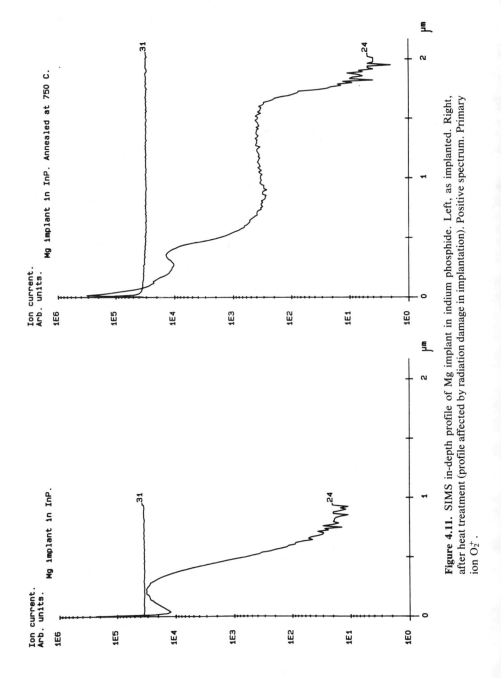

Figure 4.11. SIMS in-depth profile of Mg implant in indium phosphide. Left, as implanted. Right, after heat treatment (profile affected by radiation damage in implantation). Positive spectrum. Primary ion O_2^+.

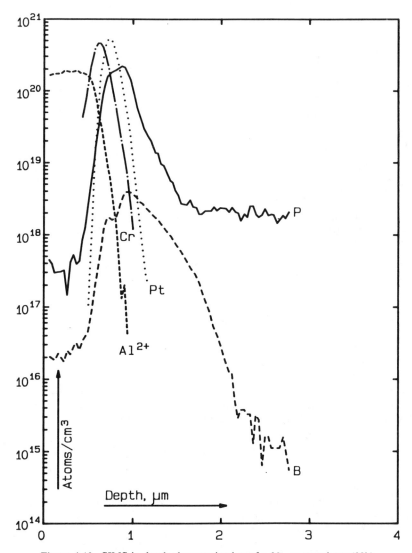

Figure 4.12. SIMS in-depth characterization of a 20-μm transistor (88b).

effects are accounted for (see Section 4.6), the arguments of Eqs. (1)–(4) apply, and it is normally a primary concern in profiling to work with relatively high specimen consumption, that is, high $v_{sp}A$. If high in-depth resolution is desired, a wide analyzed area, rather than high sputtering speed, must be utilized. A limit to this is posed by the need (see below) to analyze only the central portion of the bottom of the sputtered crater and also by the necessity to sputter through the profile layer within a reasonable period of time. At any rate, the dynamic range is furthered if the yield parameters, particularly γ and η, are high.

A limit to dynamic range is posed also at the high-concentration end of the profile, owing to the restricted receptivity of the collector and amplifier. The effective capacity of an electron multiplier assembly is normally of the order of 10^7 counts/s, and in quantitative applications account must be taken of the counting deadtime. For a modern well-adjusted collector, the intrinsic deadtime τ_0 is of the order of 10 ns. However, in actual SIMS profiling the effective deadtime $\tau_{eff} = g\tau_0$ is a function not only of instrumentation but also of the relation between the analyzed area on the one hand and the size and raster of primary beam on the other (71). The factor g (essentially the ratio of the counts with a stationary beam to those with a rastered beam) may in practical profiling amount to well over 20.

Thus, the true intensity I_M of a mass peak M is related to the recorded collector counts I_M^{eff} by

$$I_M = \frac{I_M^{eff}}{1 - g\tau_0 I_M^{eff}} \tag{13}$$

4.2. Cyclic Profiling

Although a good profile may be recorded when concentrating on only one mass number, usually reference to other mass peaks is required for reasons of stability control and quantitation. Practical profiling is as a rule performed cyclically for several masses, as illustrated in Fig. 4.13 for constituents and impurities in an alkali borosilicate glass. Within each cycle the different intensities are counted successively, each for its chosen time interval τ_M.

A pragmatic way of defining the lowest detectable concentration for M is to choose a lowest distinguishable count in each cycle, f_{min}, say (such as 4 ± 2 counts/s). Then, with $f_{min}q_e/\tau_M$ replacing fJ_M in Eq. (3),

$$(c_{min}) = f_{min}b_n^{-1}(\tau_M v_{sp}A)^{-1}K_{pM} \tag{14a}$$

$$K_{pM} = q_e K_{vM} = (m/zN_0\rho)(s_M\eta_M\gamma_M)^{-1} \tag{14b}$$

The magnitude of K_{pM} may vary about 10^{-18} by up to two powers of 10. As an example, $f_{min} = 4$, $\tau_M = 1$ s, $v_{sp} = 10^{-7}$ cm/s, and $A = 4 \times 10^{-5}$ cm^2 yields typically $c_{min} \cong 1$ ppm. Again the sensitivity is seen to be proportional to the amount of specimen consumed.

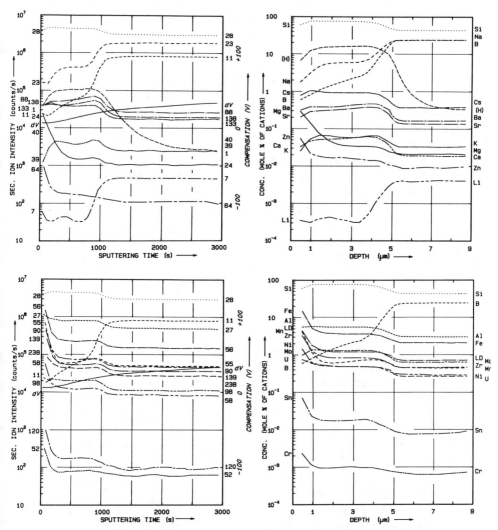

Figure 4.13. SIMS in-depth profiles of elements in a leached alkali borosilicate glass (40). 24 elements recorded cyclically. Left, raw data. Right, data processed by relative sensitivity factors, mole % of sum of cations except H, and by measured differential sputter rates. Corroded depth ~5 μm. Primary beam 0.7 μA, 50 μm diam., 150 μm scan, O⁻.

The effective in-depth resolution in cyclic profiling is given by the thickness sputtered in each cycle, $\Delta z \cong \frac{1}{2} v_{sp} \tau_M$, and so

$$c_{min} \Delta z \, A = \frac{1}{2} K_{pM} f_{min} \tag{15}$$

The total time for each cycle, τ_c, in addition to τ_M and the other peak-counting intervals, also covers the switching and peak-finding times between successive cycles and mass numbers. For profiling in insulator specimens, time in each cycle is also needed for surface-charge compensation of specimen voltage. Thus τ_c is considerably greater than τ_M, which gives considerable gaps between consecutive points of the profile and limits the available counting statistics. When very great counting accuracy is required, such as in isotope studies (55, 71, 72), this is a critical drawback. To bring down the waste time (noncounting time) in each cycle, it is desirable to effect the switching between different mass peaks by electrostatic (as in quadrupole instruments) rather than magnetic means. A fast and exact electrostatic peak-switching device, primarily for isotopic measurements, has recently been developed (73) as a complement to existing ion probes of the magnetic deflection type.

4.3.   Quantitation Aspects in Profiling

Particular difficulties of quantitation are encountered in the outermost layers of a profiled specimen, in a near-surface transition zone before equilibrium ion emission is attained. First, the various elements exhibit *differential sputter yields*; those with great S_M values are initially relatively depleted, until the changed concentrations in the analyzed zone stabilize. The ion yields from the *changed* matrix are only then, in the steady state, meaningfully related to the respective concentrations in the *original* matrix (74, 75). The zone in which preferential sputtering affects quantitation is usually of the order of 10–20 nm. Theoretical models have been proposed to describe this transition zone (76, 77).

As discussed in Section 3.6, ion yields from an initially homogeneous multielement specimen may be influenced by *implantation effects* from the primary beam. Again the analyzed matrix is changed by the presence of the implant, and a constant quantitation relation is attained only at depths beyond the projected range of implantation. Both the sputtering yields S and the elemental ionizabilities γ are affected. As discussed above, the latter effect, "chemical enhancement," may be quite dominant unless the reactive species is already abundant in the original matrix.

Close to the surface the effect of implantation is frequently dominated by chemical enhancement from other impurities. For example, when a metal is bombarded with oxygen, the implant concentration according to Eq. (11) may be far below saturation, and spurious depth profiles may therefore be obtained owing to variations in ambient gas pressure in the sample chamber.

Nonequilibrium conditions of quantitation are obtained whenever the composition of the matrix exhibits significant and fast changes with depth, and in particular at an *interface* between two matrices. The transition zones on both sides of the interface are usually given by the roughness and alignment of both the outer surface and the interface (67) and/or by cascade mixing and knock-on effects (67, 78, 79).

4.4. Atomic Mixing; Segregation; Surface Charging

Depth resolution Δz is usually defined in terms of the measured width of a sharp interface [step function (69, 67)], such as two standard deviations (2σ) of the error curve profile approximation. The ultimate practical limitation to depth resolution is provided by *cascade mixing*. The relatively large number of atoms involved in a collision cascade will "homogenize" the region to a depth typically of the order of 0.5–10 nm. The mixing range is given by the energy and direction of the primary ions and is not depth-dependent. Optimal depth resolution, if limited by cascade mixing, is obtained with low-energy ions incident at a low angle to the surface.

If the first collision between a primary ion and a target atom entails a near head-on hit, a large amount of energy is transferred, and the recoiling atom may penetrate relatively deeply into the solid, 20 nm or more. Such *recoil mixing* or *knock-on effects*, again mainly given by primary ion energy, are generally less dominant than cascade mixing (68).

In the collision cascade, very high concentrations of vacancies and interstitials are created, which affects the mobilities of surroundings atoms. The observed depth profiles may consequently be blurred by *radiation-enhanced diffusion* (80).

The implantation of primary beam species may, depending on, among other things, the direction of incidence, create layers of *lattice strain* and so cause some impurity elements to segregate either at the surface or at the inner end of the changed layer (79). Since such an effect may completely hinder the detection of the segregating impurity, care must be taken in regard of the stoichiometric products of implantation.

Another type of segregation may apply to mobile ions in the *films* of *insulators*, such as Na^+ in layers of SiO_2 on silicon substrate (81). The deposition of surface charge induces a sharp potential drop across the layer, and ion displacement may take place by electromigration. Thus if the primary ion is O_2^+, sodium is segregated at the substrate interface. If the specimen is sputtered by O^-, the displacement of Na^+ may tend toward the outer surface (81). However, since most of the charging is usually caused by emission of secondary electrons, the use of negative primary ions (and conducting surface coating, such as gold or carbon) is generally preferable for the profiling of insulators (82, 41). Positive charging may also be suppressed by the action of an electron spray gun. The particularly difficult problem of negative secondary ion extraction from insulators has only recently been rationally solved (83) by a specially designed auxiliary electron gun that renders a self-regulating potential at the specimen surface.

4.5. Surface Roughening

Even if the intensity distribution of the primary beam over the analyzed area is quite uniform, local differences in sputtering rate will arise due to heterogeneous sample composition and irregularities in the original surface. Consequently the depth resolution Δz will progressively deteriorate with increasing depth z. Ion etching around local impurities and defects may induce spectacular cones or ripples on the crater bottom, even in single crystals (84, 66). When specimens containing several phases varying in hardness, density, and so on, are sputtered, all meaning of depth resolution may be lost unless very narrow beams or very small gating areas are utilized.

In single-phase polycrystalline specimens, the local erosion rate depends on the angle between the primary beam and the crystal axes (62, 85). If the analyzed area is of the same order as the crystalline dimensions, $\Delta z/z$ may be as high as 0.5. If the field of analysis is much wider, the roughening is of the order of crystal size.

Even amorphous flat specimens will eventually exhibit uneven surface topography on sputtering (86), due to the inherent instability of local primary beam interactions (87). On very well prepared surfaces, defects on the scale of a single atomic layer (kinks, ledges, dislocations) will nucleate uneven sputtering. With optimal flatness of the original surface, $\Delta z/z$ values as low as 0.01–0.05 have been obtained.

In most systems, the relative depth resolution $\Delta z/z$ is found to be nearly constant. However, at depths much in excess of the range of primary ion penetration, the dependence of Δz on z has been observed to decrease, even approaching constant Δz at continued sputtering (85).

When an interface is being penetrated, previously developed roughening may strongly affect the depth resolution (and ionizability). Regular cone formation has been observed (85).

Crystallographic roughening as well as cone formation and preferential yield effects can often be alleviated by the use of reactive gas backfill or reactive primary ions (62, 85). Another possible remedy is to rotate the specimen. This may be particularly useful in countering the effect of interface misalignment during sputtering of multilayer sandwich specimens, but is practicable in SIMS only if relatively large areas are analyzed.

4.6. Instrumental Factors

Ion current density is not uniform across the entire primary beam; the crater produced by a stationary beam is therefore more or less rounded. To assure that simultaneously recorded ions all originate from one given depth, a *flat crater bottom* must be produced, either by using a defocused beam or, most commonly, by rastering the beam over the sample surface. Even so, both depth resolution and dynamic range are impaired if one cannot avoid the detection of species originating from localities other than the crater bottom—from crater walls, primary beams (ions and neutrals), the surface nest to the crater, residual gas, and material in adjacent instrumentation.

1. In all quantitative SIMS profiling, *gating* by mechanical or electronic aperture is used to extract ions only from a central portion of the crater into the analyzer (67).

2. Impurities in the primary beam are implanted and subsequently detected. Typically a 1-ppm impurity in the beam may be spuriously recorded as ~0.1 ppm in the specimen (11). High purification of the gas or vapor to be ionized is therefore essential. Ionized impurities can be suppressed, for example, by a Wien *mass filter*. To avoid also the neutral atom component of the beam, a *bend* in the beam column line is required; the impact of unfocused neutral species is reported to be a major limitation to the dynamic range in profiling (67). The beam may also entrain evaporated atoms or charge-neutralized ions from the specimen. Inert *coating* of the surface is usually beneficial in this context.

3. Previously sputtered or evaporated atoms from the specimen, represented in the residual atmosphere near the surface, may redeposit in the crater and then be detected at the wrong depth. Optimal *vacuum* combined with relatively high *beam density* can, according to Eq. (12), minimize this effect.

4. The deposition of sputtered specimen material on adjacent mechanical and ion-optical details may cause serious memory effects. Efficient routines of *cleaning* and baking are therefore required. The erosion of the details themselves brings selective contamination of the mass spec-

tra. For sensitive analysis of, for example, dopants in semiconductors, it has been found rewarding to introduce an easily interchangeable extraction lens design and to manufacture the instrumental parts next to the speimen from a pure refractive metal (88).

4.7. Depth Scale Determination

As the sputtering rates depend on numerous instrumental and matrix factors, they may differ even between two consecutive profiles in a given specimen. The determination is easier if a depth marker, such as an interface, is available at known depth in the specimen. Otherwise, the crater depth must be measured, which is normally possible only after a completed profile. This measurement can be made either (on light-reflecting specimens) by optical fringe interferometry or by a mechanical stylus-type surface level monitor (Talystep, Talysurf, DECTAC).

If the sputtering rate can be expected to vary because of compositional changes, several profiles to different depths, with subsequent crater measurements, may have to be performed on a given specimen.

For exact comparison of different spectral peaks in cyclic profiling, the successively observed intensities have to be carefully interpolated to the "common depth" within each cycle according to the switching-time algorithm (71).

5. IMAGING

In the discussion of surface mapping by SIMS, the notion of *lateral resolution*, Δy, may be brought into Eq. (15) when replacing the notion of a profiling cycle with that of exposure time, that is, understanding τ as the time per image or per complete x–y scan for a surface element, $\Delta A \cong (\Delta y)^2$, and so (see Fig. 4.5).

$$c_{min} = \tfrac{1}{2}K_{pM}f_{min}/\Delta z(\Delta y)^2 \qquad (16)$$

where $\Delta z(\Delta y)^2 = \Delta V_{xyz}$ is the amount of consumed material per detected image on the smallest laterally distinguishable topographic detail.

As mentioned above, from the point of view of imaging, two classes of SIMS instrumentation may be distinguished: (a) direct-image microanalyzers and (b) scanning microprobes (see Figs. 4.1 and 4.2 and Table 4.2). The lateral resolution in the former group is limited by the ion-optical aberrations in the objective (the emission lens). An increase in topographic resolution leads to a steep decrease in the transmission of the lens, which limits the practically obtainable lateral resolution of the direct image to about 0.5 μm. In scanning microprobes, on the other hand, the resolution is given by the size of the primary beam. The usefulness of this mode of imaging is conditional on the production of microbeams of sufficiently high ion-current density. The best obtainable lateral resolution is of the order of the dimensions of a sputtering cascade, a few tens of nanometers, but the current density of such a beam is not, at the present stage of development, much higher than that in the direct imaging technique (see Table 4.2), and the time-averaged flux of impinging ions in the raster mode is much lower than in direct imaging.

Consequently the large-beam imaging instruments have asserted themselves best in detection-sensitive imaging with only moderate lateral resolution and magnification, while the narrow-beam microscopes have produced excellent high-resolution micrographs, although with limited element sensitivity and dynamic range. The two concepts of SIMS imaging are illustrated in Figs. 4.14–4.17 by examples obtained with the

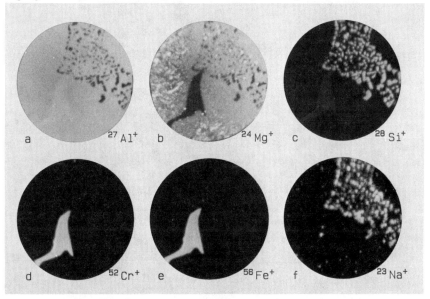

Figure 4.14. SIMS micrographs of element distributions in an Al base alloy, studied with a direct-imaging ion microanalyzer (Cameca IMS-3F). Primary beam 1.3 μA, 8 keV, O_2^+. Bombarded area 200 × 200 μm^2. Exposure 1 s; intensity adjusted by channel-plate voltage individually for each image. Full scale 150 μm. SIMS quantitation (at %): ground phase—Al 98.6, Si 0.7, Mg 0.65, Fe 0.01, Cr 0.05; brittle β phase—Al 70.0, Si 14.0, Mg 0.15, Fe 10.5, Cr 5.5; dispersed Si phase—Al <4, Si 95.4, Mg 0.04, Fe, Cr >0.002. The dimensions of the MgO particles dispersed in the ground phase are only of the order of 15 nm, but they are visible due to high ionizability. Bright shade on β phase in the Si image, $^{56}Fe^{++}$.

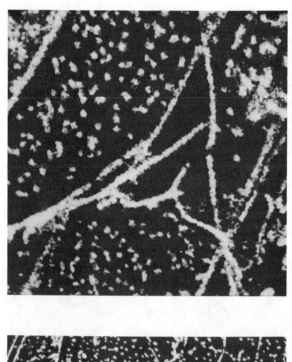
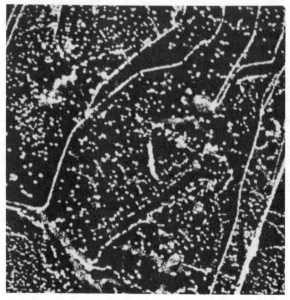

Figure 4.15. SIMS distribution map of $^{35}Cl^-$ in graphite, intercalated with $SbCl_5$, studied with a scanning ion microprobe (26). Primary ion: 1.6 pA, 40 keV Ga^+, 512s. scan. Left, 20 μm full scale. Right, 6.6 μm full scale.

162

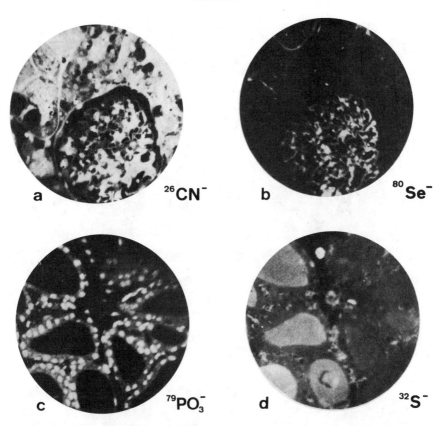

Figure 4.16. SIMS micrographs of element distribution in sections of biological soft tissues, studied with a direct-imaging ion microanalyzer (102). (*a,b*) Liver of rat intoxicated with selenium. CN^- yields a topographic view of tissue matrix; Se^-, distribution of intoxicant. (*c,d*) Thyroid gland. PO_3^- shows phosphorus mainly located in the nuclei of peripheral cells. S^- shows sulfur dominating in thyroglobulin colloids. Imaged field ~200 μm diameter. Primary ion: 1 μA, 15 keV, Cs^+. Potassium-saturated embedding medium, to stimulate negative ion emission.

instruments discussed in Table 4.2. The fine topographic sharpness achieved with the narrow-beam scanning microscope is seen demonstrated only for relatively concentrated and easily ionized elements. The topographically cruder micrographs from the direct-imaging microscope, on the other hand, allow fast distribution studies of a wide range of elements, even at impurity concentrations. Comparable sensitivity cannot possibly be obtained with a 20–40-nm beam microprobe, as this would require sputtering several microns' thickness in a single exposure. With a compromise situation of $\Delta x \cong \Delta y \cong 0.1$ μm, such sensitivity is a rea-

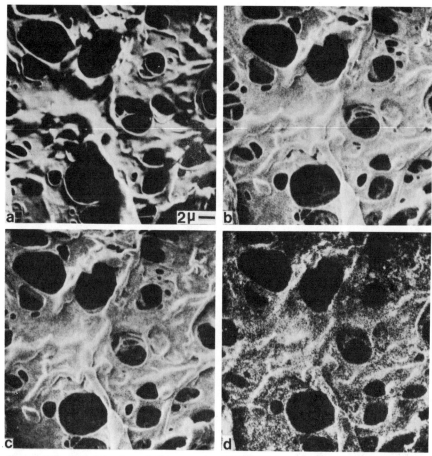

Figure 4.17. High lateral resolution scanning SIM-SIMS images of neonatal mouse skull tissue (calvariae) (103). Primary ion: 10 pA, 40 keV, Ga$^+$, ~1 min scan, 25 μm full scale. (a) Total secondary ion micrograph (SIM); (b) Na$^+$ distribution; (c) K$^+$ distribution; (d) Ca$^+$ distribution.

sonable goal, but because of the raster mode and consequently the low v_{sp}, each exposure would be quite time-consuming in comparison with the direct-imaging mode.

Imaging SIMS is primarily a microscope technique, and quantitation poses particular problems. On each point, or pixel, of the mass-resolved image, the registered brightness is dependent on the quantitation parameters discussed in preceding sections (S, s, γ, η), all with likely lateral variations. Chemical enhancement and memory effects, as well as, particulary, redeposition of sputtered material, mentioned above in connection with depth profiling, are at least as important for surface mapping (89, 90). The observed shade of gray is also influenced by particular imaging artifacts such as microscope relief contrast and depth sharpness. In direct imaging, additional artifacts are caused by local variations in sensitivity over the multichannel plate, screen, and viewing equipment. In principle, quantitation should take place at each point individually, with account taken of all these local factors. This has in recent years led to the employment of *digital methods* for evaluating ion micrographs in terms of element concentrations (90–96). Encouraging solutions have been reached, and the chief remaining limitations appear to lie in the practical combination of sufficient dynamic range of signal, realistic calibration algorithm, and reasonably sized computer infrastructure. Digital imaging is now in the commercialization stage (97).

The latest progress has been inspired by the fact that, basically, SIMS is a microanalytical technique in three dimensions. In early realizations of three-dimensional SIMS, series of mass-resolved ion micrographs, taken at successive depths of a given sputtered spot, obviously gave information on element distribution both laterally and in depth. The assembly process was cumbersome. The efficiency of three-dimensional characterization by SIMS was greatly furthered by the introduction of online image acquisition and image processing (93, 94, 98). Each point of a sputtered layer corresponds to a microvolume, a *voxel*, with coordinates x, y, and z, and can be analyzed, registered, stored, and processed (90), in theory, with only computing power as a limitation. In practice, of course, all complications and artifacts of depth profiling and imaging combine in three dimensions, and quantitation for nonhomogeneous systems entails formidable tasks. Progress in the pioneering work is due partly to the introduction of primary beam internal standards (99, 100). The particularly difficult problem of local assignment of depth coordinates has been tackled by employing sputter burnthrough maps of thin multiphase specimens on homogeneous substrate (101).

The principle of three-dimensional digital imaging (i.e., storing and processing "stacks" of images, each made as an x–y raster frame or digitalized video frame and assigned its z coordinate) is illustrated in Figs. 4.18a and b, as conceived (94) and used for scanning microprobe instrument (102). Figure 4.19 shows the x–y display at different depths of the concentration of an internal implant standard as obtained on the screen of an online computer attached to a direct-image microscope (89).

Three-dimensional digital imaging is still in a pioneering stage. Once the numerous artifacts are under control, the technique can be expected to yield fast and quantitative information according to any of the presently

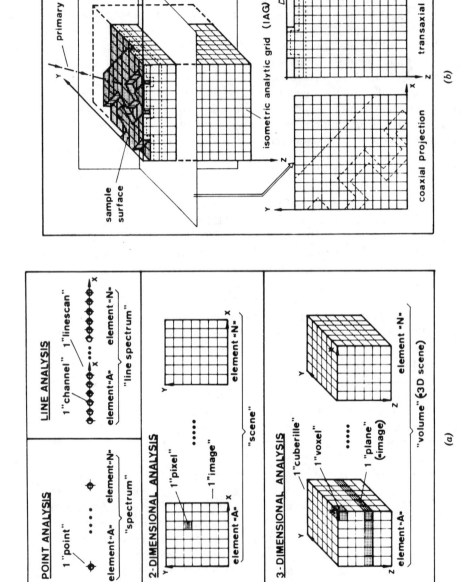

Figure 4.18. (a) Modes of multidimensional distribution analysis by SIMS microprobe (94), schematic. (b) Display of three-dimensional elemental distribution as coaxial and transaxial images (94).

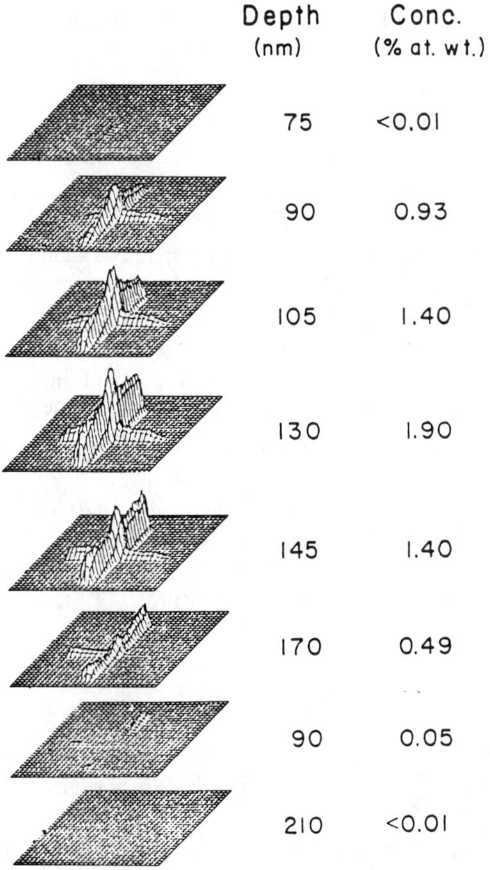

Depth (nm)	Conc. (% at. wt.)
75	<0.01
90	0.93
105	1.40
130	1.90
145	1.40
170	0.49
90	0.05
210	<0.01

Figure 4.19. Display of element distribution in three dimensions from an ion implant (89), obtained by a direct-imaging microanalyzer (Cameca IMS-3F) on screen of on-line computer.

employed modes of SIMS: point analysis, line profiling, in-depth profiling, or $x-y$ imaging. Further, one should readily be able to effect transaxial imaging ($x-z$, $y-z$, or any plane vertical to the surface); select regions of special interest for individual profiling; and carry out many other hitherto unusual systematic probing applications (90). The great challenge of these near-future SIMS concepts is evident.

REFERENCES

1. J. J. Thomson, *Phil. Mag.*, **20**, 752 (1910).
2. R. F. G. Herzog, Patent DRP H172192 IXa/42h, 1942.

3. R. F. K. Herzog and F. P. Viehböck, *Phys. Rev.*, **76**, 855L (1949).

4. R. E. Honig, *J. Appl. Phys.*, **29**, 549 (1958).

5. R. E. Honig, *Int. J. Mass Spectrom. Ion Processes*, **66**, 31 (1985).

6. H. J. Liebl and R. F. G. Herzog, *J. Appl. Phys.*, **34**, 2893 (1963).

7. G. Slodzian, thesis Univ. Paris (1963); R. Castaing, B. Jouffrey, and G. Slodzian, *C. R. Acad. Sci.*, **251**, 1010 (1960).

8. H. J. Liebl, *J. Appl. Phys.*, **38**, 5277 (1967).

9. J. M. Rouberol, J. Guernet, P. Deschamps, J.-P. Dagnot, and J.-M. Guyon de la Berge, *Proc. 5th Int. Conf. X-Ray Opt. Microanal.*, Springer-Verlag, Heidelberg, 1969, pp. 311–318.

10. A. Benninghoven, *Z. Phys.*, **230**, 403 (1970).

11. J. A. McHugh, in S. P. Wolsky and A. W. Czanderna, Eds., *Methods of Surface Analysis*, Elsevier, Amsterdam, 1975, pp. 223–278.

12. H. W. Werner, in T. L. Barr and L. E. Davis, Eds., *Applied Surface Analysis*, ASTM STP 699, 1980, pp. 81–110.

13. A. Lodding, in L. Niinistö, Ed., *Reviews on Analytical Chemistry*, Akademiai Kiadó, Budapest, 1982, pp. 75–104.

14. K. F. J. Heinrich and D. E. Newbury, Eds., *Secondary Ion Mass Spectrometry*, NBS Spec. Publ. 427, U.S. Nat. Bureau of Standards, Washington, DC, 1975.

15. A. Benninghoven, C. A. Evans, R. A. Powell, R. Shimizu, and H. A. Storms, Eds., *SIMS II* (Springer Ser. Chem. Phys. Vol. 9), Springer Verlag, New York, 1979.

16. A. Benninghoven, J. Giber, J. László, M. Riedel, and H. W. Werner, Eds., *SIMS III* (Springer Ser. Chem. Phys. Vol. 19), Springer Verlag, New York, 1982.

17. A. Benninghoven, J. Okano, R. Shimizu, and H. W. Werner, Eds., *SIMS IV* (Springer Ser. Chem. Phys. Vol. 36), Springer Verlag, New York, 1984.

18. A. Benninghoven, R. J. Colton, D. S. Simons, and H. W. Werner, Eds., *SIMS V* (Springer Ser. Chem. Phys. Vol. 44), Springer Verlag, New York, 1986.

19. H. Liebl, Ref. 14, pp. 1–32.

20. (a) V. E. Krohn, Jr., *J. Appl. Phys.*, **33**, 3523 (1962). (b) H. A. Storms, K. F. Brown, and J. D. Stein, in M. Someno and D. B. Wittry, Eds., *SIMS, Fundamentals and Applications*, Riaru-Kogeisha, Tokyo, 1979, pp. 267–272.

21. (a) V. E. Krohn and G. R. Ringo, *Appl. Phys. Lett.*, **27**, 479 (1975). (b) M. J. Higatsberger, P. Pollinger, H. Studnicka and F. G. Rüdenauer, Ref. 16, pp. 38–42.

22. A. Benninghoven, Ref. 16, pp. 438–442; Ref. 17, pp. 342–356.

23. C. A. Andersen and J. R. Hinthorne, *Science*, **175**, 853 (1972).

24. G. Bart, T. Aerne, U. Flückiger, and E. Sprunger, *Nucl. Instrum. Methods,* **180,** 109 (1981).

25. A. R. Waugh, A. R. Bayly, and K. Anderson, *Vacuum,* **34,** 103 (1984).

26. R. Levi-Setti, G. Crow, and Y. L. Wang, *Scanning Elec. Microsc.* **85-II,** 535 (1985).

27. R. Castaing and G. Slodzian. *J. Microsc.,* **1,** 395 (1962).

28. J. M. Rouberol, M. Lepareur, B. Autier, and J. M. Gourgout, in D. R. Beaman, R. E. Ogilvie, and D. B. Wittry, Eds., *X-Ray Optics and Microanalysis*, Pendell, Midland, MI 1980.

29. G. Slodzian, in Ref. 14, pp. 33–61.

30. H. J. Liebl, *Scanning,* **3,** 3 (1980); *Vacuum,* **33,** 525 (1983).

31. F. Rüdenauer, W. Steiger, M. Riedel, H. E. Beske, H. Holzbrecher, H. Düsterhöft, M. Gericke, C.-E. Richter, M. Rieth, M. Trapp, I. Giber, A. Solyom, H. Mai, and G. Stingeder, *Anal. Chem.,* **57,** 1636 (1985).

32. A. E. Morgan and H. W. Werner, *J. Chem. Phys.,* **68,** 3900 (1978).

33. F. G. Satkiewicz, in Ref. 18, pp. 63–65.

34. A. Lodding, in Ref. 17, pp. 478–484.

35. H. W. Werner, *Surf. Interface Anal.,* **2,** 56 (1980).

36. D. E. Newbury, *Scanning,* **3,** 110 (1980).

37. V. Leroy, M.-P. Servais, and L. Habraken, *C.R.M. (Centre Recherche Metallique, Liége),* **35,** 69 (1973).

38. M. Bernheim and G. Slodzian, *J. Microsc. Spectrosc. Elec.,* **6,** 141 (1981).

39. A. Lodding, *Scanning Electron Microsc.,* **83-III,** 1229 (1983).

40. H. Odelius, A. Lodding, L. O. Werme, and D. E. Clark, *Scanning Electron Microsc.,* **85-III,** 927 (1985).

41. A. Lodding, H. Odelius, D. E. Clark, and L. O. Werme, *Mikrochim. Acta Suppl.,* **11,** 145 (1985).

42. D. P. Leta and G. H. Morrison, *Anal. Chem.,* **52,** 277, 514 (1980).

43. A. M. Huber, G. Morillot, and A. Friederich, in Ref. 17, pp. 278–284.

44. G. Slodzian, in Ref. 16, pp. 115–123.

45. C. A. Andersen and J. R. Hinthorne, *Anal. Chem.,* **45,** 1421 (1973).

46. A. Lodding and H. Odelius, *Mikrochim. Acta Suppl.,* **10,** 21 (1983).

47. C. A. Andersen, in Ref. 14, pp. 79–89.

48. V. R. Deline, in Ref. 15, pp. 48–52.

49. D. E. Clark, A. Lodding, H. Odelius, and L. O. Werme, *Materials Sci. Eng.,* **91,** 241 (1987)

50. M. A. Rudat and G. H. Morrison, *Surf. Sci.,* **82,** 549 (1979).

51. E. Michiels, M. de Wolf, and R. Gijbels, *Scanning Electron Microsc.,* **85-III,** 947 (1985).

52. P. Williams, in Ref. 18, pp. 103–104.

53. (a) H. Oechsner, W. Rühe, and E. Stumpe, *Surf. Sci.*, **85**, 289 (1979). (b) H. Oechsner, in Ref. 17, pp. 291–295.

54. (a) C. H. Becker and K. T. Gillen, in Ref. 18, pp. 85–88. Also *Anal. Chem.*, **56**, 1671 (1984). (b) F. M. Kimick, J. P. Baxter, D. L. Pappas, P. H. Kobrin, and N. Winograd, *Anal. Chem.*, **56**, 2782 (1984).

55. J. C. Lorin, A. Havette, and G. Slodzian, in Ref. 16, pp. 140–150.

56. (a) N. Shimizu, in Ref. 18, pp. 45–47. (b) N. Shimizu and S. R. Hart, *J. Appl. Phys.*, **53**, 1303 (1982).

57. A. Wucher and H. Oechsner, in A. D. Romig and W. F. Chambers, Eds., *Microbeam Analysis*, San Francisco Press, 1986, pp. 79–81.

58. (a) P. Sigmund, *Appl. Phys. Lett.* **25**, 169 (1974). (b) P. Sigmund *Nucl. Instr. Meth.*, **B18**, 375 (1987).

59. (a) U. Södervall, H. Odelius, A. Lodding, G. Frohberg, K. H. Kraatz, and H. Wever, in Ref. 18, pp. 41–44. (b) U. Södervall, H. Odelius, A. Lodding and E. U. Engström, *Scanning Microsc.*, **1**, 308 (1987).

60. S. A. Schwarz, *J. Vac. Sci. Technol*, **A5**, 308 (1987).

61. H. Oechsner and Z. Sroubek, *Surf. Sci.*, **127**, 10 (1983).

62. M. Bernheim and G. Slodzian, *Surf. Sci.*, **40**, 169 (1973).

63. (a) K. Mann and M. L. Yu, in Ref. 18, pp. 26–28. (b) M. L. Yu and N. D. Lang, *Phys. Rev. Lett.*, **50**, 127 (1983).

64. R. T. Lareau and P. Williams, in Ref. 18, pp. 149–151.

65. V. R. Deline, C. A. Evans, Jr., and P. Williams, *Appl. Phys. Lett.*, **33**, 7 (1978).

66. E. Zinner, *Scanning*, **3**, 57 (1980); *J. Electrochem. Soc.*, **130**, 199C (1983).

67. C. W. Magee, R. E. Honig, and C. A. Evans, Jr., in Ref. 16, pp. 172–185.

68. W. O. Hofer and U. Littmark, in Ref. 16, pp. 201–205.

69. S. Hofmann, *Surf. Interface Anal.*, **2**, 148 (1980).

70. K. Wittmaack, *Vacuum*, **33**, 119 (1983); *Nucl. Instrum. Methods*, **168**, 343 (1980).

71. H. Odelius and U. Södervall, in Ref. 17, pp. 311–316.

72. U. Södervall, U. Roll, B. Predel, H. Odelius, A. Lodding, and W. Gust, in F. J. Kedves and D. L. Beke, Eds., *Diffusion in Metals and Alloys*, Trans-Tech, Aedermannsdorf, 1983, pp. 492–499.

73. G. Slodzian, J. C. Lorin, R. Dennebouy, and A. Havette, in Ref. 17, pp. 153–157.

74. J. W. Mayer and J. M. Poate, in *Thin Films—Interdiffusion and Reactions*, Wiley, New York, 1978, pp. 119–160.

75. Z. L. Liau, W. L. Brown, R. Homer, and J. M. Poate, *Appl. Phys. Lett.*, **30**, 626 (1977).

76. M. Arita and M. Someno, in R. Dobrozemsky, F. Rüdenauer, F. P. Viehböck and A. Breth, Eds., *Proc. 7th Int. Vac. Congr. & 3rd Conf. Solid Surf.*, Vienna, 1977, pp. 2511–2514.

77. H. W. Winters and J. W. Coburn, *Appl. Phys. Lett.*, **28**, 176 (1976).

78. I. S. T. Tsong, J. R. Monkowski, and D. W. Hoffman, *Nucl. Instrum. Methods*, **182**, 237 (1981).

79. P. R. Boudewijn, H. W. P. Akerboom, and M. N. C. Kempeners, *Spectrochim. Acta*, **39B**, 1567 (1984).

80. P. Blank and K. Wittmaack, *Radiat. Eff. Lett.*, **43**, 105 (1979).

81. H. L. Hughes, R. D. Baxter, and B. Phillips, *IEEE Trans. Nucl. Sci.*, **19**(6), 256 (1972).

82. H. W. Werner and A. E. Morgan, *J. Appl. Phys.*, **47**, 1232 (1976).

83. G. Slodzian, M. Chaintreau, and R. Dennebouy, in Ref. 18, pp. 158–160.

84. P. Dorner, W. Gust, M. B. Hintz, A. Lodding, H. Odelius, and B. Predel, *Acta Met.*, **28**, 291 (1980).

85. W. O. Hofer and P. J. Martin, *Appl. Phys.*, **16**, 271 (1978).

86. I. H. Wilson, *Radiat. Eff.*, **18**, 95 (1973).

87. P. Sigmund, *J. Mater. Sci.*, **8**, 1545 (1973).

88. (a) H. N. Migeon, C. LePipec, and J. J. LeGoux, in Ref. 18, pp. 155–157.
    (b) H. N. Migeon and A. E. Morgan, in Ref. 17. pp. 299–301.

89. A. J. Patkin and G. H. Morrison, *Anal. Chem.*, **54**, 2 (1982).

90. F. G. Rüdenauer, *Surf. Interface Anal.*, **6**, 132 (1984).

91. J. D. Fasset, J. R. Roth, and G. H. Morrison, *Anal. Chem.*, **49**, 2322 (1977).

92. J. H. Schilling and P. A. Büger, *Int. J. Mass. Spectrom. Ion Phys.*, **27**, 283 (1978).

93. B. K. Furman and G. H. Morrison, *Anal. Chem.*, **52**, 2305 (1980).

94. F. G. Rüdenauer and W. Steiger, *Mikrochim. Acta Suppl.*, **2**, 375 (1981).

95. R. W. Odom, B. K. Furman, C. A. Evans, Jr., C. E. Bryson, W. A. Petersen, M. A. Kelly, and D. H. Wayne, *Anal. Chem.*, **55**, 574 (1983).

96. S. R. Bryan, W. S. Woodward, D. P. Griffis, and R. W. Linton, *J. Microsc.*, **138**, 15 (1984).

97. C. A. Evans & Associates, "Digital Image Processing System," new product announcement, 1983.

98. S. R. Bryan, D.P. Griffis, W. S. Woodward, and R. W. Linton, *J. Vac. Sci. Tech.*, in press, (1987).

99. H. Gnaser, F. G. Rüdenauer, H. Studnicka, and P. Pollinger, in H. O. Andrén and H. Nordén, Eds., *29th Int. Field Emission Symp.*, Almquist & Wiksell, Stockholm, 1982.

100. H. E. Smith, M. T. Bernius, and G. H. Morrison, in Ref. 18, pp. 121–123.

101. A. J. Patkin, S. Chandra, and G. H. Morrison, *Anal. Chem.*, **54**, 2507 (1982).

102. J. P. Berry, F. Escaig, and P. Galle, *J. Microsc. Spectrosc. Electron*, **9**, 475 (1984).

103. D. A. Bushinsky, R. Levi-Setti, and F. L. Coe, *Amer. J. Physiol.* **250**, F1090 (1986) See also R. Levi-Setti, G. Crow, and Y. L. Wang, in Ref. 18, pp. 132–138.

CHAPTER

5

LASER MICROPROBE MASS SPECTROMETRY

A. H. VERBUEKEN, F. J. BRUYNSEELS,
R. VAN GRIEKEN, and F. ADAMS

University of Antwerp, Antwerpen-Wilrijk, Belgium

1. INTRODUCTION

Soon after their development in the early 1960s, lasers were applied in mass spectrometry (1) and atomic emission spectroscopy (2). The introduction, at that time, of the focused laser beam as an improved sampling device and a new versatile ion source not only permitted highly effective analysis of bulk samples but also initiated a breakthrough in microanalysis (3–5). Indeed, the laser makes it possible to vaporize, excite, or ionize a small and well-defined volume of solid material. Hence, the laser offers an alternative means to electron and ion beams for localized chemical analysis (6, 7).

1.1. Laser Microprobe Mass Spectrometry

The first laser mass spectrometry studies involved large spot sizes (≈ 150 μm) leading to limited spatial resolution. The long laser pulse lengths and the common reflection geometry resulted in broad kinetic energy spreads of the laser-produced ions, necessitating double-focusing mass spectrometers with their inherent low ion transmission (3). Later research concentrated on reducing laser pulse lengths (by Q switching) and spot diameters (by focusing the laser beam and reducing its wavelength), and on the use of a transmission configuration for ion production and extraction. Additional improvement was gained by using a light microscope system for further focusing of the laser beam down to a diffraction-limited spot size of about 0.5 μm (at 266 nm, Ref. 8), a time-of-flight (TOF) mass spectrometer allowing a high ion transmission, and a time-focusing ion reflector in the TOF drift tube to considerably improve the mass resolution to about 850 at $m/z = 208$ (9).

It is essentially this instrumentation that constitutes the laser microprobe mass analyzer or LAMMA-500 (Leybold-Heraeus, Cologne, FRG) which has been available for applied research for a few years now (10). Application of the LAMMA-500 is limited to microscopic samples thin enough to be perforated by the laser beam: histological sections, films, and powdered materials on a thin supporting foil. To extend the utility of the instrument for the chemical microanalysis of bulk specimens, the LAMMA-1000 has recently been developed; its sample chamber and ion source are modified to have the laser irradiation and ion extraction on the same side of the sample (reflection geometry) (11–16).

Recently, a laser-induced ion mass analyzer called the LIMA-2A has been manufactured by Cambridge Mass Spectrometry Ltd., Cambridge, U.K.; it is capable of performing microanalysis in either the transmission mode for thin samples or using a reflection geometry for thick specimens (17–22). This instrument, like the LAMMA instrumentation, uses TOF mass spectrometry and is equipped with a caroussel carrier that can accommodate up to eight samples. A high vacuum can be maintained in the main analysis chamber to provide compatible conditions for physico-chemical studies of surfaces.

Another approach, intensively followed by Eloy and coworkers (23, 24), is represented by the laser probe mass spectrograph (LPMS), a single-focusing magnetic-sector instrument originally designed for photoplate detection. In this apparatus the laser–solid interaction occurs in the reflection mode, which allows for microchemical analysis of large industrial, geological, and biological samples (23–29). In its earliest version, using a photographic plate ion detector, several mass spectra could be accumulated by repetitive sampling before being evaluated with a recording densitometer (23). The instrument has recently been modified substantially, with respect to not only the laser probe system itself (25) but, in particular, the mode of ion detection. The new version combines the magnetic deviation with TOF measurements using a panoramic electro-optical detector: the photoplate has been replaced by a device combining an ion–electron converter with a scintillator, which allows photons emitted from the whole image plane to be transmitted to a fast photomultiplier through a convenient light guide (28, 30). In this way, as in LAMMA and LIMA, high sensitivity and fast analytical response are combined.

The development of the scanning laser mass spectrometer (SLMS) began in 1976 when Conzemius and coworkers (31–34) adapted a commercial laser to a double-focusing mass spectrometer of the Mattauch–Herzog type, equipped with both electrical and photoplate ion detection. The system was shown to possess a spatial resolution of 25 μm and excellent analytical capabilities for measuring concentration profiles for trace-level analytes in metal systems and solid films. Jansen and Witmer (35) also reported on the successful coupling of a pulsed laser ion source in a reflection geometry with a double-focusing mass spectrometer, while maintaining an easy conversion to a more conventional rf spark ion source. The method permits semiquantitative analysis of both conducting and nonconducting inorganic materials without the use of reference specimens, not only for a wide variety of bulk samples, but also in microanalysis and thin-film analysis. The application of different kinds of lasers in combination with double-focusing mass spectrometers, enabling a high mass and energy resolution, was reported in earlier literature, for example, by Dietze and Zahn (36), Bykovskii et al. (37), and Bingham and Salter (38). A complete detailed survey can be found in excellent review articles by Kovalev et al. (39) and by Conzemius and Capellen (3). They cover the literature on early developments and compare various system designs.

Furman and Evans (40) combined a laser source for ion formation with a Cameca IMS-3F ion microanalyzer (with its high-quality ion optical performance). The instrument provides direct-imaging laser mass analysis (DILMA) and has successfully produced the first ion images resulting from photon irradiation of various materials. The ability to directly image ions produced by laser desorption with about 1 μm lateral resolution is most important. The simultaneous use of ion beam bombardment and laser irradiation (pulsed or continuous-wave laser) on the same surface area further provides a unique way for actual thermal processing of the material in the sample chamber of the ion microanalyzer (41). The laser irradiation is used for both ionization of surface contamination and in situ laser annealing.

During the past 20 years, laser mass spectrometry has matured to a powerful analytical technique with a wide variety of applications (3, 39). A bibliography covering this topic for the years 1963–1982 has been compiled by Conzemius et al. (42), who classified all relevant publications up to 1982. Recently, Michiels et al. (6) reviewed the use of laser microprobe mass analysis (LAMMA) as a molecular microprobe, and Verbueken et al. (7, 43) reported on applications of LAMMA in medicine, biology, and environmental research, while Cotter (44) has focused attention on two new areas using laser mass spectrometer combinations—the laser desorption and multiphoton techniques.

1.2. Related Laser Microprobe Techniques

Laser-induced plasmas have been exploited by emission spectroscopists for analytical purposes since 1962 (45). Indeed, laser microspectral analysis (LMA), with or without electric spark cross-excitation for the intensification of the spectral emission (46, 47), has proved to be useful for the local chemical analysis of a variety of samples of different types (48). Although it is not a mass spectrometric technique, LMA is mentioned here because of its possible use as an analytical microprobe in, for example, biomedical applications (49).

Radziemski et al. (50) reported on the direct detection of aerosols in ambient air using the emission spectra generated by the repetitive electrodeless laser spark in so-called time-resolved laser-induced breakdown spectroscopy (LIBS). In another approach, laser-induced resonance flu-

orescence is utilized by the combination of laser ablation and laser selective excitation spectroscopy (TABLASER) (51).

The combined use of laser ablation and resonance ionization spectrometry (LARIS) was successful for the detection of extremely low impurity
levels in solid materials (52). In this experimental setup, a gas proportional
counter measures the photoreleased electrons of the atomic species of
interest. Recently, the combination of resonance ionization and time-of-
flight mass analysis has been proposed as a powerful analytical tool for
monitoring trace pollutants (53). This technique consists of selectively
exciting and ionizing atoms of the chosen elemental impurity from an un-
ionized gaseous sample by means of tuned laser radiation. After a certain
time delay the resulting ions are analyzed by a TOF mass spectrometer.
Only the neutral atoms are thus used for the resonance ionization process.
In a similar way, the direct action of a vaporizing laser pulse on solid
matter may yield the involved neutrals as well.

2. INSTRUMENTATION

Although hundreds of publications have appeared in the past decade on
the development and applications of laser ionization mass spectrometers,
few have dealt directly with the design and performance of the instruments, while the manufacturers' operating instructions give little attention
to the analytical problems arising from the instrument's performance. For
the sake of brevity, we will limit our description to the design and performance of the commercially available LAMMA-500.

2.1. General Outline of the LAMMA-500

The principle of the laser microprobe mass analyzer LAMMA-500 (Leybold-Heraeus, Cologne, FRG) is based on the excitation of a microvolume
of the sample to an ionized state by a focused laser beam. The analytical
information is derived from mass spectrometry of these ions. The special
feature of LAMMA is that the analysis is performed under microscopic
control.

A detailed description of the instrument can be found in the literature
(8, 10, 54, 55). A schematic diagram of the system is shown in Fig. 5.1.
The sample is mounted in vacuum at an operating pressure of usually
10^{-4} Pa on a movable x–y stage; it can be observed in a binocular microscope through a thin quartz glass window (0.2–0.3 mm), which simultaneously serves as a vacuum seal. Either transmitted or reflected
light illumination can be chosen for specimen observation. The optical

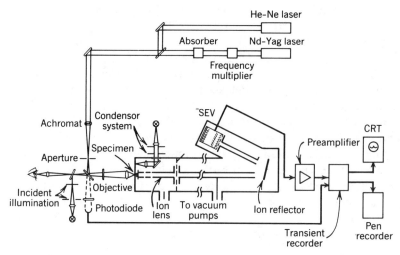

Figure 5.1. Schematic diagram of the LAMMA-500 instrument. A time-focusing ion reflector has been introduced into the time-of-flight (TOF) mass spectrometer to compensate for any spread of initial ion kinetic energy (9).

microscope (magnifications of $100\times$ up to $1250\times$) is equipped with ultraviolet-transparent immersion (glycerol) objectives and is simultaneously used for focusing the red spot of a continuous low-power (2-mW) pilot He–Ne laser onto the region of analytical interest (e.g., a few square micrometers of a thin histological section or a single microscopic particle).

The vaporization and ionization of the selected sample volume is accomplished by a single pulse ($\tau = 15$ ns) of a high-power Q-switched and frequency-quadrupled Nd:YAG laser ($\lambda = 266$ nm, power density in the range of 10^7–10^{11} W/cm^2 for a 1-μm laser focus) that is collinear with the visible He–Ne search laser. The laser intensity can be attenuated to 2% of its initial value by a 25-step UV-absorbing filter system. The energy of each laser pulse is monitored and displayed directly for reference purposes. The best spatial resolution of analysis is obtained at low laser energy. The laser-generated ions are almost exclusively of unit charge. They are accelerated and collimated by an Einzel-type ion lens into the drift tube of a 1.8-m TOF mass spectrometer with a high ion extraction and transmission efficiency (up to 50%, within a limited kinetic energy range). The initial kinetic energy spread of the extracted ions is further strongly reduced by passing through a time-focusing electrostatic reflector (ion mirror) in the folded flight path (9). This ion-reflecting element thus allows bunching of ions of the same mass but different velocities, and thereby considerably improves the attainable mass resolution of the instrument. The ions are postaccelerated toward the first dynode of an open secondary electron multiplier with 17 copper–beryllium dynodes and a gain of about 10^6. The output signal is stored in a fast 2048 channel and 100-MHz transient wave-form recorder of 8-bit resolution. The recorded spectrum is displayed on a fast cathode-ray-tube screen. It can be plotted as hard-copy output by a strip-chart recorder or transferred to a computer for data processing, for example, mass scale calibration, ion signal integration, spectral averaging, and further statistical data treatment.

The microplasma produced by the interaction of the focused laser beam with the sample contains electrically neutral atoms and molecular fragments, as well as atomic and molecular ions of either charge, which are all derived from the constituents of the analyzed volume. The TOF mass spectrometer can be used for the detection of negative as well as positive ions by simply switching the polarity of the electric fields. In this way a complete mass spectrum can be recorded for each laser pulse. Table 5.1 summarizes some technical data and specifications of the apparatus.

Since the LAMMA-500 instrument operates in the transmission mode (ion extraction at 180° relative to the incident laser beam), the sample usually consists of thin specimens (≤ 2 μm)—for example, biological microtome sections or a thin polymeric supporting foil loaded with micrometer-sized particles.

2.2. Ionization Source

2.2.1. Description of the Laser and Focusing System

The LAMMA-500 employs a Nd:YAG solid-state laser in an oscillator–amplifier combination. The laser medium consists of a crystalline YAG host of high thermal conductivity (YAG = yttrium aluminum garnet, or $Y_3Al_5O_{12}$) doped with Nd^{3+}, as the paramagnetic lasing ion, in a concentration or ion density of about 10^{20} cm^{-3}. (Initially, in the LAMMA prototype a ruby laser was installed.) The selected lasing transition is from the $^4F_{3/2}$ state to $^4I_{11/2}$, giving 1064-nm laser light. The laser rod is located at one focus of an elliptical cavity (a reflecting gold-plated elliptic cylinder as a common pumping geometry), where a linear (xenon-filled) flashlamp centered at the conjugate focus efficiently provides the optical pumping. Energy stored in a capacitor bank (high voltage around 1 kV) is dumped into the flashlamps of the pumping chamber, thereby exciting a broad range of Nd^{3+} electronic states. This discharge is triggered by the release of the laser-firing pushbutton of the LAMMA apparatus. Cooling of the laser material, which is in general necessary for a high repetition rate (and cw) laser operation, is provided by a self-contained closed loop

Table 5.1. Summary of Technical Data and Specifications of the LAMMA-500[a]

Microscope	Eyepiece 10×
	Objectives 10×, 32×, 100× (UV objectives)
	Incident and transmitted light illumination
	Phase contrast (option)
Laser	Pilot laser: He–Ne, 2 mW
	High power: Nd–YAG, TEM_{00}, Q-switch
	Energy: 80 mJ at λ 1060 nm, $\tau = 15$ ns
	Frequency doubled, tripled, and quadrupled (at $\lambda = 530$, 353, and 265 nm, resp.)
Time-of-flight mass spectrometer	Length of the drift tube 1.8 m including the ion reflector
	Drift voltage 0 to ±5 kV
Electronics	Open 17-dynode Cu–Be multiplier, preamplifier 100 MHz (10×)
	Transient recorder (2048 channels, 100 MHz, 8-bit resolution, sensitivity range 100 mV–10 V
	Display 10 × 12 cm, strip-chart recorder, laser pulse energy monitor
Vacuum system	
Sample stage	Forepump D 16 A, turbomolecular pump
	Automatic valve system, vacuum gauge
Mass spectrometer	Ion getter pump
Power	3 × 220 V ± 10%, 50 Hz, 16 A
Dimensions	Spectrometer length 1.8 m; width 0.8 m, height 1.5 m
	Electronics 2 × 19″ racks, 28 chassis units high
Weight	800 kg
Sensitivity	Signal height for ^{208}Pb >0.1 V (S/N >10^2) with a multiplier gain of 1 × 10^6 measured on a standard section (10 mM Pb)
Lateral resolution (minimum perforation diameter)	1 μm
Mass resolution of the time-of-flight mass spectrometer	500 FWHM at mass 208 (lead isotope ^{208}Pb)

Table 5.1. Summary of Technical Data and Specifications of the LAMMA-500^a
(*continued*)

Mass range		
	0–100 amu 100–400 amu 400–1000 amu	with 0.02 μs sampling time, depending on delay of transient recorder (2048 channels, min. sampling time 0.01 μs)
Laser energy		20 μJ on entrance aperture of the microscope objective with 100% filter transmission, λ = 265 nm

^{a} Registered trademark of Leybold-Heraeus, Cologne, FRG.

of circulating distilled water containing 8% $NaNO_2$. The oscillator, in LAMMA, has a stable-resonator design (55). Its primary function is to provide the optical feedback mechanism needed to return the radiation emitted by the laser medium back to it (gain medium) for repeated amplification. In the LAMMA apparatus, a concave rear mirror (totally reflecting) and a plane, partially reflecting output mirror are used. Also included in the cavity is a mode-selecting aperture (1.8 mm diameter) to restrict the laser output to a single transverse mode, namely the fundamental TEM_{00} mode. The intensity profile of such a laser beam is shown to follow a Gaussian distribution (55). The exclusion of the higher-order transverse modes is necessary to finally achieve diffraction-limited focusing of the laser beam.

In order to control the laser output and to concentrate all of the available energy into a single intense, short pulse (down to the nanosecond range), the oscillator is Q-switched (56). In this way, high and reproducible peak powers can be obtained that may subsequently be doubled and quadrupled in frequency (the frequency-doubling process requires a high intensity of the incoming light). The Q-switching technique is thus often introduced whenever analytical methods are considered that utilize short laser pulses with high output intensities. In LAMMA, a very fast Q switch is provided by an electro-optical shutter employing a combination of a Pockels cell and a thin associated polarizer positioned in the optical cavity. The light emitted from the laser rod becomes linearly polarized in passing the plane polarizer. The electro-optic crystal (KDP: potassium dihydrogen phosphate) or Pockels cell is maintained at such a voltage (~3.5 kV) that the plane-polarized light incident on the crystal is converted into circularly polarized light by transmission through the crystal. The induced phase shift in the Pockels cell depends directly on the high voltage applied, and hence the effect is often referred to as the *linear electro-optic effect* as opposed to the Kerr effect (56). The laser mirror at the back reflects this circularly polarized light, and in doing so it reverses the direction of polarization. Hence, on reemerging from the Pockels cell, the light is again plane polarized but now oriented at 90° to its original direction and therefore not transmitted by the polarizer. As a consequence, the shutter is closed and laser action will not occur; hence the population inversion can reach very large values. If the bias voltage is reduced to zero, the electro-optic crystal will have no effect other than a small transmission loss within the cavity (minimized by index-matching fluid), as it no longer interferes with the polarization state of the light. When the shutter is suddenly opened, the laser will have a gain much in excess of the losses, and the stored energy will be released in the form of a short (about 15 ns) and intense light pulse. The change in voltage is synchronized with the optical pumping and is triggered at a suitable delay time (~280 μs) after the ignition of the flashlamp.

By a set of external nonlinear crystals (two thermostatted frequency-doubling crystals), the original frequency is quadrupled to the UV wavelength of 266 nm (providing 400 μJ to 2 mJ of laser energy). The unconverted lower harmonics of the laser beam are filtered out each time from the output by an absorbing-glass wavelength separator. At this stage, the beam of a common continuous He–Ne laser (red light with λ = 632.8 nm; typical power of 0.5–5 mW) is introduced in the optical pathway and is aligned collinearly with the invisible UV light of the high-power Nd:YAG laser. This red pilot laser eases the alignment of the laser beam optics and serves for aiming and focusing at a selected area of the specimen. In essence, a quartz plate acts as a beam combiner. Additional mirrors (prisms) are required for correct alignment of both laser beams relative to the optical axis of the microscope.

An intermediate optical system is used to demagnify the laser beam waist to the useful aperture of the objectives of the light microscope and to focus the laser exactly onto the intermediate image plane of the microscope objective. It consists of an achromatic lens with a circular pinhole at its focus. This small aperture is used as a spatial filter to eliminate higher-order contributions, yielding a more homogeneous and circular intensity distribution. The lens pinhole system is axially adjustable for correct focusing of both laser beams onto the sample. Thereafter, a constant fraction of the UV-laser light is transmitted through a beam splitter and is detected by a calibrated photodiode meter for laser energy monitoring. The splitter reflects part of the incident laser light and introduces it into the optical microscope. Before the light enters the microscope, a set of tandem UV-absorbing filters can be used to reduce the laser intensity. A folding mirror inside the microscope nosepiece reflects the laser beam and orients it into the selected objective lens (Zeiss Ultrafluar 100 × or 32 ×, both of the oil immersion type, and a 10 × objective for use at longer working distance). These UV-transmitting objectives are thus used to focus the laser beam onto the thin specimen or particle, and simultaneously they serve for the light microscopic observation of the sample with either incident or transmitted light (total magnification up to 1250 ×).

2.2.2 Laser Power at the Sample

In general, for multimode beams, as for pulsed solid-state lasers, the approximation (57):

$$D = fQ_0$$

is used, where D is the diameter of the focused spot and Q_0 the divergence angle of the beam approaching the lens (with focal length f). If the power in the beam is W watts [$W = P/\tau$; P is the measured quantity of energy delivered during the laser pulse of duration τ (T_{FWHM}), which equals 15 ns], then the power density E at the focused spot is given by

$$E = \frac{4W}{\pi f^2 Q_0^2}$$

For a beam of uniform intensity distribution and radius R, one has

$$Q_0 = \frac{1.22\lambda}{R}$$

Hence,

$$E = \frac{0.86WR^2}{f^2\lambda^2}$$

In these equations, the power densities are derived on the assumption that the light is concentrated uniformly over the focused spot. In practice, however, the intensity at the center will be greater than at the edge; the beam produced by a resonator operated in the TEM_{00} mode has a radial amplitude profile, which is Gaussian, and a characteristic diameter at which the amplitude is $1/e$ times the maximum and consequently the intensity $1/e^2$ times the maximum. If the laser is operating, as we have seen above, in the uniphase mode and if light of intensity less than $1/e^2$ of that at the center is excluded, then

$$Q_0 = \frac{2\lambda}{\pi R}$$

And so in this case

$$E = \frac{\pi W R^2}{f^2 \lambda^2}$$

Calculations based on idealization and geometrical approximation readily provide approximate values of laser intensity, focal diameter, and energy available at the sample. With commercially available equipment it is possible to achieve focusing of the laser beam to nearly the limit imposed by the laws of optical diffraction. As a first approximation, the diffraction-limited spot size may be used and the radial amplitude profile estimated as Gaussian. Estimates of the minimal laser-induced perforation radii are 0.2–0.3 and 0.3–0.5 μm at $100\times$ and $32\times$ objectives, respectively, in polymer films (55) and 1.8 μm in silver foils (58), comparable to 0.48 μm with the $100\times$ objective as derived from diffraction theory and geometrical optics (63).

The energy actually deposited in the sample, being the time integral of the power available at the sample and the absorbance of sample and plasma, is not directly measurable. The energy delivered is estimated by integration of the current through a photodiode, which, by means of a beam splitter, measures part of the laser beam. The fraction absorbed is not directly known.

A pulse of 15 ns delivering 1 μJ to a spot with a Gaussian profile and a radius parameter of 1 μm would have a peak intensity of 1.06×10^9 W/cm². In practice, absolute laser energies between about 0.001 and 10 μJ may be available at the sample surface, leaving a power density in the range of 10^7–10^{11} W/cm² for a spot of about 1 μm diameter. The latter figure is subjected to considerable uncertainty, since an absolute power density range and the pulse energies stated by different laboratories are not strictly comparable. Indeed, interlaboratory standardization is uncommon, nor has the method of calculating focal intensities been agreed upon. Uncertainty in estimates of power density at the sample is aggravated by chromatic aberration of the microscope between the UV focus of the pulsed laser, the red laser for sampling, and the visual focus, which may easily differ from each other by a few micrometers. Short-term variability of the laser energy is estimated to be 8–15% RSD (Fig. 5.2).

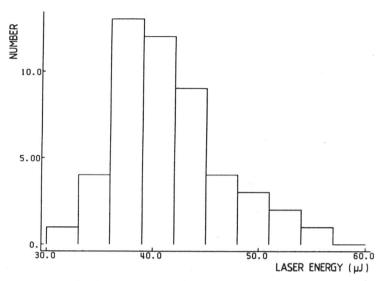

Figure 5.2. Histogram of laser energy (μJ) readings illustrating the pulse-to-pulse variability. A relative standard deviation (RSD) of 12.4% is obtained (N = 49, mean = 41.43, standard deviation = 5.141, variance = 26.43, lowest value = 33.00, highest value = 55.90).

The mean energy available to the sample molecules is the time integral of laser intensity minus reflection and transmission losses in sample and plasma, divided by the number of excited molecules or atoms. Intensity, transmittance, and reflectance are not constant over time or space, or from sample to sample. This energy is partitioned into vaporization, ionization, and plasma expansion. Hence, inevitable variations in laser intensity or laser focusing quality will lead to differences in the volume analyzed, the degree of molecular fragmentation, and the ionization efficiency.

2.3. Mass Spectrometer

The microprobe is based on the timed ionization of solids by the application of a short laser pulse. Hence, a continuous instead of gated ion extraction into a TOF mass spectrometer is allowed. However, as a result the mass resolution directly depends on the time definition of the generated ion bunch. For the sake of simplicity, we will assume that the ion formation follows a pulse-shaped time profile.

Since the ions are extracted from the side of the sample that is opposite the laser beam, the spectrometer configuration of the LAMMA-500 is referred to as a *transmission type*. This contrasts with the frequently used reflection geometry, in which the laser and ion optics are on the same side of the sample. The latter configuration is utilized in the LAMMA-1000, introduced in 1982, which was developed for the analysis of larger bulk samples.

2.3.1. Ion Extraction and Mass Separation

The ions formed upon laser impact are extracted from the source to the drift tube by a potential difference of typically 3000 V, and then they pass through an ion lens that focuses them toward the detector. The ions are thus accelerated toward the drift region of the TOF mass spectrometer by a "homogeneous" electrostatic field formed between the sample itself and the entrance electrode of the spectrometer (since the gap covers a distance of 5.75 mm, the electric field is about 5200 V/cm). The LAMMA-500 can be used to analyze negative as well as positive ions, depending on the polarity of the potentials applied to the electrodes. In the drift region of the TOF tube the ions separate according to their velocities. The principle is outlined in Fig. 5.3. The mass discrimination and dispersion of a TOF mass spectrometer derives from the classical energy–velocity relationship $E_k = \frac{1}{2}mv^2$, or $v = \sqrt{2E_k/m}$, where v represents the velocity of the ion, m its mass, and E_k the kinetic energy. In the first part of the mass spectrometer the bunch of ions has been accelerated over a short distance in order to gain a sufficient and ideally identical kinetic energy. Let us assume that the contribution of any initial kinetic energy E_0 due to the laser ionization process can be neglected. The kinetic energy obtained by the ions after passing the whole acceleration region equals

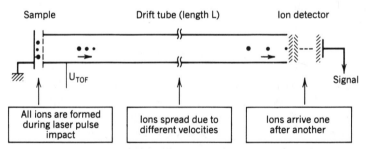

Figure 5.3. Principle of a time-of-flight (TOF) mass spectrometer (55). A bunch of ions is accelerated and then projected through a drift space toward a detector. The flight time for each ion is a function of its mass, so the detector receives a series of separate ion bunches, one for each mass present. A mass spectrum is obtained by recording the detector current versus time.

the potential difference U_{TOF} multiplied by the charge of the ion q. So we obtain

$$v = \sqrt{2qU_{TOF}/m}$$

The ion mass can be expressed in atomic mass units (1 amu $= 1.6604 \times 10^{-27}$ kg), and its charge by z times the electron elementary charge ($e = 1.6021 \times 10^{-19}$ C), so that

$$v = 1.39 \times 10^4 \frac{\sqrt{U_{TOF}}}{\sqrt{m/z}}$$

In the case of LAMMA, z is almost exclusively equal to 1.

From this point on, the ions are allowed to drift for a distance L sufficient for the velocity differences to spatially separate the various mass species from each other. These spatially resolved groups of ions arrive in succession at the detector, presenting a series of ion pulses with specific arrival times indicative of their mass-to-charge ratios. As a first approximation, the arrival time t of the ions will then be

$$t = L \sqrt{\frac{m}{2qU_{TOF}}} = \frac{7.2 \times 10^{-5} L\sqrt{m}}{\sqrt{U_{TOF}}} = C\sqrt{m} \tag{1}$$

where L represents the length (in meters) of the effective drift region, implied mainly by the geometry of the instrument. The proportionality of time versus $\sqrt{m}$ is a crucial property for the mass calibration of time-of-flight spectra (see further below). The constant C can be calculated from the geometry and the electric potentials of the TOF mass spectrometer. In practice, however, it is easier to derive the value of C by identifying several mass peaks of spectra obtained from reference samples.

A practical TOF spectrometer will have not only a field-free drift region but also accelerating and decelerating regions whose transit time will contribute to the total flight time. In the LAMMA-500, an ion lens of the immersion type is introduced to increase the acceptance of the spectrometer and to collimate and focus the ion beam. Furthermore, a time-compensating ion-reflector system and a postaccelerating region near the detector cathode supply additional electric fields that influence the velocity and pathway of the ions. A thorough calculation of the ion flight time for mass calibration purposes will have to deal with all these contributions. In the earlier designs of TOF mass spectrometers, ions were focused in the plane of the detector by a double acceleration of the ion bunch through two grids (constituting a kind of slit lens). Once the geometrical dimensions were fixed, the electric field strengths could be optimized for focus conditions, independent of the ionic mass and kinetic energy after acceleration (59).

In general, the resolving power of a mass spectrometer will be better if the components of the ion beam are simultaneously well separated in space and well focused to small images. The spectrometer completely resolves adjacent masses if the spread of flight times for ions of a single mass does not exceed $t_{M+1} - t_M$. If the spectrometer just resolves an ion bunch of mass M from another differing in mass by ΔM, then the mass resolution R of the instrument is defined as

$$R = \frac{M}{\Delta M}$$

Two ion types are just resolved when the distance Δt between their centers becomes equal to their mean width W. Hence, the following expression for the resolving power RP, as stated by Vogt et al. (55), is obtained:

$$RP = \frac{\Delta t}{W} \left(\frac{M}{\Delta M} \right)$$

For a good estimation, Δt may be considered to be proportional to $L/\sqrt{U_{TOF}M}$, as obtained by differentiation of Eq. (1). The peak width W is influenced by a number of factors: the distribution of the initial kinetic energies E_0 of the ions, differences in ion optical pathways, the length of the ion acceleration and drift path, and the response time C_A of the ion detection system. The initial energy E_0 results from the laser ionization process itself and depends on the excitation state of the sample, which in turn is influenced by the applied laser intensity and the absorption characteristics of the irradiated sample area. Due to the insertion of the ion reflector, the resolution of the TOF mass spectrometer is considerably improved by eliminating the ΔE_0 effect to a first-order approximation. In this way, the possible dependence of the mass resolution on the sample excitation conditions is strongly reduced. Eventually, the contributions from the instrumental response time C_A, essentially determined by the transit-time spread of the electron multiplier and the frequency bandwidth of the preamplifier and transient recorder (typically around 10^{-8} s) (55), become dominant and lead to

$$RP = C \sqrt{\frac{M}{U_{TOF}}} \frac{1}{C_A}$$

where C is a constant.

Kaufmann et al. (9) improved the mass resolution to about 850 by inserting an ion reflector in the TOF mass spectrometer to compensate for the spread of initial ion energies. This value is certainly sufficient for the identification and measurement of elemental ions. At the high mass range in particular, any deterioration of the instrument's performance will have a pronounced effect on the spectral quality and the obtainable mass resolution. Anyway, it is difficult to accomplish a baseline separation of full-scale peaks at adjacent masses in the range from $m/z = 600$ onwards (Van Vaeck, personal communication). Moreover, comparison of instruments with and without ion reflector reveals that this element reduces the peak width by overlapping several possible contributions from isobaric ions with different initial velocities. As a result, the reflector allows for a better peak shape at the cost of some specific information.

An electrostatic TOF mass spectrometer provides some particular advantages over other types of mass analyzers. Since there are no deadtimes of scanning, a whole mass spectrum of one polarity can be recorded from a single laser pulse. Since the TOF mass spectrometer essentially consists of cylindrical tube electrodes, a high transmittance of the ions is obtained, leading to a high intrinsic ion-detection capability. The latter is also determined by the ion-extraction efficiency. The quality of the ion extraction will depend on both the geometry of the system (instrument design and applied voltages) and the specific distribution of radial ion kinetic energies. Despite its importance, a fully quantitative treatment of the extraction efficiency as a function of general physical parameters is not available. The fraction of the ions that can be extracted efficiently will thus be characterized by the distribution of the ions in space and the distribution of their respective velocity vectors. If such distributional differences occur in single or sequential measurements, then ion discrimination during the extraction process is to be expected (60). It has been suggested that atomic ions are formed mainly in the central core region (of a higher ionic temperature) of the laser-induced microplasma, while molecular ions may be formed by a so-called laser desorption process in the lower-intensity fringe of the beam. During the ion formation, initial kinetic energy distributions may be generated that differ from species to species (61). The geometry and position of the sample, the laser beam intensity profile and focus, and the deposited laser energy can vary, being changed either by the operator or by random shot-to-shot variation. All these factors can have a pronounced influence on the extraction efficiency and, as a consequence, also on the finally obtained experimental signal variability.

2.3.2. Ion Lens

In the LAMMA-500 mass spectrometer an ion lens is positioned in the drift tube, centered about 2 cm behind the accelerating electrode. It consists of a tubular three-element (with the same radius R) constant-voltage immersion ion lens. The central segment is held at a different and lower potential (U_{lens}), in absolute value, with respect to the two outer electrodes, which are kept at the same potential, the U_{TOF} value. Such an ion lens is classified as an Einzel (single) lens. The lens length L is 13 mm, and its geometrical parameters are $L/R = 2.6$ and $D/R = 0.4$, where D represents the spacing between the segments. It serves to collimate the ion beam (Fig. 5.4) and thereby to improve the transmittance of the ions through the spectrometer by focusing them onto the detector surface. Otherwise, most of the ions would be lost due to collision with the tube walls. Such lens effects are governed by the physical laws of ion optics. An ion beam crossing a potential discontinuity suffers a change in its normal velocity component, while its velocity parallel to the equipotential surface remains constant. Since the velocity of the ions is proportional to the square root of their energy, which in turn is determined by the local potential, an equation is obtained similar to the ordinary law of optical refraction:

$$\sqrt{U_1} \sin i_1 = \sqrt{U_2} \sin i_2$$

where U is the local potential and i the angle with respect to the normal of the surface line separating the two regions. In this way, it is possible to visualize how the ion beam is traced across a region of varying electrostatic potential. The higher the voltage change, the more pronounced its focusing effect.

Two objectives are to be attained in TOF mass spectrometry. First, monoenergetic ions with the same mass must all have the same flight time from the source to the detector, whatever the angles and positions of emission are. Second, the flight times must remain constant even if the ion energy varies slightly around a central value. Thus, the ion optical design should have as few spherical and chromatic aberrations as possible. In practice, this is not achievable with a single electrostatic converging lens. Ion lenses, as used in LAMMA, inevitably suffer from some chromatic aberration, a dependence of the focal length on the ion kinetic energy. A different ion lens potential would be required to focus ions of different kinetic energies onto the same point. Also, at a certain voltage setting of U_{lens}, a varying transmittance is observed for ion species with different kinetic energy distributions. It has been shown that in LAMMA, ions of different chemical origin show different distributions of initial ion kinetic energy (60–64).

For optimal ion transmission, the zone of ion formation has to be imaged onto the sensitive diameter of the detector. Therefore, any defo-

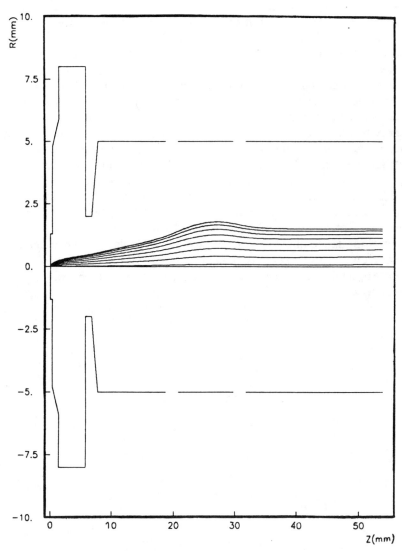

Figure 5.4. Focusing effect of the ion lens on the ion trajectories: $U_{\text{TOF}} = 3000$ V, U_{lens} = 900 V for ions with an initial kinetic energy of 5 eV, starting from the sample surface on the optical axis and for specific angles (2.5°, 12.5°, 22.5°, 32.5°, 42.5°, 52.5°, 62.5°, 72.5°) (62).

cusing condition of the ion lens will influence the spectral intensity. Not only can the total spectral intensity vary, but also the relative signal intensities of the different chemical species. Degradation of spectral information and even masking of peaks can occur due to incorrect selection of the ion lens potential.

Careful focusing of the ion lens allows the ion signal detection to be maximized. By measuring the signal intensities at various lens potentials while keeping all other instrumental parameters constant, the effective energy bandpass of the system can be estimated as well as the extent of spectral attenuation due to ion lens defocusing. Figure 5.5 gives an example of such a study for Pt^+ ions derived from a 200-Å-thick Pt/C coating. The resulting signal intensities show an optimal value around $\mid U_{\text{lens}} \mid$ = 1020 V. A typical feature of the intensity profile is that a sharp falloff can be observed at the low-voltage side (near $\mid U_{\text{lens}} \mid \approx 900$ V), while at the other side a tailing effect is seen. Quantitative examination of the spectral influence of the lens settings for a particular sample at specified experimental conditions may indicate the typical order of magnitude of the ion optical discrimination effect (60). Variation in sample position and ionization conditions may require optimizing the ion lens potential for each specimen, and if various ion species with different kinetic energy

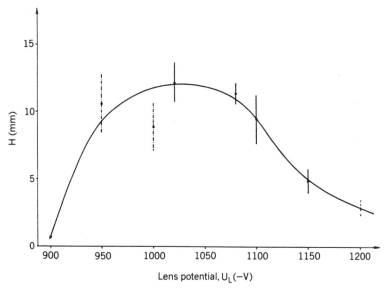

Figure 5.5. Effect of the ion lens potential on the signal intensity of $^{194}Pt^+$ in the positive-ion mass spectra of a 200-Å-thick Pt/C coating (95% Pt) onto a Formvar film (at -3000 V accelerating potential). The error bars represent the RSD for up to eight measurements.

distributions are to be compared, then the lens potential should be optimized and specified for each type of ion.

The ability to essentially extinguish the ion beam by overfocusing the ion lens was used by Mauney (63) to estimate the chromatic aberration of the ion lens. The lens potential was varied until the ion signal sharply decreased with increasing lens voltage differential $| U_{TOF} - U_{lens} |$. Thus, at the low-voltage side (low $| U_{lens} |$) a threshold lens potential (threshold for overfocus) was defined and measured as a function of ion kinetic energy determined by the ion accelerating potential. On the basis of the negative-ion spectra of carbon foil, the potential required to focus ions was shown to be proportional to the ion kinetic energy (the ratio of threshold lens potential to TOF potential was measured to be 0.327). A 150-eV shift in ion kinetic energy requires a 100-V shift in the lens potential (63). In practice, a ratio U_{lens}/U_{TOF} close to 1/3 is used for proper focusing of the lens. The exact ion lens voltage for the maximum detected signal will, however, be dependent on the specific kinetic energy and the spatial position of emission of the ions.

2.3.3.   Ion Reflector

The LAMMA-500 design includes a single-stage electrostatic ion reflector that compensates for possible differences in the ion flight times resulting from different ion kinetic energies. The reflector causes ions with high kinetic energy to be retarded by their longer flight into the retarding field, and so their arrival time at the detector can be matched with that of the slower ions of the same mass. The principle is outlined in Fig. 5.6. The

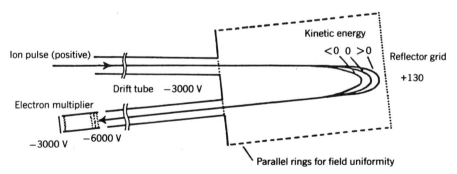

Figure 5.6. Schematic representation of the effect of the ion reflector of the type employed in the LAMMA-500. Ions of higher kinetic energy penetrate more deeply into the reflecting field and are delayed with respect to ions of lower kinetic energy. The delay compensates for differences in transit time of the field-free regions.

ion transmittance of the reflector has been studied by Mauney and Adams (65) by means of a simple thermal ion source that emits Na^+ ions from a tungsten filament.

As stated above, the mass resolution of the TOF mass spectrometer is considerably improved by employing the ion reflector system in the drift path. This is due not only to its energy- and time-focusing action, but also to the fact that it simultaneously enlarges the effective drift path (and thus the mass dispersion). Compared to a linear TOF mass spectrometer, the folded drift tube nearly doubles the available length of flight path without consuming much more space. On the other hand, it can influence the relative signal intensities in the spectrum by acting as a filter that cuts off all ions having more kinetic energy than $(U_{TOF} - U_{refl}) \cos^2 \theta$, where θ is the reflector incidence angle. It also attenuates ions when the reflected image is not centered well on the detector (63).

The spectral resolution and intensity are only weakly affected by the reflector voltage. Resolution is optimized when the derivative of arrival time with respect to the mean ion kinetic energy (KE) which is expressed as a function of the potential difference in V, is zero. This condition holds for the LAMMA-500 when $U_{TOF} - U_{refl} = 1.0832KE$, where the kinetic energy is nominally equal to U_{TOF}. Intensity should be maximal when the beam centered in the first drift tube is reflected to the center of the detector. This occurs when $U_{TOF} - U_{refl} = 1.07KE$ for $\theta = 3°$. Both these functions are relatively insensitive to changes of kinetic energy by ~50 eV, and they are almost simultaneously optimal.

The reflector's influence as an instrumental parameter is not important when applied in the normal working range around $U_{refl} = 125$ V, where all signal intensities have reached a plateau. However, its energy cutoff property may be utilized to obtain cumulative ion kinetic energy distributions (60–67). Such an operation requires the recording of repetitive spectra from homogeneous samples under uniform ionization conditions, while the cutoff level can be stepped across the energy range by adjusting the reflector potential.

2.3.4. Mass Calibration

For the LAMMA-500 spectrometer, the following experimental flight time equation is established:

$$t = a + b\sqrt{m/z}$$

The parameter a represents the time delay between the nominal starting time of the experiment and the actual time of ion emission. It includes the delays in propagation of the electronic signals, the various triggering events, and the ion production process itself, which is induced within a time interval of the order of the laser pulse duration (10 ns). The value for the parameter a was determined experimentally by Mauney (63) to be about 3.0 μs. The coefficient b comprises the flight equation coefficients for all of the various regions of the spectrometer. Its value can be calculated from instrumental geometry and electronic parameters and compared to the value obtained experimentally.

The apparatus can be divided into the following regions: acceleration, ion lens, drift tube 1, reflector, drift tube 2, and postacceleration at the cathode. The latter region refers to a gap of about 3 cm between drift tube 2 and the electron multiplier cathode in which the ions receive an additional acceleration in order to minimize mass discrimination of the detection efficiency. Thus,

$$t_{flight} = t_{accel} + t_{drift} + t_{lens} + t_{refl} + t_{cath}$$

Using some geometrical approximations and basic assumptions on the electrostatic potentials along the spectrometer axis to simplify the calculations, the coefficient b can be expressed as (63)

$$b = 1.972 \times 10^{-6} + \frac{6.42 \times 10^{-3}}{|U_{TOF} - U_{refl}|}$$

$$(U_{lens} = \tfrac{1}{3}U_{TOF}; \ U_{cath} = 2U_{TOF})$$

for an acceleration voltage U_{TOF} of 3000 V and with the time t expressed in seconds. Variation of the ion reflector potential thus changes the drift time for ions of a particular mass and energy: Ion signals are shifted to shorter delay times (lower b) at higher (more repulsive) reflector voltages due to a decreased effective drift length. As an example, the flight time of $^{208}Pb^{+}$ ions under routine conditions ($U_{TOF} = -3000$ V; $U_{refl} = 125$ V) is calculated to be 58 μs. The relative contribution of the time, in seconds, spent by the ions in the first acceleration region, for $U_{TOF} = -3000$ V, is given by

$$t_{accel} = 15.1 \times 10^{-9}\sqrt{m/z}$$

For the $^{208}Pb^{+}$ ions this would be 218 ns, which is indeed small compared to the actual ion drift time.

Mauney (63) demonstrated a good correspondence between the calculated flight-time coefficients and the experimental calibration coefficients obtained by making two-point evaluations of sets of mass spectra. The relative discrepancy between the values was not statistically significant.

2.4. Ion Detection

Since ions striking metal surfaces cause the emission of secondary electrons, open electron multipliers are often used in mass spectrometers to detect small ion currents. The ion current is converted to an electron current at the first dynode of the multiplier and is magnified by further dynodes, the total amplification or gain being given by

$$G = G_1 G_2^m$$

where G_1 is the ion–electron conversion coefficient (number of secondary electrons released per incident ion at the conversion dynode) and G_2 the multiplication factor of each of the m remaining dynodes (at each of the multiplier stages, further secondary electrons are emitted). Understanding of the ion-induced electron emission at the first dynode is important for the use of an open electron multiplier in mass spectrometry, because at this particular stage, mass discrimination complicates the measurements; that is, isotopes of smaller mass bombarding the metal surface produce a heavier emission of electrons than isotopes of greater mass with equal kinetic energy. In general, the ion–electron emission parameter G_1 is dependent on the composition and purity of the target dynode, the relative angle of ion impact, the charge and polarity of the ions, their electronic configuration, and the ion kinetic energy. Discrepancies in ion detection occur as a function of the chemical nature and structure of composite ion species and, mostly, of the ion mass (68–71). In practice, it is often assumed that ions with the same velocity also produce the same number of secondary electrons. For a constant ion kinetic energy, the detector response will then decrease with the square root of ion mass. However, Fehn (68) established the relationship

$$G_1 \sim v^b$$

where the exponent b decreases with increasing ion kinetic energy and increases in a nonmonotonous periodic way with increasing atomic number of the impacting ions. In particular for the heavier elements, a simple signal correction by a factor proportional to the inverse square root of the isotope mass is no longer valid. Molecular and cluster ions have been studied less extensively than atomic ions (70, 71). Positively charged dimers (M_2^+) and monoxide ions (MO^+) result in a higher ion–electron conversion coefficient than the corresponding elemental ion (M^+). Also, the conversion efficiency appears to be a factor of 2 higher for positive ions than for negative ions. The average secondary electron yield increases with the number of atoms per ion, and for constant mass it is larger for polyatomic ions than for single atomic ions. The difference in yield between cluster ions in homologous series (of the type M_{n+1}^+, M_n^+) decreases for larger values of n. Such experimental data extrapolated to ions of 6 keV energy (see below) may be useful for ion signal correction in LAMMA. However, a comparison with published data may be troublesome or even unjustified, because the conversion efficiency of the cathode will depend on the type, age, and previous history of the electron multiplier and the experimental conditions.

At the end of the drift tube the ion bunch is postaccelerated toward the first dynode (cathode) of the detector (U_{cath} is usually -6000 V for positive-ion detection). The nonlinear field near the cathode induces a slight lens effect that tightens the ion beam and minimizes angular losses due to collisions outside the sensitive area of the detector. The detector consists of an open 17-stage CuBe venetian blind secondary-electron multiplier (Model 9643/4B, EMI, Middlesex, UK), which provides a gain of about 10^6 (for a multiplier voltage $U_{mult} = 3000-3250$ V). The spectrum may be recorded after an additional preamplification ($\times 10$) of the output of the electron multiplier. Signal pickup can be performed at the sixth, twelfth, or seventeenth (last) dynode. By using these different outlet connections and by varying the respective voltages U_{mult}, the signal amplification can be varied over a wide range and adapted to the signal intensities. This is particularly useful to reduce the overall spectral intensity and to avoid overloading of intense peaks (see further below). A maximal voltage of 3.5 kV is recommended when using the total dynode series. This value has to be lowered when fewer dynodes are used.

An important property of the detector, in particular for quantitative measurements such as isotope ratio determinations, is the experimental linearity of response, that is, the proportionality between the input ion current and the signal output. Substantial nonlinearity for output peaks greater than 0.2–0.3 V has been observed by Simons (72) for ion detection in LAMMA. A decrease in gain of the electron multiplier with increasing output pulse amplitude restricts the analytical accuracy (Fig. 5.7). Within the experimental error, a simple saturation model fits the response curve, so its effect can be calibrated and corrected. In the case of large ion currents, for which saturation of the last electron multiplier stages would distort the spectra, the signal can be tapped at the sixth or twelfth dynode

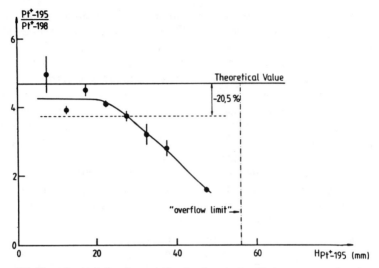

Figure 5.7. Experimental abundance ratio of major to minor Pt isotopes as a function of the Pt signal intensity (input range 0.5 V). The graph illustrates the extent of saturation and signal suppression when intense ion signals are measured.

stage. In the low-signal regime, the statistical nature of the ion-counting process becomes important and must be duly considered.

2.5. Spectrum Recording

In any form of TOF mass spectrometry, the time-dependent ion signal must be recorded as the series of masses arrive over a short period of, usually, microseconds.

In the original design the pulsed ion signal from the detector is fed into a fast transient recorder (Model B-8100, Gould-Biomation, Cupertino, California) that incorporates a 100-MHz, 8-bit resolution analog-to-digital converter and a total of 2048 memory channels (2 kbytes). The input sensitivity of the transient recorder (TR) ranges from 50 mV to 5 V and allows matching of the signal peak heights according to the nominal "8-bit" dynamics of the TR. Digitizing of the signals is usually performed in sample intervals of 10, 20, or 50 ns. The selected delay time and sampling rate determine the recorded mass range and to some degree the mass resolution achieved. Using an acceleration voltage of 3000 V, the principal portion of the mass spectrum, between 20 and 200 mass units, arrives in the time interval 20–60 μs after the laser pulse. The TR is triggered by the Pockels cell controller when the Q-switched laser pulse is released. For adequate representation of the peak shape, several time intervals should be sampled within each peak profile. If the routinely used sampling interval of 20 ns is selected, about seven measurements are made per mass peak in the lower mass range (below $m/z = 120$, at $V_{acc} = 3$ kV).

The resolution and dynamic range of the TR are limited by the 8-bit word size (static resolution). At high-signal slew rates, the precision of digitization decreases and the resolution of measurement is severely reduced for large peaks. The resolution of the TR is limited by the aperture uncertainty and signal frequency. Simons (72) stressed the importance of this decrease in resolution, from nominally 1 in 256 down to about 1 part in 5 (2.3 bits) for large mass spectral peaks at 10 MHz, which impairs the precision of measurement. Solutions to minimize this problem are based on the use of transient recorders of a higher performance such as the TR 8818 B from LeCroy (LeCroy Research Systems Co., New York), whose hybrid circuitry allows for an aperture uncertainty of only 25 ps. It allows the retention of 8 significant bits at 100 MHz signal frequency. According to Simons (72), an effective dynamic range of 6.5 bits (1 part in 90) is achievable in practice. An additional advantage is the memory size of 32 kbytes, sufficient to record a properly resolved mass spectrum up to $m/z = 600$ on each laser shot. By installing two recorders operating at different signal attenuations, either working in the same mass range or in two sequential, possibly overlapping, mass regions, one can attempt to overcome the limitations imposed by the preset input range. Nevertheless, in addition to the digitization errors, the nonlinearity of the electron multiplier must be considered when large relative abundance ratios have to be measured accurately, for example, for quantitative isotope ratio measurements.

2.6 Spectral Data Processing

The nature of the instrument allows the acquisition of a complete mass spectrum in less than a millisecond, and the complete cycle of selecting a target, focusing the laser, and acquiring and displaying the spectrum can be completed in less than a minute for simple samples. Hence, the use of a computer to rapidly store the spectra considerably simplifies the overall operation. Further, a computer can conveniently perform many tasks that are slow and tedious manually.

The program LAM, designed in the authors' laboratory (63, 73), is briefly discussed as an example. It readily allows calibration of the mass spectra, integration or peak height measurement of the ion signals, logging of specific instrumental parameters and informative comments, and calculation of the laser energy available at the sample. It also provides a record of the details of calibration of the TOF mass spectra (the calibration coefficients are reported and can be checked). A particularly interesting feature of the program is the ability to perform a channel-by-channel averaging of the spectral data over a large set of spectra recorded under identical conditions, so that an averaged mass spectrum is obtained that can be plotted by the computer to provide a representative mean. In conjunction with the program LAM, an additional program for simple statistical analysis and tabulation can be used to process the output from LAM, for example, to select sets of spectral data that can subsequently be averaged and/or compared with respect to a specified reference mass signal. Together, these programs allow the analyst to acquire, store, integrate, report, and process a large number of mass spectra per day. More sophisticated manipulation of the LAM output can be performed by applying elaborate statistical programs, either available from the literature or developed in house.

3. LASER–SOLID INTERACTION

3.1. Mechanism of Laser-Induced Ion Production

LAMMA has been used to study the influence on laser-induced ion formation of physicochemical parameters such as type of bond, crystal structure, heat of formation, and ionization potential for a number of foils (74–76). The laser energy threshold values for foil damaging are significantly different for metals (Al, Fe, Ag, Au) and semiconductors (Si, Ge) on the one hand, and insulators (such as epoxy resin or SiO_2) on the other hand. The former show no significant wavelength dependence, whereas dielectric material exhibits considerably higher threshold values and a decrease in threshold with increasing photon energy. The different behavior of these materials can be understood if the mechanism leading to damage is treated in terms of Joule heating. This means that energy dissipation from the electromagnetic field to the lattice is dominated by interaction with free electrons in the conduction band. Furthermore, the main features of the observed cluster-ion intensity distributions can be explained in terms of molecular stability and/or thermodynamic properties of the evaporation process. Nevertheless, some results led to the assumption that nonequilibrium properties might also be important in the laser-induced phase transition. Therefore, a model has been developed to describe the influence of crystal structure on the cluster distributions, provided nonequilibrium properties dominate laser-induced phase transitions and cluster stability effects are of minor influence. This model describes evaporation of clusters as a two-step process. The first step consists of the formation of near-order structures during the heating and before transition into the gas phase. The near-order structures are defined as regions of lower vibrational excitation than the surrounding lattice. The second step then describes evaporation of the near-order structures of clusters into the vacuum, with the assumption of negligible recondensation and gas-phase reactions. It has been concluded that both equilibrium and nonequilibrium properties of the laser-induced evaporation process may determine the occurrence of inorganic molecular structures, depending on the target as well as on the irradiation conditions.

As for the ionization, the plasma can be approximated by a local thermal equilibrium (LTE) model, which implies energy relaxation through collision processes, without any radiation. In this case, ion concentrations are described by a chemical equilibrium yielding the well-known Saha–Eggert equation:

$$\frac{n_+}{n_0} = \frac{2Z_+}{Z_0 n_e} \frac{(2\pi m_e kT)^{3/2}}{h^3} \exp\left(-\frac{I - \Delta I}{kT}\right)$$

where n_+, n_0, n_e are concentrations of ions, atoms, and electrons, respectively; Z_+, Z_0 are partition functions of ions and atoms, respectively; m_e is electron mass; T is absolute temperature; h and k are the Planck and Boltzmann constants; I is ionization energy; and ΔI is the lowering of I depending on n_e. A second ionization model was also considered, since high plasma densities indicated that properties of the solid state possibly play a role in ion formation. Ionization equilibrium at a solid surface is described by the Saha–Langmuir equation, where the solid is characterized by its electronic work function Φ:

$$\frac{n_+}{n_0} = \frac{Z_+}{Z_0} \exp\left(-\frac{I - \Phi}{kT}\right)$$

The results of both equilibrium ionization models appear to be in reasonable agreement with experimental values. However, a very dense plasma state would be required to explain the experimentally observed significant contribution of large cluster ions. Their formation would, moreover, be inconsistent with a dissociation equilibrium in the gas phase at $T \approx 5 \times 10^3$ K. Therefore, it is assumed that ionization processes in contract with the solid phase play a dominant role.

Ion formation processes of alkali halides were studied in a systematic way by Jöst et al. (77) and Schueler et al. (78, 79). The fast dissipation of energy at solid surfaces resulting in ion production was treated extensively by Krueger (80, 81). It is thought that the ions (including clusters) are formed directly from the solid state, with several uppermost atomic layers being involved (77). Gas-phase interactions are negligible. The absorption of light by the transparent halide crystals is assumed to be initiated by multiphoton absorption with production of free electrons, which are excited and transfer energy to the lattice. The further energy transfer is maintained by rapid polaron (individual or collective electron–phonon interaction) Joule heating. If enough energy is offered to the solid, a breakdown can be initiated with formation of free ions. The dependence of the alkali cation yield on the static sample temperature was discussed by Schueler et al. (78, 79) for the LAMMA-1000 instrument (reflection geometry). These results show that the fast-ion formation process induced by pulsed laser irradiation cannot simply be related to a heating process. The influence of the static sample temperature on the mass spectra of alkali halides and in particular the irreversible breakdown of the cluster-ion detection at higher static sample temperatures indicate that the physical and chemical properties of the sample itself are important for the ion formation process. The action of the laser irradiation on the sample seems to be fundamentally different from and independent of static heating of the sample. Different laser irradiation wavelengths do not significantly affect the experimental results.

A large number of compounds from different chemical classes were analyzed by Heinen (82) using LAMMA in the laser desorption (LD) mode. In this mode, sample excitation must be kept as low as possible so that a minimum number of ions (roughly $\geq 10^3$ ions) are produced to yield a few significant mass peaks. These ions are, in most cases, directly related to the structure of the material analyzed, often with a high yield of the parent molecular ions (M^+), in competition with protonated ($M + H)^+$ and cationized $(M + Na)^+$ and $(M + K)^+$ species, in addition to some specific fragment ions. Heinen (82) demonstrated a characteristic influence of the chemical nature of the compounds on the types of ions observed in the LAMMA mass spectra. It was suggested that molecular ion formation from organic compounds and inorganic salts by laser irradiation is primarily a thermal process. In particular, in the case of formation of molecular ions from aromatic compounds, an electronic process was evident (82). As to the positive ions, Van Vaeck et al. (83) observed a striking similarity between LAMMA and electron impact mass spectrometry with a direct probe. A major role was attributed to the formation and subsequent fragmentation of odd-electron molecular ions upon laser microbeam irradiation. In the negative-ion detection mode, the LAMMA spectra revealed a ready disintegration of nonionic organic compounds: The major signals were due to carbon cluster-type ions (C_n^- and C_nH^-), which do not contain specific molecular information.

Hercules et al. (84) and Novak et al. (85) noted that a variety of processes contribute to laser-induced ion emission from solids, including vaporization, shock waves, and gas dynamics. In the central region of the laser impact zone, a direct interaction takes place between the laser and the sample, resulting in ionization. This is a region characterized by extensive fragmentation and formation of atomic species. At high laser power densities, this region of direct laser impact can be classified as a microplasma. The adjacent region is an area of high thermal gradient. This region, which has been referred to as the selvedge, can be looked upon as some sort of condensed yet mobile phase between the solid state and the gas phase. It is an area where collisions can readily occur, some of which may result in chemical reactions. The surface of this region is rapidly expanding, and here the quasi-liquid sample becomes a gas. Finally, there is a "cloud" produced by expulsion of material into the vacuum, where ion–molecule reactions (gas-phase reactions) could occur.

Seydel and Lindner (86, 87) proposed a model for laser desorption of organic compounds based on experimental results obtained for chemically different compounds. The analysis of thermolabile nonvolatile high-molecular-weight compounds such as phospholipids was successful only at high laser power densities. Useful mass spectra were obtained by irradiating the back surface of the copper grid, which carries the sample and faces toward the laser. They used relatively thick samples (~30 μm) that were not perforated at laser impact. The organic solid absorbs the energy from the impacting UV-laser pulse by means of electron excitation. The electronic energy is converted to vibrational energy of the molecules, which leads to fast heating via the liquid state of the small volume of direct laser–solid interaction. After superheating of the formed liquid, a "phase explosion" caused by an avalanche-like generation of vapor nuclei may take place. This phase explosion leads to the generation of a shock front that traverses the remaining solid with energy dissipation. When the shock front reaches the back surface of the sample, intact molecules and fragments are released due to a momentary perturbation of the binding potentials. Quasi-molecular ion formation should then take place in the selvedge by the attachment of alkali ions.

An excellent review has been published by Hillenkamp (88, 89) on laser-induced ion formation from organic solids, with emphasis on LD mass spectrometry. Various experimental and instrumental approaches are discussed; various mechanisms of ion formation are given, which contribute in varying degrees to the mass spectra obtained with the different techniques. These processes are

Thermal evaporation of ions from the solid,

Thermal evaporation of neutral molecules from the solid followed by ionization in the gas phase,

"True" laser desorption, and

Ion formation in a laser-generated plasma.

The first two processes are referred to as thermal, because they can also be induced by classical Joule or nonlaser radiative heating, usually in conjunction with heat conduction to the sample surface. In the case of "true" laser desorption, the laser-induced ions show properties that exclude a thermal process of generation. The ions have initial kinetic energies of more than 10 eV and often even 50–100 eV. The mechanisms that lead to such "true" laser desorption are believed to be collective, nonequilibrium processes in the condensed phase, as discussed by Jöst et al. (77). At high laser intensities ($\geq 10^{10}$ W/cm^2), dense plasmas are formed from the solid. Molecules will be broken down to their atomic constituents, multiply charged atomic ions will grow in abundance with increasing laser power, and initial ion energies will typically be high. The electrons, ions, and multiply charged particles that are initially present in the plasma recombine quickly to yield un-ionized atoms and clusters. The ions and electrons in the outer layer of the initial plasma undergo Coulomb repulsion; they escape recombination and are detected. It is most likely that the experimental results have to be explained by more than a single process. One can simultaneously create a plasma in the center of the laser-irradiated area and get "true" laser desorption from the periphery and even thermal ion emission from the heated substrate for times longer than the laser pulse duration.

Wieser and Wurster (90) demonstrated that cationization of sucrose by Li$^+$ attachment also takes place when the sucrose and LiCl are spatially separated. Under these conditions, the cationization process requires an instantaneous and sufficiently fast mixing process, which may occur in the region of transition from the solid to the gas phase. They also showed that part of the analytical material leaves the interaction region in the solid–liquid state (91). Due to the spatial distribution of laser energy, ion formation in the central part of the laser spot will take place under quite different energetic conditions than in the periphery of the zone of laser impact. Laser pyrolysis may characterize the ion formation in the central part of the laser spot. In contrast, in the outer regions of the high-power laser interaction zone, soft ionization such as laser-induced desorption may dominate. Wurster et al. (58) observed $Ag_nCu_m^+$ cluster ions coming from physically separated silver and copper double foils. The mass spectra of ^{12}C and ^{13}C double foils separated from each other by a copper grid contain C_n^+ cluster ions consisting of both ^{12}C and ^{13}C atoms (92). Signals from Si_2^+ ions appear in the LAMMA mass spectra of SiO_2, although each silicon atom is surrounded by four oxygen atoms in the crystal lattice of SiO_2 (93). These experiments demonstrate that recombination reactions occur during the laser irradiation of solids. Consequently, not only are ions formed at the position of the sample, but various rearrangements and/or chemical reactions may take place at certain distances from the sample surface.

The influence of laser wavelength on the ionization process is particularly significant for organic materials. Hillenkamp (89) proved that aromatic amino acids have a much higher ionization probability at 266 nm than at 355 nm. Antonov et al. (94) showed that the ionization energy threshold of anthracene was dependent on the irradiation wavelength. On the other hand, Hurst (95) proved that photoionization of free atoms was selective for ionization wavelengths related to specific electronic transitions. Krier et al. (96) obtained a significant improvement of elemental sensitivity by coupling selected radiation from a tunable dye laser to the Nd:YAG laser of the LAMMA-500.

3.2. Energy Distributions of Laser-Produced Ions

Ions formed in LAMMA have been reported to have kinetic energies of 5–30 eV, which are influenced by chemical and laser intensity effects (8). Mauney and Adams (66) developed a method for ion energy determinations in the LAMMA-500, where the ion reflector precludes obtaining such measurements directly from the TOF mass spectra. Cumulative distribution curves can be obtained by varying the cutoff level over the energy domain of interest while repetitive spectra are acquired on a homogeneous specimen. These curves reveal both the shape of the ion energy distribution and its center relative to the accelerating voltage.

The observed kinetic energy distributions show substantial contributions from ions with kinetic energy less than the total accelerating potential, defined as "negative" initial kinetic energies. In the example of Fig. 5.8, these values are shown for negative ions derived from a carbon foil measured at 1.1 μJ and 5.3 μJ per laser pulse. Not only do the distributions include "negative" energy tails (compared to the charge times accelerating potential), but the median of the distributions also lies at negative values. Moreover, this deficiency of energy is increased with

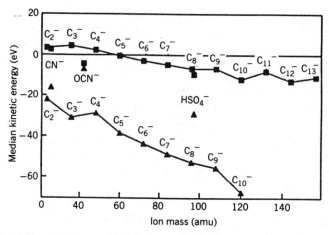

Figure 5.8. Median kinetic energies (relative to the accelerating potential) of negative ions derived from a carbon foil. Laser energies per pulse are approximate. (■) 1.1 μJ/pulse; (▲) 5.3 μJ/pulse. U_{TOF} = 3000 V, U_{lens} = 1070 V, 32× objective. CN^-, OCN^-, and HSO_4^- result from impurities.

increasing laser intensity. This conflicts with the concept that ions are ejected from the sample with substantial velocities that are added to the electrostatic acceleration.

The fact that distinct species possess different median energies indicates that the observed median is not the result of any process applying equally to all ions. Also, species such as CN^-, OCN^-, and HSO_4^-, which deviate from the trend of the carbon clusters C_n^-, demonstrate that the effect is not purely mass related but, at least in some way, chemically influenced. A number of possible explanations for ion formation in LAMMA are shown schematically in Fig. 5.9.

In some cases, atomic ions, including Na^+, K^+, Al^+, and Ti^+, may be extracted with low positive initial kinetic energies (0–15 eV), in agreement with the early results of Hillenkamp et al. (8). In other cases, even atomic ions are deficient in kinetic energy (64, 67). Polyatomic ions have predominantly kinetic energy deficiencies, sometimes differing by over 20 eV for ions with different composition but separated by only one mass unit. The observation that many of the dominant species in the LAMMA spectra have distribution medians at negative values indicates that their occurrence is not the result of processes that could produce only minor species, such as counterflow collisions producing ions with negative net velocity. This leaves as reasonable possibilities (Fig. 5.9) incomplete acceleration due to the locus of ion formation as well as energy-reducing collisions. For the actual dimensions, a deficit of 10 eV would imply a

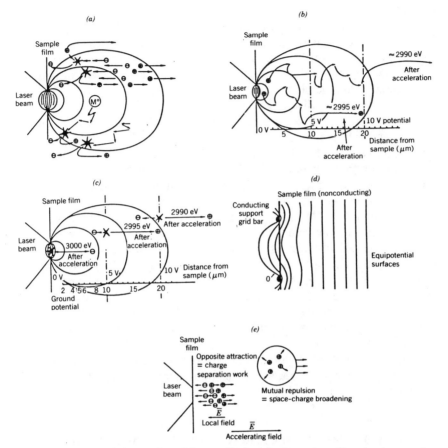

Figure 5.9. Schematic illustration of five mechanisms for formation of ions having negative (deficient) initial kinetic energy values: (*a*) Collisions lead to ions formed with velocity away from the spectrometer. These include thermal collisions and collisions with undetected ions of opposite polarity, which accelerate toward the grounded sample. (*b*) Ions collide with neutral vapor and lose part of their energy gained from electrostatic acceleration. (*c*) Ions formed from neutral vapor after gaseous expansion by a few micrometers are accelerated through a reduced potential difference. Vapor-phase ion formation may occur by many processes, including collisions mentioned in (*b*). (*d*) Accelerating potential is perturbed by a few volts (<3 V) by the grid spacing, and by unknown amounts if electrostatic charges accumulate on the sample. (*e*) Charge–charge electrostatic interactions (space charge effects, plasma screening) alter the accelerating field experienced by the ions.

drift of only 20 μm through the acceleration field prior to ionization or effective escape from the dense region. The effect of different positions of ion formation cannot at present be distinguished from general collisional features. Nevertheless, the trend toward more negative energy values for larger clusters suggests that the larger clusters are formed by recombination as the plasma cools, rather than being formed initially and fragmented as the collisions proceed.

The present technique of measuring kinetic energy distributions is laborious and is limited to an energy resolution around 1 eV. However, the information revealed is as important to the investigation of ion formation processes as that provided by other techniques such as mass-analyzed ion kinetic energy spectrometry (MIKES). In addition to the study of ion formation processes, the kinetic energy measurements may have practical applications in LAMMA, such as in the discrimination of chemically different species or the removal from the spectrum of certain undesirable ions. Substantial differences in initial ion kinetic energy such as those that occur in secondary-ion mass spectrometry (due to the discrimination of the ion extraction lens) produce deviations in spectral intensity by as much as two orders of magnitude. In the case of LAMMA, estimates of the ion temperature based on the LTE model have implicitly assumed that all ions are transmitted with equal efficiency. Deviation from this equality may have contributed to the reported lack of fit with such models (64). Hence, it will not be possible to make reliable calculations until the efficiency of ion extraction, as well as other aspects of the spectrometer, are accurately known.

4. ANALYTICAL CHARACTERISTICS OF THE LAMMA TECHNIQUE

The LAMMA technique was originally developed for highly sensitive microanalysis of thin histological sections in applications of biological and medical relevance (97). In particular, it was intended to be a complementary method to other existing microanalytical techniques such as electron probe X-ray microanalysis (EPXMA), secondary-ion mass spectrometry (SIMS), and Raman microprobe analysis (see Table 5.2), for determining the intracellular distribution of physiological cations such as Na^+, K^+, and Ca^{2+}, as well as toxic constituents, in biological tissues. During recent years it has increasingly been applied to a large variety of biological, technical, organic, and inorganic specimens of interest in biology and medicine, environmental research, geology and mineralogy, forensic science, materials research, and general chemistry (98, 99). The main advantage of LAMMA is its ability to perform a rapid and highly

Table 5.2. Summary of Characteristics of Four Types of Microprobes[a]

	X-Ray Microanalysis	Ion Microprobe (Ion Microscope)[b]	Laser Microprobe (Transmission Geometry)	Raman Microprobe
Probe	Electrons	Ions	Photons (laser)	Photons (laser)
Detection method	Characteristic X-rays: WDS or EDS	Ions ("+" or "−") Double-focusing MS	Ions ("+" or "−") TOF MS	Photons (Raman) Double monochromator, PM
Resolution of detection	WDS, 20 eV; EDS, 150 eV	$M/\Delta M = 200\text{–}10{,}000$	$M/\Delta M = 800$	$0.7\ cm^{-1}$ (spectr.), $8\ cm^{-1}$ (image)
Lateral resolution (analyzed area)	WDS, 1 μm; EDS, 500–1000 Å[c]	1–400 μm	1 μm	1μm
Imaging (spatial resolution)	SEM, 70 Å; STEM, 15 Å	0.5 μm (SII)	1 μm	1 μm
Information depth	≤1 μm	Tens of Å	—	—
Detection limits	WDS,100 ppm; EDS, 1000 ppm	ppm[d]	ppm	Major comp.
Elemental coverage	WDS, $Z \geq 4$; EDS, $Z \geq 11$	H–U	H–U	—
Isotope detection	No	Yes	Yes	No
Compound information	No	Yes[e]	Yes	Yes
In-depth analysis	No	Yes	Difficult	No
Destructive	No	Yes	Yes	No
Quantitative analysis	Yes	(Yes)	(Yes)	Yes

SEM: scanning electron microscope
STEM: scanning transmission electron microscope
WDS: wavelength-dispersive spectrometer
EDS: energy-dispersive spectrometer

SII: secondary-ion image
TOF-MS: time-of-flight mass spectrometer
PM: photomultiplier

[a] Depends very much on element of interest and on the chemical environment including the nature of the primary ion beam (Ar^+, O_2^+, O^-, Cs^+).
[b] Ion microscope Cameca-3f.
[c] High concentration deposit.
[d]
[e] More in "Static SIMS".

sensitive mass analysis of both atomic and molecular species present in a microregion of the sample, in either the positive- or negative-ion detection modes.

4.1. Sample Requirements

Inherently limited by its transmission geometry, the LAMMA-500 requires samples such as thin sections (usually 0.1–2 μm thick) or small particles (0.5–5 μm) that can be perforated or vaporized by laser impact. In particular cases, however, some ingenuity may circumvent this obvious limitation. For specific applications in LD experiments, thick specimens may even be of some advantage (87, 100).

Also, the sample should, of course, be stable in the vacuum of the mass spectrometer ($\sim 10^{-4}$–10^{-5} Pa). Only a special technique allows compounds of high volatility to be analyzed under normal atmospheric conditions (101, 102). In this procedure the vacuum seal of the LAMMA-500 is not a quartz window but a thin polymer film that carries the deposited material to be analyzed on its atmospheric side; the laser pulse will then generate tiny perforations in the vacuum seal, through which the ions from the analytes formed under atmospheric conditions can enter the instrument. The spectra have been found to be identical with those obtained from the same compounds by ionization in the conventional way inside the vacuum chamber. Although this mounting procedure has the potential to create new opportunities for mass spectrometric microanalysis, further experiments are needed to demonstrate its practical applicability in the study of, for example, aerosols and biological material at ambient pressure and relative humidity.

For the analysis of biological and clinical samples mounted conventionally under vacuum, microtome sections can be prepared as for routine transmission electron microscopy (TEM). The absence of low-temperature facilities in the sample chamber (cryostage) necessitates the use of common embedding procedures (103). Cryosectioned shock-frozen material (without embedding) supported by a very thin Formvar film (polyvinyl formaldehyde) may also be analyzed, with or without prior lyophilization (104). Similarly, thin sections of other materials can be mounted directly provided that they have sufficient mechanical stability. Thin foils of metals and semiconductors (58, 74–76), thin sections of soils (105–107), and even very thin sheets of minerals (mica) (108) can be studied. An advantage of LAMMA (and of laser mass spectrometry in general) is that electrically nonconducting samples may be analyzed.

A common approach in LAMMA of particles is to mount the particulate matter on a TEM grid coated with a thin organic or metal supporting foil

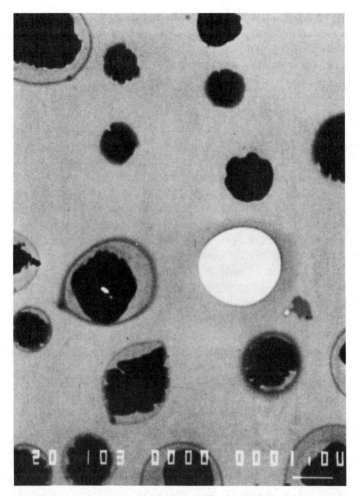

Figure 5.10. Electron micrograph (original magnification 10,000×, bar represents 1 μm) illustrating atmospheric aerosol particles directly impacted on a thin Formvar foil. The hole shown in the center of the picture originated from the pulsed laser interaction with a particle.

(≤0.1 μm thick). This can be done by bringing small particles, fibers, or finely pulverized material into close contact with such a Formvar-coated electron microscope grid to allow the particles to adhere sufficiently. Aerosols can also be collected by direct impaction in a number of current collector devices, with or without size classification (Fig. 5.10) (109–112). Larger particles or fibers may sometimes be attached to the metal grid bars without any film coating, though such samples are usually mechanically unstable. Particles with larger diameters (0.1–0.5 mm) may well be mounted in so-called sandwich grids, while macroscopic samples up to some millimeters in size can be clamped in a modified sample holder, which holds the particle with a special tweezer on the rear side. In such a case, one should pay attention to the orientation of the sample and make sure that interesting areas lie near the edges visible in the light microscope, so that they can be analyzed with laser light at grazing incidence (72, 113–115). In principle, this allows the recording of mass spectra free from any additional background. Another approach is to use multiple laser shots until complete perforation is reached, producing a mass spectrum from the final shot. However, preferential vaporization and migration of some species may occur and lead to nonrepresentative analytical results. This laser drilling on, for example, a multilayered sandwich of different materials impairs the obtainable lateral resolution of analysis by irreproducible crater shape formation (9, 97). Large bulky specimens may also be analyzed in a "two-step" procedure in which regions of the sample are vaporized by laser impact in the absence of the accelerating electric field. Under these conditions the evaporated material is recondensed and deposited on the quartz window vacuum seal; it is reionized by subsequent laser shots and further analyzed in the normal way (110). This procedure is applicable only for those specimens that can recondense from the plasma state without change of composition. If the sample is transparent to the laser wavelength and suitable for use as a vacuum seal of the mass spectrometer, it is sometimes possible to focus the laser beam onto the back of the specimen and produce ionization from that surface. This method was applied, for instance, to mineral quartz samples (116).

Soluble compounds or suspensions may also be investigated. A microliter droplet of a dilute solution is brought onto a thin film suspended on a grid. After evaporation of the solvent, small crystallites or particulate residues remain on the foil that can then be visualized in the microscope and analyzed (117). In some special cases, as for anionic surfactants (wetted with, e.g., chloroform) or viscous fluids, the material can be prepared directly for analysis by simply dipping a grid into the sample (118).

4.2. Instrumental Features

The laser intensity of the UV laser will determine both the elemental detection efficiency and the spatial resolution of the analysis. It can be controlled to a certain extent by using the set of filters situated at the entrance of the light microscope. In this way, the laser power can be attenuated to provide relatively low laser energies that will produce sufficiently small perforation diameters (Fig. 5.11). Optimal lateral resolution

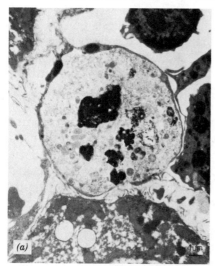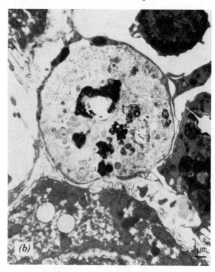

Figure 5.11. Electron micrograph (TEM) of a thin histological liver section (thickness 0.2 μm) taken before (*A*) and after (*B*) perforation by the laser beam. It demonstrates the obtainable spatial resolution of analysis: the diameter of the laser-induced hole is about 1–1.5 μm.

is then reached at the cost of reduced sensitivity. A spatial resolution of analysis of about 1–2 μm is practically obtainable in LAMMA of soft tissue sections, which may, however, be larger than some cellular organelles of interest. In aerosol research, particles as small as 0.5 μm may be spotted and analyzed, though an increased background contribution of the supporting foil is inevitable. The actual analytical volume (usually around 10^{-13} cm^3) is a function of laser focusing, morphology, local mass density, and chemical composition of the sample, which influence the relative absorption and dissipation of the laser energy. For particulate matter, target geometry and individual particle size are also important. For thicker samples, higher laser power is required for perforation, which will affect the perforation diameter and ionization efficiency.

With every laser shot, a complete mass spectrum (positive- or negative-ion mode) is recorded with a mass resolution of $M/\Delta M = 850$ at $m/z = 208$ (9). Under these conditions, unit mass identification is clearly feasible, and individual isotopes can be detected and discriminated. LAMMA is thus capable of providing isotopic information for tracer experiments using stable isotopes (isotope dilution) or radionuclides. The mass resolution in LAMMA is, however, insufficient to avoid mass interference between elemental signals and contributions from both organic fragments and inorganic cluster ions.

One major advantage of LAMMA is the short analysis time in which the multielement spectra can be obtained. A sample can easily be mounted and the appropriate vacuum conditions established within minutes. Detailed analysis often requires initial optimization of instrumental settings, such as the selection of laser power and ion lens voltage. Moreover, the sample should sometimes be analyzed in both positive- and negative-ion modes using different laser powers and multiple mass spectra recorded for signal averaging. The shot-to-shot spectral reproducibility is limited by variations in instrumental parameters and specimen properties. The former concern the laser fluence, focusing conditions, ion extraction, and ion detection. Inherent chemical or physical inhomogeneities of the sample as well as the sample preparation itself may influence the overall analytical reproducibility. Since the LAMMA technique is destructive, it is obviously impossible to record both positive- and negative-ion spectra at a single location within a section or from a single particle. Moreover, a number of possible interesting sample areas or particles are destroyed while optimizing the operational conditions.

With biological samples, in particular, the time-limiting step in the overall procedure is the selection of the analytical area of interest. The moderate quality of the light microscopic image and the fact that unstained histological sections are usually used encourage the LAMMA user to rely on electron microscopy (EM) to define the ultrastructure of the specimen so that it can be correlated with the analytical results. One drawback of LAMMA is that, in contrast to EPXMA, the method is destructive, because the laser beam perforates the thin section. TEM micrographs are taken before and after LAMMA of the cellular regions of interest to provide records of the analyzed subcellular fragments and to measure the evaporated area (Fig. 5.11) (119–121). Tissue sections may be slightly colored, for example, with toluidine blue, for light microscopy (LM). Reflected light illumination may sometimes improve the visual differentiation of tissue structure (122). Incorporation of UV transillumination can give satisfactory images by enhancing the absorption contrast (123).

The high speed at which spectra can be acquired and transferred to a computer for storage and further data processing allows screening, by multiple laser shots, of the distribution of physiological cations and toxic trace metals at the subcellular level. It permits the averaging of mass signals recorded from the same type of organelles and cell structures to obtain representative spectra. On the other hand, in particle studies it makes it possible to analyze large populations of individual particles for the compilation of statistically meaningful data (e.g., classification, source identification, and structural morphology in combination with EM). In most LAMMA laboratories, software for data evaluation has been developed and applied (73).

4.3.   Detection Limits and Sensitivity

As in any microprobe technique, the detection limits of LAMMA depend on operational conditions as well as on properties of the specimen under investigation. A number of factors need to be considered: the first ionization potential for the formation of positive ions and electron affinity for negative ions, the binding state of the elements with the surrounding matrix, the bond strength of organic molecules, the stability of the generated ions, and the probability of recombination processes in the laser-induced microplasma. In addition, instrumental parameters such as laser wavelength, pulse duration, and irradiance in the focus, as well as specific ion optical properties, will affect the overall analytical sensitivity.

The high overall ion transmission through the TOF mass spectrometer of the LAMMA instrument enables very low detection limits for a wide range of elements (9, 91, 97, 110, 123–131). LAMMA is particularly sensitive for those elements with a low first ionization potential, such as physiological cations. It has been stated that absolute amounts down to about 10^{-20} g are detectable in the microvolume analyzed, resulting in relative detection limits in the ppm range (Table 5.3). On the other hand, some essential elements such as zinc, chlorine, and phosphorus; the protein constituent atoms sulfur and nitrogen; and toxic heavy metals such as mercury, cadmium, platinum, and gold show poor detection limits in the positive-ion-mode LAMMA spectra. All these elements have high ionization energies. LAMMA can easily be operated to detect negatively charged ions. In the negative-ion mode, nonmetallic elements such as the halogens with high affinities for electron capture can be detected with high sensitivity, either as their elemental anions or as cluster ions, of the type XO_n^- (X = P, S, N), for example.

The detection limits of LAMMA are markedly lower than the values obtained in EPXMA (Table 5.2). A detailed comparative study of laser and ion microprobe sensitivities for the detection of lead in biological material has been made by Linton et al. (128). At a comparable lateral resolution of analysis of about 1 μm, a relative detection limit of 5 μg/g was found in LAMMA, which is superior by 2 orders of magnitude to that obtained with the Cameca IMS-3F ion microscope. Also, the useful ion yield for lead was 2 orders of magnitude better in LAMMA than the IMS-3F value (3×10^{-3} versus 5×10^{-5}). Although sensitivity factors for the alkali and alkaline earth metals show similar trends in SIMS and LAMMA, the sensitivity factor for lead is orders of magnitude lower (0.5

Table 5.3. Relative Detection Limits (ppmw) for Elemental Analysis (Positive Ions)

Element	Glass Matrices[a]	Organic Matrices[b]	Element	Glass Matrices[a]	Organic Matrices[b]
Li	3	0.4 1.6[c]	Nb	7	
Be	25		Ag		1
B	9		Cd		20
Na		0.2	In	7	
Mg	8	0.4	Cs	6	0.5
Al		0.3 5[d] 4[e]	Ba	6	0.5 2.5[c]
Si	10		La	4	10[c]
K	0.8	0.1	Ce	5	
Ca		1 1.6[c]	Pr	3	
		0.4[f]	Ho	14	
Ti	1		Tm	11	
V		0.5	Lu	6	
Co		10	Ta	6	
Cu		15	Pt		20[d]
Rb		0.5	Pb	10	0.4 10[c]
					5[g]
Sr		0.5 2.5[c]	Th	6	
Y	4		U	6	2.0

[a] NBS Standard Reference Materials of the SRM 610–616 series. Limits defined by mass signals exceeding 3 times standard deviation of the background noise (124).

[b] Epoxy resin (Spurr's low-viscosity medium) standards containing trace amounts of organometallic complexes (10^{-13} g evaporated material, irradiance set at 5 times the damaging threshold, signal intensities averaged over 5 analyses must exceed 3 times the background noise) (110).

[c] Data obtained in either natural pigment granula or epoxy resin matrices (0.3-μm thick sections) (123).

[d] Metal-loaded Chelex-100 beads embedded in Epon resin (0.2-μm thick sections) (127).

[e] Lyophilized serum (129, 130).

[f] Standard specimen of CaF_2 deposited on a 0.3-μm epoxy resin section (123).

[g] Epoxy resin (Epon) thin sections (0.5 μm) doped with known concentrations of lead acetate (128).

in LAMMA versus 3×10^{-4} in SIMS when normalized to potassium). Other LAMMA studies also showed an enhanced lead sensitivity compared with LTE calculations (126). A specific reason for the large enhancement of the lead ion yield in LAMMA was, however, not given.

Krier et al. (96) reported the coupling of a tunable dye laser to a LAMMA-500 instrument and found that when the wavelength of the laser beam corresponds to a particular electronic transition of the element being analyzed, resonance absorption phenomena may be observed. The detection thresholds fell from 50 ppm at 266 nm to 1 ppm at 228.80 nm for cadmium and from 20 ppm at 266 nm to 50 ppb at 324.75 nm for copper. The irradiation wavelength at the lowest detection threshold corresponds to an electronic transition of the metal with a high transition momentum. The same authors also reported that the energy required to ionize anthracene is 100 nJ at 286.5 nm, 50 nJ at 266 nm, and only 3 nJ at 225.7 nm.

In LAMMA, polyatomic ions are produced by fragmentation and/or recombination processes. These ions can be derived from the biological tissue itself, the embedding material, and/or the supporting foil. Organic ions and major inorganic cluster ions can interfere with the atomic mass signals to be measured. For this reason the detection limit of the element of interest may be worse in practical analysis. In "real" specimens, the majority of potentially interfering background peaks appear in the mass range m/z = 20–140, with maximum ion intensities near the lower mass end; the spectral pattern of these background peaks is strongly influenced by the laser power used. This may create a problem during the analysis of trace amounts of the transition metals whose atomic weights fall in this mass range. Small ion signals from iron or copper, for example, may be hidden by larger peaks of organic fragments with the same nominal m/z number. Only if these background mass peaks are reproducible and result in typical fingerprint spectra can background subtraction procedures by computer become applicable to some extent, on the additional condition that the ion formation for the total sample and for the background alone are the same. The laser power density at the focus of the sample plays a crucial role in the process of matrix fragmentation and ion formation. Therefore, the power has to be controlled to avoid excessive spectral background. Since the transmission of the TOF mass spectrometer for complex molecular ions may be lower than for simpler atomic ions, due to differences in initial kinetic energy distributions, appropriate selection of ion optical parameters has been able to suppress part of the mass spectral interferences (60, 63, 64, 132).

4.4. Ionization Characteristics and Speciation Capabilities

A considerable advantage of the laser ion source is the fact that the laser intensity can easily be varied to provide a wide range of ionization conditions (89). Experience has shown that irradiances close to the threshold of ion formation at low laser power (laser desorption) will preferentially desorb ions from the target surface facing the spectrometer (surface characterization) and promote parent ion formation (84). On the other hand, laser irradiances three- to tenfold above this threshold will lead to evaporation of the full target volume and yield information on the bulk composition. In the latter mode, the high-power ionization causes extensive fragmentation (laser pyrolysis) and molecular rearrangement. Exploiting these special features of LAMMA, a two-step procedure can be applied for the analysis of micrometer-sized particles:

1. A laser shot at low irradiance with a slightly defocused beam is aimed at the particle surface facing the mass spectrometer. Selective evaporation of the surface material is obtained, while the particle appears to remain intact and no gross morphological changes are observed.
2. A second laser shot with a focused laser beam is aimed at the center of the particle; this results in complete destruction of the particle and the registration of a fragmentation spectrum.

This procedure has proved to be very useful for the surface characterization of asbestos fibers (133) and aerosols (134).

Lindner and Seydel (87) showed that laser irradiation of a thick sample (≈ 20 μm), which does not lead to perforation, also offers soft ionization conditions. They described a desorption mechanism based on a nonthermal, shock-wave-driven process, leading to the release of mainly intact molecules from the solid sample surface (see Section 6.3).

An interesting feature of LAMMA is its ability to produce molecular ions (e.g., for aromatic compounds) or quasi-molecular peaks (e.g., for polysaccharides) from many nonvolatile and/or thermally labile compounds. Frequently, in LAMMA, a cationization or simple protonation of nonionic molecules—attachment of a metal, alkali ion, or proton to a neutral organic molecule—occurs. In this way, molecular information is obtained from labile species that may be difficult or impossible to obtain by conventional mass spectrometry (with an electron impact source). Moreover, organic salts and acids, free organic bases, organometallic complexes, organic polymers, and inorganic compounds ranging from simple halides or oxides to complex inorganic salts all give typical mass spectra suitable for identification of molecular species (82, 83, 135–143). In addition to the signals that are characteristic of molecular structure and functional groups (quasi-molecular ions or counterions in salts), typical cluster ions, which are formed by fragmentation and/or recombination, are also observed. Cluster ions show characteristic intensity distributions that can be described by theoretical or empirical models of ion formation such as the "valence model" of Plog (74–77, 93, 116, 132, 142–144).

Organic fingerprint mass spectra can also be obtained by analyzing entire biological entities such as isolated bacteria. This can be done either at low laser power, to produce specific molecular fragment ions, or at high laser power, comparable to laser pyrolysis ionization mass spectrometry (98, 99). In the latter mode, organic polymers such as the embedding media routinely used in EM, are fingerprinted. In single-particle analysis, this approach can be used to deal with chemically complex particulate matter in classification studies (112).

Limited but valuable information on inorganic species and organic components may thus be extracted from the LAMMA spectra, although so far only for the major constituents within the sampled area.

4.5. Quantification

In the present state of the LAMMA technique, accurate quantification is still problematic because of the lack of a detailed theory describing the physical processes of the laser-induced ion formation. Absolute quantitative analysis requires data concerning the ion yield, the evaporated volume, the overall ion transmission, and the ion–electron conversion efficiency (dependent on ion mass, structure, and energy) of the secondary electron multiplier. Since no theoretical model at the moment can predict the ion yield for a specific specimen as a function of target and laser beam parameters, empirical procedures must be applied for quantification.

One such approach is the use of suitable calibration standards (123, 127). Such homogeneous standards are necessary, not only for direct comparison with a variety of samples but also for use as simple model systems to study the general analytical features of the laser microprobe. Ideal elemental calibration standards should fulfill the following requirements:

1. Well-defined chemical composition and structure
2. Matrix composition similar to that of the specimen
3. Homogeneity at the level of spatial resolution
4. Controllable elemental doping possible
5. A wide variety of elemental dopants possible
6. Elemental concentrations adjustable within a range typical for the specimens to be analyzed.
7. Elemental concentrations assessable by several analytical techniques

In LAMMA, additional properties should be similar for standards and specimens:

1. Target geometry
2. Matrix mass density
3. Optical characteristics and surface texture
4. Instrumental conditions

Standards can be made by dissolving metal-containing organic compounds such as crown ether complexes (145) into the epoxy resins that are routinely used for the embedding of biological specimens. After polymerization, microtome sections are cut from the cured blocks for use as thin-film standards (9, 97, 126, 131, 146). Thin films of vacuum-deposited metals or dielectric materials were also proposed for standardization purposes; a grid mask is used to obtain a regular pattern of coated next to uncoated areas on the surface of the specimen (120, 146). Good results were obtained by coating inorganic compounds (e.g., fluorides or oxides) containing elements of the same sensitivity to LAMMA as the element to be determined. It would be very interesting to use a thin coating of a stable isotope of the element of interest itself as an internal standard, thus providing the possibility of the quantitative application of "isotope dilution" analysis. However, instrumental effects of LAMMA have to be considered carefully to allow a meaningful interpretation of the results for quantitative purposes because of the inhomogeneous distribution of the analyte over the sample thickness. One has to take into account the geometric effect of ion extraction; it favors the detection of ions generated from the side of the sample that is faced toward the extraction lens of the TOF mass spectrometer (92). Moreover, the coated element is present in the sample in a different chemical environment than the analyte, so that differences in sensitivity factors as a result of matrix effects are to be expected.

Proteinaceous standards consisting of an organic matrix such as gelatin or albumin doped with a variety of elements have been investigated by LAMMA for use as thin cryosections to be compared with shock-frozen freeze-dried biological material (104, 147). Another interesting possibility of calibration is the use of anionic surfactant films containing several metal cations (118). Mixtures of purified surfactants each prepared with a specific alkali metal ion are made by weighing suitable quantities and dissolving them in chloroform. By bringing an unfilmed TEM grid (of a high mesh number) into close contact with a beaker wall previously wetted with the surfactant solution in chloroform and after evaporation of the solvent, a single surfactant film is obtained. The film thickness can be varied up to 1–2 μm by adjustment of the surfactant concentration. A similar approach consists in dissolving a metal-containing organic complex in Formvar (in chloroform solution at about 0.3% w/w), which is routinely used as a thin supporting aid in TEM. Grids may then be coated by the thin polymer film containing the dopant of interest (64). A novel calibration system for LAMMA analysis of biological microtome sections is based on ion-chelating resin beads homogeneously loaded with the metal ions of interest (122, 127, 148). These beads with their immobilized metals may be coembedded with the soft biological material to be investigated and cut to the desired thickness. The standard is thus present in a similar chemical environment and in the same section as the analyte.

Only a few methods for incorporating internal standards into thin sections for quantitative analysis of organic compounds have been reported. Interesting approaches include the dissolution of pure organic compounds (e.g., aromatic monomers or polar molecules) in polyethylene (3% w/w) or in a polymeric resin for the identification of molecular or quasi-molecular mass signals by selective laser desorption (82, 123). Another possibility is to use labeled organic compounds; any covalently bound atom of high ion yield and minimal mass interference may be a suitable candidate. In a preliminary attempt, fluorinated derivatives of cardioactive drugs were dissolved in epoxy resin as standard specimens (9). Also of interest might be the use of Nafion, a perfluorinated cation-exchange polymer, as an immobilizing adsorption medium for simple cations and large biological molecules (149). Thin films of this material can be prepared on suitable backings using the electrospray method.

The LAMMA analysis of particulate matter obviously requires another type of calibration: particle standards. For this purpose, micrometer-sized particles may be synthesized directly in the suitable dimension. The generation of particles with a well-defined size and chemical composition by atomization of liquid solutions with the desired particulate material as the solute provides a suitable method of calibration for LAMMA in, for example, aerosol research and fundamental studies (150, 151). This method was used by Kaufmann et al. (110) to calibrate the dependence of the ion signal on the particle size for NaCl particulates. Heavy-metal-doped organic particulate matter has also been proposed as a standard (91, 125); flat particles were produced by nebulizing aqueous solutions of sucrose containing known amounts of metal salts and subsequently drying the generated droplets. For the preparation of salt aerosols coated with polynuclear aromatic hydrocarbons (PAH), Niessner et al. (152) used a gas-phase condensation apparatus coupled to the outlet of a Collison nebulizer equipped with a diffusion dryer. For the preparation of multielement salt standards, mixtures of inorganic salts can be dissolved in water with some ethanol added, pipetted in thin lines onto a supporting film, rapidly frozen, and finally freeze-dried to obtain a structure of small granules less than 1 μm in size (153).

Bulk homogeneous material can be crushed to micrometer-sized particles and serve as standard reference samples. Such materials include silicate minerals (augite, hornblende, lepidolite) (154), silicate glass microspheres and glass fibers composed of various element oxides (110, 124, 155–157), and glass films and uranium oxide microspheres (72, 158).

Another approach is the use of larger carrier particles consisting of a polymer matrix (159) or synthetic highly porous silica spheres (Spherisorb particles), about 3 μm in diameter, which can be loaded by soaking with solutions of various salts (160). Submicrometer-sized Polybead polymeric spheres—monodisperse latex particles available with carboxy, hydroxy, or amino functionality (Polysciences Inc., Warrington, Pennsylvania)—may also provide an alternative means for standardization (64). It has been reported that the spark of the emission spectrometer may be a suitable and efficient source for the production of metal and metal alloy particle standards in the 5 nm to 10 μm range (161). Several hydrous metal oxide sols consisting of spherical particles with narrow size distribution have also been described (162); they might be useful as a uniform calibration system.

A basic prerequisite to the quantification of LAMMA results is the ability to establish a straightforward relation between the ion signal intensity and the local concentration of the analyte. Using different types of standard specimens, linear calibration curves over a broad range, even up to 3 orders of magnitude, have been experimentally obtained with a fair reproducibility of measurement (10–30% RSD, in practice, for homogeneous thin sections) (54, 118, 125, 127, 129, 163, 164). These calibration curves represent absolute signal intensities or show relative data with respect to a reference mass peak.

The amplitude of the ion signal is susceptible to a number of sources of variability that influence the ion yield and the evaporated volume. These include laser intensity fluctuations, differences in target geometry, local mass density, and physical and microchemical properties of the particulate matter or biological sample to be analyzed. Correction for variation in, for example, the evaporated volume can be obtained by normalizing the signals to one or several suitable internal reference mass peaks. Reference candidates include an element with known and homogeneous concentration, such as a matrix element, a doped or coated element (including stable isotopes), or an organic mass fragment. Only under carefully controlled conditions can a direct sample-to-standard comparison provide successful quantification.

In practice, element-specific relative sensitivity coefficients (RSC) obtained from multielement standards may empirically permit quantification. Elemental RSC values generally do not vary by more than a factor of 10 (at high laser intensity) (9, 91, 97, 104, 110, 123–128, 154, 155). This kind of standardization has been applied with different degrees of success. However, it remains limited to semiquantitative results because of analytical discrepancies, in particular when the matrix composition and the sample geometry are varying (123–126). A theoretical approach is to apply the LTE model to interpret the relative sensitivity factors of the ion emission. Satisfactory analytical results have been shown to ensure from this model in SIMS for a variety of materials (165, 166). With the assumption that a local thermal equilibrium exists in the laser-induced microplasma, the degree of ionization of an element may be estimated to a first approximation, provided that realistic estimates can be made of the relevant parameters of the model such as electronic partition functions, electron density, and plasma temperature.

Phenomenological laws such as the Saha–Eggert ionization equation have been applied to some extent to metal alloy (123) or sandwich foils (75), glass particles (124, 155), anionic surfactant films (118), sucrose particulate matter (125), and doped epoxy resins (126). The calculated parametric temperature values have been shown to range from about 5000 to 13,000 K. The LTE model applied to homogeneous silicate minerals was tested in a comparative study of SIMS and LAMMA, using both the transmission and reflection types of instruments (154). The relative intensities of the positive ions in the LAMMA spectra followed a dependence with ionization potential similar to that observed in SIMS (Fig. 5.12). In the case of the transmission-type LAMMA-500 instrument, considerably larger deviations from the Saha–Eggert ionization equation were noted. This deviation might be associated with the inherent lack of reproducibility of the method for these samples compared to SIMS rather than with the inapplicability of the LTE approach, since excellent results were obtained with the reflection-type LAMMA-1000 instrument (Fig. 5.12). Ion signal discrimination in the mass spectrometer and mass interference should be avoided or at least corrected in order to allow a better estimation of the relative ion concentrations produced by the laser impact.

The ratios of the different isotopic signals of an element can be used as a sensitive measure to recognize nonlinearities in the ion detection and recording system (Fig. 5.7). Not only are these relative isotopic abundances relied upon for element identification in the LAMMA spectra, but they may also provide quantitative chemical and physical data through isotope dilution and tracer kinetics (120, 146, 167). The determination of isotope ratios in LAMMA is inherently more precise than the determi-

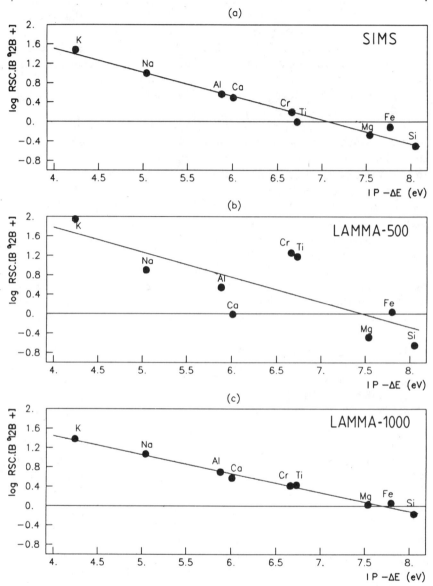

Figure 5.12. Fitting of the Saha–Eggert ionization equation for positive ions obtained from a hornblende sample. RSC = $[I_{M^+}/C_M]/[I_{Al^+}/C_{Al}]$, B = internal partition function, IP = ionization potential (eV), ΔE = ionization potential depression (eV) due to Coulomb interactions of the charged particles. (*a*) SIMS, O$^-$ bombardment, 0–20 eV secondary ions; (*b*) LAMMA-500; (*c*) LAMMA-1000. For SIMS and LAMMA-1000 analyses, selected fragments of hornblende were mounted with Sn–Bi eutectic in a metal ring and polished with

224

nation of relative elemental abundances. A thorough investigation of the performance and accuracy of LAMMA for such measurements is, however, still needed for full exploitation of this potential (72). One instrumental limitation of precision is the rapidly degrading 8-bit resolution of the transient waveform recorder used to digitize and store the mass spectra (see Section 2.5). The accuracy of the isotopic measurements is restricted by a decrease in gain of the electron multiplier as the output pulse amplitude increases (see Section 2.4). This detector nonlinearity can be corrected via a calibration procedure (72). To overcome to some extent the limited dynamic range of the instrument, one can use independent recording channels operating at different input sensitivities or insert a pulse preamplifier with logarithmic compression.

5.   APPLICATIONS OF LAMMA IN BIOLOGY AND MEDICINE

Bulk analytical methods such as atomic absorption spectrometry can be applied only to a small weighed mass of a particular organ or to a large number of cells or isolated organelles, so that in situ localization of elements at the subcellular level is not feasible (168). In current biomedical research of electrolyte metabolism and human pathology, for example, microanalysis is used to reveal the local variation in concentration of a specific constituent within the cell. In this way a concentration-dependent spatial distribution of multiple elements is provided. Since its initial development (169–171), LAMMA has become an important research tool, finding increasing application in the life sciences as a technique complementary to analytical electron microscopy, for element identification, localization, and quantification at the ultrastructural level in human, animal, and plant tissues. In the following sections a comprehensive compilation of a wide variety of applications of LAMMA in biology and medicine is presented. Some selected applications of other laser microprobe techniques of interest are also discussed. Only the most significant results and implications are included. For more specific information concerning such topics as specimen preparation, the reader is referred to the proceedings of LAMMA symposia (98, 99) and to review papers by Kaufmann (123) and Verbueken et al. (7, 43).

---

diamond paste. After ultrasonic cleaning with CCl_4, an electrically conducting gold layer (30–40 nm) was vapor-deposited onto the surface. For analysis with the LAMMA-500, hornblende was crushed in an agate mortar to micrometer-sized particles, which were attached to a Formvar-coated TEM grid (154).

5.1. Specimen Preparation

In microanalysis of biological tissue, sample preparation plays an important role. The preparative procedure must preserve ultrastructural detail while ensuring that the elements to be analyzed have neither been lost nor displaced from their original sites. In most instances perfect preservation of the in vivo intracellular distribution of the analytes is difficult to achieve and to check. One usually tries to minimize artifacts. In some cases, even conventional techniques of wet fixation, dehydration, and embedding as used for routine EM may prove satisfactory (103). However, for the local analysis of highly diffusible, water-soluble electrolytes such as sodium, potassium, and chloride, the latter procedures are inappropriate. Cryotechniques must then be employed using snap-freezing and vacuum-drying procedures. After shock-freezing, the specimen can be further processed in different ways: (a) it can be sectioned in a cryoultramicrotome and analyzed either in the hydrated state (cryostage in the sample chamber) or after freeze-drying; (b) the bulk specimen can be freeze-dried *in toto* at low temperature followed by epoxy resin embedding under vacuum and cutting of dry sections; or (c) freeze-substitution by an apolar medium can be carried out before embedding and sectioning. The loss or displacement of any element in the tissue will depend critically on the degree of binding to immobile cellular structures or constituents. For tightly bound elements and insoluble deposits, these tedious cryopreparative procedures are unnecessary. Another possibility is to combine wet processing of the sample with ion-capture cytochemistry to prevent dislocation of the element to be analyzed by its forced precipitation. Such combined histochemical techniques have been used for both diffusible and nondiffusible elements. In addition, the possible loss or displacement of the organic material to which the elements of interest may be bound must be taken into consideration. Staining procedures, either to improve the morphological identification in concomitant TEM investigations or to enable easy recognition of the sites of analytical interest in the LAMMA light microscope, must be checked for contamination and possible leaching effects, or else they must be avoided completely.

5.2. Applications in Plant Biology

LAMMA has been applied successfully to problems in wood research. These concerned the detection of ions in single cell wall layers of wood treated with inorganic preservatives and involved the cell wall of softwood samples treated in the laboratory by impregnation with selected electrolyte solutions (172) as well as pinewood poles commercially treated with

CCF (chromium–copper–fluorine) or CCB (chromium–copper–boron) wood preservative solutions on an industrial basis (173, 174). Of special interest was the detection of an accumulation of boron and fluorine in the cell wall of thin-sectioned wood at the microscopic level. More recently, the distribution of some elements in individual cells of the fine roots of healthy and diseased conifers from a few sites was investigated (175, 176). A lack of the major nutrient elements calcium and magnesium was demonstrated in the case of diseased trees growing on acidified soils. The supposed key position of aluminum being the toxic element could not be confirmed from these results, but clearly the altered interaction with other elements might intensify any given damage.

LAMMA has been applied in studies on the accumulation of heavy metals by microorganisms. The intracellular distribution of lead in artificially exposed cultures of an algal specimen (Chlorophyta) was compared to the localization obtained by general histochemical precipitation methods (177, 178). LAMMA has also been used for the elemental analysis of inclusions in thin-sectioned halotolerant algae (179). The uptake of uranium and its deposition within the algal cells was reported in a later experiment (180). The amount of uranium in the single dried cells was determined after total evaporation of the cells by LAMMA under light microscopical control.

LAMMA might also offer new perspectives for lichenology. It allows the analysis of minute amounts of products present in microlichens. Individual organic microcrystals located in the fluorescent cortex have been identified in situ as lichexanthone (181, 182). This unique example of characterizing organic substances on a microscale illustrates the capacity of LAMMA as a molecular microprobe. In the dark carbonized area of a cryosectioned microlichen, unexpected barium has been detected (183).

Fingerprint mass spectra of the inorganic cation and anion content of the trapping slimes of carnivorous plants have been obtained with LAMMA (184). This information might help in elucidating the mode of secretion and lead to further species classification.

In single cells of soybean cotyledons that had been infected with a fungus, the phytoalexin glyceollin could be detected by LAMMA at the site of infection (100). A steep rise in glyceollin content toward the infected area was observed. This demonstration of a highly localized glyceollin accumulation at the cellular level again illustrates the suitability of LAMMA for the detection of organic molecules in biological tissues with high lateral resolution.

Chamel and Eloy (25) have reviewed some applications of the laser probe mass spectrograph (LPMS) in plant biology. These studies included the in situ determination of the distribution, in the treated lamina, of elements applied to the plant leaves: copper (23, 185), boron (186), and zinc (contained in fungicides) (187) located in the treated zone after the foliar application. LPMS is capable of analyzing microstructures when the sample size is very small as for root nodules (188), in cases where it is rather difficult to gather enough material for most other analytical methods. It is also useful when a specific zone inside a biological structure has to be analyzed without being isolated, as for intraroot fungal vesicles in endomycorrhizas (188). Other application examples are the qualitative analysis of the local mineral composition in transverse sections of woody stems of the walnut tree (189) and the analysis of the vessel wall of a piece of wood impregnated with a commercial salt solution containing chromium, copper, and arsenic (25, 26).

5.3. Applications in Biomedicine

De Nollin et al. (190) investigated the influence of antimycotic drugs on the ultrastructure of a human pathogenic fungus and on the cytochemical localization of Ca ions. Throughout this study a combined oxalate–pyroantimonate precipitation technique was used to immobilize intracellular Ca^{2+}. LAMMA confirmed that the measured amounts of calcium corresponded to the electron-dense precipitate formed in the different cellular compartments. In mycological research, elemental analysis of the inorganic composition of fungal tissues has also been performed by laser microscopy using emission spectroscopy, LMA (191, 192).

Siebert et al. (193) evaluated the potential use of minced krill meat for food supply. They could not recommend it for human consumption due to its high fluorine content, which might cause disturbances in the mineralization of teeth and bones.

The cation distribution in developing ovarian follicles of the fruit fly has been studied using pyroantimonate precipitation and LAMMA (194, 195). Of particular interest were the stages of oogenesis when ionic asymmetries are being built up. The results suggested that the electrical polarity observed in polytrophic ovarioles may be based on differences in the cation distribution along the anteroposterior axis of the follicie.

LAMMA has provided useful information in bacteriology. The possibility of quantitative measurements of the sodium and potassium contents in single mycobacterial cells has been tested and evaluated against conventional "integrating" analytical methods (196, 197). The influence of temperature during washing of the bacteria on the distribution of the Na^+/K^+ ratio within the cell population was investigated. In these experiments one drop of a suspension of the washed bacteria was placed on Formvar-coated copper EM grids and dried. Single (myco)bacteria could then be selected and analyzed by LAMMA. The fate of isonicotinic acid hydrazide (a tuberculostatic drug) or its fragments was traced in individual treated mycobacteria (196, 197). Encouraging results were obtained using LAMMA for the classification (taxonomy) of mycobacteria by mass spectrometric fingerprinting of single cells. LAMMA fingerprinting of different strains provides an alternative approach to Curie-point pyrolysis mass spectrometry for the characterization of microorganisms (198). Differentiation of closely related strains of bacteria has been possible (199). Time-dependent changes in the intracellular Na^+/K^+ ratio of single bacterial cells induced by an antibacterial agent (a nitrofuran derivative) have been measured with LAMMA (200). The Na/K ratio was shown to be a sensitive indicator of cell viability. Similar shifts in cation concentrations have been observed with cephalosporins and aminoglycosides and in specific culture media (201). The statistical distribution of the Na^+/K^+ ratio within a bacterial cell population may also be useful for monitoring its physiological state and alterations induced by drug treatment (chemotherapy). Cation ratio distributions in bacterial samples treated with drugs at minimal inhibitory concentrations were compared to the respective untreated controls at different times during the growth cycle (202). LAMMA was also used to assess the influence of growth conditions and drugs (isonicotinic acid hydrazide, ethambutol) on the mass fingerprint patterns and Na/K ratios in individual mycobacterial cells (203) and to detect cellular impairments externally induced by heat, X-ray irradiation, and various drugs in the (myco)bacteria (204). An extensive computerized statistical procedure has been developed to allow the differentiation of various species of the same genus or between drug-sensitive and drug-resistant strains from a limited number of single-cell mass fingerprints (202). In kinetic studies of the transport properties of the cell membrane, intracellular concentrations of exchangeable elements or stable isotopes added at different temperatures can be measured (205). Furthermore, LAMMA may serve as a so-called soft-ionization source for mass spectrometric detection and structural analysis of isolated bacterial components (e.g., lipids from the cell wall). Recent work in this area includes the identification of complex organic compounds, components of the outer membrane of Gram-negative bacteria, such as phospholipids, lipid A–like molecules, and natural lipid A (86, 205, 206). These results illustrate the application of LAMMA for the analysis of high-molecular-weight, nonvolatile, synthetic and natural bioorganic compounds by exploiting their prominent quasi-molecular ion formation originating from alkali ion attachment.

LAMMA has also been used successfully to study the inner ear (207). Orsulakova et al. (208) measured the intracellular ion concentration gradients in the lateral wall of the cochlear duct, that is, in the transport-active epithelia of the inner ear. The local K^+/Na^+ ratio was determined by profiling across the spiral ligament and stria vascularis. This concentration ratio was shown to follow a gradient. After anoxia, this gradient flattened, which suggested an energy-dependent active transport mechanism. The specimen had been shock-frozen, freeze-dried at low temperature, embedded in epoxy resin under vacuum, and sectioned dry. In order to control the quality of sample preparation, striated skeletal muscle was cryofixed in an identical way and checked for structural preservation and ionic composition (208–210). A good correlation was found between the K^+/Na^+ ratio and the preservation of the tissue fine structure in the analysis of tissues with different degrees of freezing damage. The influence of the ionic composition of various perfusion liquids on the LAMMA measurements of inner ear tissue was evaluated to allow the selection of an appropriate medium (209, 210). In studies on ion transport in the endolymphatic space, the K/Na ratio in the epithelial cells and subepithelial tissue of the endolymphatic sac has been measured under a variety of experimental conditions, such as ethacrynic acid injection (211, 212).

The LAMMA technique has been applied extensively in vision research. Of major interest in this field is the analysis of the calcium distribution in vertebrate and invertebrate retinas. It has been suggested that calcium plays an important role in photoreception by acting as an internal neurotransmitter (120). In the search for Ca^{2+} stores in the photoreceptor cells, various specimen preparation procedures have been employed that should prevent diffusion of the analyte and preserve tissue ultrastructure. These methods consist of shock-freezing of the isolated retinas or wet chemical fixation with aldehyde either in tertiary phosphate buffer or in the presence of oxalate and/or pyroantimonate precipitating agents. The major portion of calcium was found in black pigment granules within the photoreceptor cells. In vertebrate photoreceptor pigments, significant amounts of barium were found to be associated with calcium: this was the major dissimilarity compared to the invertebrate species (213). When analyzing cow retina with LAMMA, we also demonstrated the presence of barium next to high amounts of calcium (7). In crayfish, the accumulation of Ca^{2+} in intracellular compartments and its kinetics of uptake by the photoreceptor cells from extracellular sources have been studied using the ^{44}Ca stable isotope (120, 146, 214–216). The potential for such isotope labeling studies represents a major advantage of LAMMA. Stable isotopes allow measurement of the release and/or redistribution of Ca at the (sub)cellular level in ion transport studies. In shock-frozen and freeze-dried specimens of cat and frog retina, distribution profiles within the photoreceptors have been established for the leading physiological cations

(217). The presence of large amounts of barium within pigment granules in the retinal pigment epithelium and the choroid was confirmed in frog, cat, and human and was also revealed in human melanoma cells (217, 218). The combined use of LAMMA and ultrastructural cytochemistry was described for the verification of the localization of calcium in rat retina (219). The fruitful combination of both techniques may be helpful in further elucidating the physiological role of calcium in visual transduction. Recently, the light-dependent release of calcium from photoreceptors of isolated toad retinas has been measured (220–222). Bleaching of visual pigment molecules by continuous illumination caused a graded release of calcium from the outer segments of the vertebrate rods. The rate of this calcium efflux increased with light intensity. Since no effect of exposure to light on the rate of calcium influx could be detected, it was concluded that the total calcium content of the rods diminished rapidly in bright light. The analytical results were shown to be consistent with the principal postulates of the "calcium hypothesis" (221).

Considerable efforts have been made in the microanalysis of hard tissues, such as teeth and bone. Preparation methods for LAMMA of hard dental tissue have been described, especially for the analysis of the trace element fluorine (223–226). A preparation method was sought that would permit micrometer areas to be examined in whole tooth sections. Gabriel et al. (226) applied large-area grinding of the dental material, and later on they also used a groove milling technique. After treatment of teeth with TiO_2-containing dental root-filling material, an intracellular accumulation of titanium was revealed in cells of the dental pulp and tissue surrounding the treated teeth (227, 228). TiO_2 in the pulp tissue may lead to morphological alterations and to impairment of the cell function.

The elemental composition of human, amphibian, and shark teeth has been determined by local topographic and in-depth analysis using LPMS (23, 229). Undecalcified samples of bone and of calculi have been analyzed by laser microspectral analysis (LMA) to optimize the applicability of the technique to the study of calcified tissues (230–233). Also, LIMA is used in medical research releated to calcium metabolism in osteoporosis or muscular dystrophy patients (21).

During recent years, LAMMA has been applied successfully to the microanalysis of bone in clinical nephrological investigations (234). Aluminum toxicity has been implicated in the pathogenesis of a number of clinical disorders in patients with chronic renal failure treated by long-term intermittent hemodialysis, which may lead to severe bone disease (dialysis osteomalacia) (235). The accumulation of aluminum and its ultrastructural localization, together with that of possibly associated elements such as iron, have been studied using LAMMA on human bone

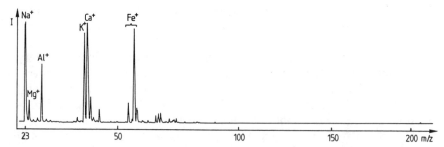

Figure 5.13. Positive-ion LAMMA spectrum of an undecalcified bone biopsy taken from a chronic dialysis patient with Al and Fe overload and osteomalacic bone disease. The mass spectrum was obtained from a region at the osteoid–calcified bone interface. Significant Al and concomitant Fe signals are observed (237).

biopsy sections (119, 121, 236–241). Different regions of the trabecular bone have been analyzed, including the osteoid–calcified bone boundary and the unmineralized osteoid (Fig. 5.13). Also, the local aluminum content within bone cells has been determined in patients with dialysis osteomalacia and compared to patients without kidney disease. LAMMA investigations have disclosed the presence of aluminum in bone marrow macrophages (240) and in single osteocytes and fibroblasts (228, 242–245).

In dynamic studies of osteogenesis associated with calcium phosphate ceramic implants containing TiO_2 as a composite material, it was shown that titanium liberated from the implant was located in the surrounding bone marrow (246, 247). No titanium was detected in the mineralized bone. Preliminary results have also been obtained by LAMMA in comparing the trace element concentrations of normal and pathological bone in the case of osteosarcoma bone tumor (243). Also of interest is the localization of lead in different cell types of bone marrow from a person who had been severely poisoned by lead nitrate over a period of several months (164, 243).

It is important to gain knowledge about the distribution and local concentration of heavy metals in soft tissues and to correlate the microanalytical findings, with clinical expressions such as atherosclerosis and nephropathy. The topochemical distribution of lead across the human arterial wall in normal and sclerotic aortas has been determined using LAMMA (128, 164, 228, 243, 248). The toxic element lead could also be detected in placental tissue and in fetal liver after acute maternal lead intoxication (163). Other LAMMA studies include the identification of lutetium in mouse liver (249) and the detection of cadmium in rabbit kidney (250). After administration of the anticancer drug cisplatin to a dog, the heavy metal platinum could successfully be localized within the proximal tubular cells of the kidney of the treated dog (119, 122, 148). In a clinical case of Wilson's disease, an inhomogeneous distribution of copper within the myocardium was demonstrated by LAMMA (163, 248, 251). Using LMA, it was possible to measure gold in cultured mouse fibroblasts treated with gold thioglucose and also in human skin biopsies after treatment with gold salts for rheumatoid arthritis (49).

In studies of cellular physiology, there has been considerable interest in the application of LAMMA to muscle specimens (cardiac, skeletal, or vascular smooth muscles). The function of muscle as chemicomechanical energy converters depends on subcellular and transmembrane movements of inorganic ions in a complicated and highly compartmentalized system. Hirche et al. (252, 253) developed a technique for sample preparation without altering tissue ultrastructure or intracellular ion concentrations. It was shown that the K/Na ratio as measured by LAMMA can be considered a sensitive criterion for the quality of the muscle preservation. A decrease of the intracellular K/Na ratio has been found in muscle cell damage (250), including the ischemic myocardium (253). In freeze-dried embedded frog sartorius muscle, a regional selective accumulation of alkali metal ions has been demonstrated (147, 254). The Ca^{2+} distribution in vascular smooth muscle preparations of the rat and in dog myocardial cells was visualized by combined oxalate–pyroantimonate precipitation and confirmed by parallel LAMMA measurements (255, 256). The effect of flunarizine, a selective Ca^{2+} entry blocker, and of noradrenaline stimulation of the heart were illustrated (255, 256). A large variation of the calcium signal intensity in LAMMA was found within the same skeletal muscle cells, in contrast to the rather homogeneous sodium and potassium signals (250, 253). This suggests a highly compartmentalized intracellular calcium distribution, in agreement with the hypothesis of a calcium-dependent control of contractile activity. The cardioactive drug verapamil was traced in preincubated cardiac tissue (papillary muscle) by means of its fluorinated derivative (9, 97). This exemplifies the unique potential of LAMMA in the study of subcellular pharmacokinetics through appropriate labeling of organic compounds. A covalently bound atom of high ion yield in LAMMA may be a suitable candidate if its location remains representative for the presence of the compound or its main metabolites. Using LMA, element changes in human heart biopsy tissues after heart transplants or in various heart diseases could be monitored, and correlations were suggested with pathological events (49).

In rat pancreas, the cellular uptake of Ca^{2+} into the acini has been studied by Wakasugi et al. (257). The accumulation of calcium in the dark areas containing cytochemical oxalate precipitates, mainly in the rough endoplasmic reticulum, was confirmed by LAMMA. It was suggested that during stimulation with secretagogues, Ca^{2+} is taken up in intracellular ATP-dependent Ca^{2+} pools. One of the first applications of LMA described the chemical analysis of rat pancreas, in addition to brain, elastin and calculus (2).

Inorganic lithium salts are potent drugs in the treatment of some mental disorders, mainly those of the depressive type. Small amounts of Li^+ could be detected by LAMMA in brain cells of a rat after oral administration of a single therapeutic dose of LiCl (97, 131, 258–260).

LAMMA confirmed the accumulation of iron within granules of macrophages and uterine glandular cells (decidua) of pregnant Tupaia, which had been suggested by histochemistry (131, 258). In a bacterial species (pedomicrobium) known to concentrate particular divalent cations, the elements manganese and iron could be detected (131, 258). In particle-laden alveolar macrophages from a heavy cigarette smoker, LAMMA could readily disclose the presence of aluminum together with iron and calcium in the cellular phagolysosomes (261). The sample had been prepared by conventional fixation and embedding procedures. LAMMA has also been used to establish a mass spectrometric chemical classification of coal-mine dust particles taken up by cultured human lymphocytes after in vitro incubation (9, 97, 131). Particles of a predominantly organic (pit coal) composition were preferentially phagocytosed compared to inorganic particles (containing mainly Na, K, Al, Fe, and Ca).

A review by Verbueken et al. (234) dealt with the ultrastructural localization of the clinically toxic element aluminum in a variety of tissues and organs in normal and pathological conditions. The finding of a concomitant storage of aluminum and iron in the hepatic lysosomes of patients on chronic hemodialysis, as revealed by LAMMA, well illustrates the potential of a sensitive, multielemental microanalytical technique to reveal possible element associations at the microscopic level (Fig. 5.14) (119, 121, 148, 262, 263). Using LMA, unique profiles of groups of elements (including Al and Fe) have been established in human cancerous liver, lung, and colon and compared to the nonmalignant tissues (49).

Some LAMMA studies have been performed on blood samples, in particular of patients suffering from renal insufficiency. The trace element content of individual human erythrocytes was investigated, with special emphasis on magnesium and aluminum (163, 164, 243, 248, 264). In such patients on regular dialysis, aluminum has been detected in lyophilized samples of serum (129, 130, 163, 243) and in synovial fluid (129). Laser microanalysis may also be used to determine the elemental composition of bloodstains in medico-forensic work (265).

Using LAMMA, the skin of patients on long-term chronic hemodialysis was shown to contain very low concentrations of aluminum (266, 267).

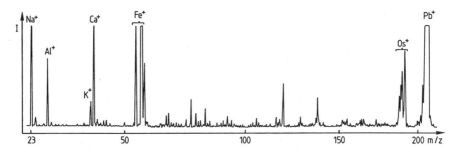

Figure 5.14. Positive-ion LAMMA spectrum obtained from a lysosome of a human liver cell, demonstrating a concomitant Al and Fe accumulation. The liver biopsy was taken from a chronic dialysis patient. The Os and Pb signals are due to the fixation and staining procedures, respectively (237).

The skin biopsy had immediately been frozen, cut, and recovered from a water surface. In addition, the potential of LAMMA as a molecular microprobe was illustrated by the tentative identification of $CaHPO_4$ and of phthalate plasticizer in these human skin samples (266, 267). In skin biopsies of a patient with salicylate-elicited urticaria, pyroantimonate precipitates within cutaneous mast cells and the surrounding connective tissue have been analyzed for their calcium content. LAMMA revealed an interference between the antimonate stain and histamine (268, 269). By means of LMA, an elevated amount of calcium could be demonstrated in localized lesions in a case of calcinosis of the skin (270). In freeze-dried frog skin, the leading physiological cations Na^+ and K^+ have been probed by LAMMA over different layers of the unembedded specimen (171).

The potential use of LAMMA to study the distribution of specific isotopes ("isotopography") may be of great interest. The uptake of ^{125}I in animal thyroid gland was investigated by LAMMA, which revealed the presence of tellurium originating from nuclear decay of the tracer (163). This technique permits localization of radioactive isotopes when the mass of the radionuclide or the masses of the disintegration products do not coincide with originally present elements.

Increasing attention is being paid to the in situ probing of specific organic constituents and inorganic crystals in biomedical samples, with a high lateral resolution of analysis. The laser microprobe has been used, for example, to identify calcium oxalate deposits in human and rat renal histological material (64, 244, 271). Other microanalytical techniques such as EPXMA, which are limited to mere elemental analysis, are not informative for molecular compound identification. Such investigations are imporant to verify diagnostic histochemical methods for the demonstration of crystalline deposits in histological sections. LAMMA has also revealed calcium oxalate crystals in the brain (basal ganglia) of patients who suffered from encephalitic reactions to polyol-containing infusions during intensive care (272). The chemical nature of spheroliths in human cataract lenses was explored by LAMMA; calcium carbonate appeared to be the most likely constituent (273). LAMMA has also been employed to identify amino citric acid in biological peptides (calf thymus ribonucleoprotein) (274).

Further development of LAMMA will certainly focus on the microprobing of biomedically relevant organic constituents and crystalline deposits within histological sections, either by mass fingerprinting or by direct mass spectrometric identification.

6. APPLICATIONS OF LAMMA IN ENVIRONMENTAL RESEARCH

Most of the applications of LAMMA in environmental research make use of the single-particle analysis capabilities for the investigation of ambient aerosols, asbestos fibers, fly ash, and dust of industrial origin. Another important field of application is the study of geological samples.

The major advantages offered by LAMMA in these research fields are, on the one hand, its capability for simultaneous detection of nearly every element of the periodic system with a favorable detection limit and, on the other hand, its potential for the fingerprinting of organic compounds and for the speciation of inorganic constituents, all on a microscopic scale. In what follows, a survey of the LAMMA applications in environmental research is given, with brief comments, on the results obtained. Special attention is drawn to review papers that deal with the same subject (6, 7, 110, 275).

6.1. Applications in Aerosol Research

The analytical techniques most frequently used provide accurate data about the bulk chemical composition of airborne particulate matter but do not give information about its internal heterogeneity. Such information is especially useful in the study of airborne particulate matter for detailed interpretation of the data with respect to the origin and formation mechanism of the individual particles (e.g., marine, continental, or pollution-derived) and to their behavior during transport—that is the physico-chemical interactions with the transporting medium (e.g., gas-to-particle conversion) or with other particulates (e.g., coagulation processes). Microprobe techniques such as LAMMA can be helpful in the study of these phenomena because they allow the analysis of individual aerosol particles for their organic and inorganic composition. Moreover, other relevant data concerning trace elements and the dispersion of a chemical component within a single particle, such as enrichment at the particle surface, can be assessed. When combined with the particle size distribution, these data can lead to a better understanding of atmospheric chemistry, which is necessary for the determination of the effects of particle inhalation on human health.

Airborne particulate matter can be divided into the so-called natural background aerosol (e.g., soil erosion, sea spray, biological and volcanic emissions) and man-made pollution (e.g., fly ash and dust produced by industrial activities). Figure 5.15 shows LAMMA spectra of both anthropogenic and natural aerosol particles.

For the sample preparation of particulate matter collected in bulk, such as steel furnace dust (276), oil shale retort (277), fly ash (278–280), and refinery source emissions (281), the particles are usually dispersed on a polymer film (e.g., Formvar or Pioloform) by bringing the filmed grid into contact with the bulk sample. This sample preparation procedure is not advisable for aerosol particles, because the formation of agglomerates changes the particle size distribution and can result in the chemical transformation of reactive components that were originally distributed over specific particle types. The particles would not retain their identity, so that individual particle analysis is no longer meaningful. Therefore, these particles are often sampled directly from the air with impactors (112, 117, 160, 282). The collection substrate must be sufficiently thin to allow perforation by the laser (e.g., Formvar films). If the filmed grids are then mounted in the impaction surfaces, size-fractionated aerosol samples are obtained, and no further sample manipulation is needed for the analysis (Fig. 5.10). The sampling time must be optimized to assure an appropriate loading on the impaction surface; that is, the particles have to be spatially separated from each other so that particle interactions are negligible.

A general procedure for quantification in LAMMA particle analysis is not available at present (see Section 4.5). This must be considered a drawback, although in several cases a valuable interpretation is possible for qualitative or semiquantitative data.

The first LAMMA studies on atmospheric aerosols were reported by Wieser et al. (109, 150, 160, 283) and Kaufmann et al. (110, 117). Ambient aerosol particles in the giant and large size range were sampled with inertial impaction devices. Individual aerosol particles were analyzed with LAMMA for their major and minor elemental composition. The spectra showed the occurrence of trace contaminants, and the first preliminary investigations on the chemical state of sulfur and nitrogen were reported.

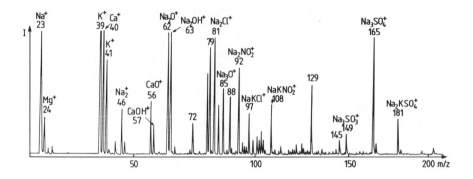

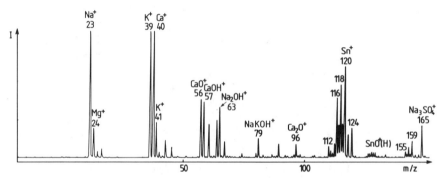

Figure 5.15. (*A*) LAMMA spectrum (+) of a coastal aerosol particle (2 μm diameter) of marine origin. (*B*) LAMMA spectrum (+) of an anthropogenic aerosol particle (1 μm diameter).

Seiler et al. (284) described the differences in the spectra for solid silicon-rich aerosols and vacuum-dried sulfur-rich particles.

Adams et al. (282) and Surkyn et al. (285) used the LAMMA instrument for aerosol source identification. It was possible to explain the anomalous enrichment of sodium as $NaNO_3$ and calcium as $CaSO_4$ in aerosols collected in remote continental regions of the southern hemisphere.

Bruynseels et al. (111, 112) applied the LAMMA technique to study a set of samples collected with a low-volume cascade impactor on a trajectory from a beach site under constant influence of steady and clean onshore tradewinds, toward and through a heavily polluted industrialized area. Six types of both positive and negative mass spectra with different inorganic and organic signals could be distinguished in the different particle size ranges. It was possible to retrieve different types of chemical composition and to evaluate the influence of the different sources—sea spray, soil erosion, and anthropogenic pollution. LAMMA was also used to study the speciation and distribution of sulfur and nitrogen in individual aerosol particles as a function of particle size and origin and for the detection of sulfate and nitrate layers on individual marine aerosols (134).

In a combined study with LAMMA and EPXMA, Bruynseels et al. (286) characterized the airborne particulate matter sampled above the North Sea from two masses of different origin as determined by air-mass back-trajectory analysis. The combination of the results of both analytical tools has considerable advantages because they can provide complementary analytical data.

Artificial PAH-coated particles, generated in the laboratory by gas-phase condensation of PAHs (anthracene, perylene, phenanthrene, pyrene, and chrysene), were investigated by Niessner et al. (152). After mixing of the PAH-coated aerosols with gases such as O_3, NO_2, and Br_2, subsequent LAMMA analysis revealed information about the chemical reactivity of the PAHs with the trace gases.

The residual particles formed after dispersion and drying of rainwater samples were analyzed by Kühme (287) to study the scavenging of trace substances from the atmosphere as well as their relative solubility.

Denoyer et al. (278, 279) used LAMMA for the characterization of coal and oil fly ash that was collected in bulk with an electrostatic precipitator and mounted on a Formvar-coated grid by touching the grid to the bulk fly ash. The mass spectra of the oil fly ash provided data on the chemical composition such as its carbonaceous nature and the presence of sulfur and vanadium compounds and trace elements. On the basis of their LAMMA spectra, the coal fly ash particles were classified into three particle types: two types of aluminosilicates showing a statistically meaningful difference in the $^{54}Fe/^{44}Ca$ intensity ratio and carbonaceous particles containing PAHs. In a multi-instrumental study of fly ash particles with respect to the morphology and surface enrichment of matrix and trace elements, Gay et al. (280) used the laser microprobe for the detection of trace elements and anions in the individual particles. Van Craen et al. (276) investigated the surface and possible leaching characteristics of steel furnace dust with LAMMA and SIMS. The negative-ion spectra from both the SIMS and LAMMA techniques provided information from which the chemical form of certain elements such as iron, sulfur, and phosphorus could be inferred. Soot from an experimental oil shale retort has been examined by Mauney et al. (277), who showed that LAMMA can detect a wide molecular weight range of PAHs simultaneously with a multiele-ment inorganic analysis of individual particles with a 1–10 μm diameter.

Dutta et al. (281) reported on the characterization of refinery emissions and ambient refinery samples. The source emissions were described as vanadium-containing carbonaceous particles and lanthanum-containing aluminosilicate particles. The source emissions were also identified in the ambient samples.

Verdun and Muller (288) used LAMMA for desorption studies of PAHs adsorbed from toluene solutions onto activated coal, silica gel, and asbestos.

6.2. Applications in Asbestos Studies

It is well known that fine respirable asbestos fibers and man-made mineral fibers are biologically active in producing lung fibrosis, cancer, and other diseases. Although the particle geometry is of primary importance for the carcinogenic effect, such chemical properties as the leachability of spe-cific elements and the surface adsorption of pollutants can be of consid-erable pathogenic importance.

LAMMA analysis of asbestos fibers relies on their light microscopical (visual) detection and on the ability to distinguish asbestos from nonas-bestos fibers and to discriminate different asbestos varieties on the basis of the mass spectra obtained. Spurny et al. (157) utilized the possibility of differentiating between the serpentines and the amphibole classes of fibers and also between amosite and crocidolite on the basis of fingerprint LAMMA spectra. De Waele et al. (133, 289) systematically studied the LAMMA fingerprint spectra of the five UICC (International Union against Cancer) asbestos standards and pointed out that the positive-ion-mode spectra are especially indicative for the different asbestos varieties. The same authors (290) also described the striking difference in spectral ap-pearance between chrysotile originating from Zimbabwe and the same material mined in Canada. The laser desorption fingerprint spectra, in-dicative for the surface material of the asbestos fiber, showed an elemental peak of iron at least a factor of 10 more intense in the Canadian chrysotile. This could be explained by the presence of magnetite microcrystallites as surface impurities in the Canadian asbestos.

The asbestos fibers were treated with several organic products such as benzidine, benzo[a]pyrene, aliphatic alkylammonium salts, N,N-di-methylaniline (DMA), and orthophenylene diamine (OPDA) (289, 291–294). It appeared that when DMA and OPDA were adsorbed onto asbestos and sepiolite fibers, reaction products such as methyl violet and crystal violet for DMA (293) and 2,3-diaminophenazine and higher-molecular-weight compounds for OPDA (294) could easily be detected on the individual fibers.

De Waele et al. (295) also reported the detection of organic surface contaminants such as phthalate plasticizers originating from their storage in polyethylene bags with a detection limit of 500 μg/g and a shot-to-shot reproducibility of ~20%.

LAMMA was also used for the characterization of chemically modified fiber surfaces, such as triethyloxyoctylsilane-coated chrysotile fibers (296), organosilane-grafted sepiolite fibers, and chrysophosphate fibers (133).

Spurny et al. (157, 297–300) studied UICC asbestos standards and glass fibers before and after treatment with acids and bases, blood sera, and animal tissues. They observed a chemical instability of the fibers in leaching experiments—for example, for the elements magnesium, iron and silicon in chrysotile and for sodium, potassium, calcium, and zinc in a man-made mineral fiber.

6.3. Application to Geological Samples

Dutta and Talmi (114) applied LAMMA to optically distinct regions of a coal surface. The edge regions of polished sections of the coal samples were investigated. Since coal is a heterogeneous mixture, several model compounds such as clays, polyethylene, polystyrene, and phenoxy resin were also studied. By comparison with these model compounds, they determined areas on the coal surface that were carbonaceous in nature and other regions that were claylike. They also identified aromatic polymers with oxygen groups. From their study of the edge regions of small chips of coal and shale samples, Vanderborgh and Jones (115) concluded that rapid heating, as by high-power laser irradiation, selectively directs energy into the hydrocarbons that are firmly sorbed onto a polymeric matrix and that the majority of the molecular fragments are neutral compounds that react with the available positive and negative ions to form detectable charged clusters. The results emphasize the formation of metallic cation adducts in the positive spectra and chlorine adducts in the negative spectra.

Henstra et al. (105–107) analyzed components from thin sections of soils developed in weathered granite. LAMMA spectra of secondary titanium compounds and other constituents were obtained. The characteristic analytical possibilities of the ion microprobe mass analyzer (IMMA), EPXMA, and LAMMA-500 were compared.

Weinke et al. (301) investigated troilite (stoichiometric iron sulfide, FeS), of meteoritic origin and pyrite of terrestrial origin. The trace element content was calculated by a method using the LTE model, and results were compared with those obtained by other instrumental techniques, such as neutron activation analysis, EPXMA, and SIMS.

Freund et al. (302) reported the detection of atomic carbon in arc-fusion-grown MgO and CaO single crystals. Similar measurements were done on synthetic forsterite and natural olivines (113) and led to the conclusion that these minerals can take up atomic carbon in the solid solution.

Osmium-containing specimens from different deposits were analyzed by Englert and Herpers (167). The samples were prepared by spreading the powders on filmed grids or by fixing larger fragments between two grids. They found isotopic anomalies for ^{187}Os (daughter of the long-lived radionuclide ^{187}Re) corresponding to the sample's age and the ^{187}Re content of the ores from which the osmium was separated. In a combined LAMMA and analytical EM study of zirconium deposits, Steel et al. (303) measured the content of the trace elements uranium, thorium, and lead as well as the relative abundance of the lead isotopes.

7. CONCLUSIONS

Commercial instruments for laser microprobe mass spectrometry have been available for applied research for only a few years. However, the usefulness of this new microtrace analysis technique has already been demonstrated in numerous and extremely diverse applications in medical, biological, and environmental science.

The most important claims made for LAMMA are its high detection efficiency (down to about 10^{-20} g), its speed of operation, its microbeam character (spatial resolution of about 1 μm), its capabilities for inorganic molecular speciation and for organic mass fingerprinting (even for nonvolatile and thermally labile organic compounds), and its potential for separate analysis of the surface layer and core of microparticles. On the other hand, the technique is destructive, the quality of the light microscopical observation is poor, and quantification of the analytical data is still under investigation.

LAMMA has indeed proved its merits as a qualitative tool and has shown some promise for quantitative measurements. However, the hopes of early investigators for straightforward quantitative protocols have not completely been fulfilled. Although the instrument can yield suitably reproducible spectral intensity data for selected samples, subtle changes in instrumental conditions can result in radical changes in the ion signal, altering both absolute and relative mass spectral intensities. It is certainly possible to make quantitative measurements if the instrumental limitations and the chemical and physical effects are acknowledged and carefully handled. We have every reason to expect that routine quantitative LAMMA analysis will develop reasonably successfully within the next few years. Important in the achievement of quantitative accurate analysis will be the necessary fundamental studies of the processes involved in the measurement, including laser–matter interactions, plasma chemistry, and physics.

ACKNOWLEDGMENTS

This work was supported in part by the Belgian Ministry of Science Policy through research grants 80-85/10 and 84-89/69. A. Verbueken thanks the Belgian National Foundation for Scientific Research (N.F.W.O.) for providing a position as senior research assistant. We thank L. Van Vaeck for his valuable comments and critical reading of the manuscript. The skillful secretarial work of N. Van Hoeymissen is highly appreciated.

REFERENCES

1. R. E. Honig and J. R. Woolston, *Appl. Phys. Lett.*, **2**, 138 (1963).

2. R. C. Rosan, M. K. Healy, and W. F. McNary, Jr., *Science*, **142**, 236 (1963).

3. R. J. Conzemius and J. M. Capellen, *Int. J. Mass Spectrom. Ion Phys.*, **34**, 197 (1980).

4. D. Glick and R. C. Rosan, *Microchem. J.*, **10**, 393 (1966).

5. F. Hillenkamp, E. Unsöld, R. Kaufmann, and R. Nitsche, *Nature*, **256**, 119 (1975).

6. E. Michiels, L. Van Vaeck, and R. Gijbels, *Scanning Electron Microsc.*, **III**, 1111 (1984).

7. A. H. Verbueken, F. J. Bruynseels, and R. E. Van Grieken, *Biomed. Mass Spectrom.*, **12**, 438 (1985).

8. F. Hillenkamp, E. Unsöld, R. Kaufmann, and R. Nitsche, *Appl. Phys.*, **8**, 341 (1975).

9. R. L. Kaufmann, H. J. Heinen, M. W. Schürmann, and R. M. Wechsung, in D. E. Newbury, Ed., *Microbeam Analysis—1979*, San Francisco Press, San Francisco, 1979, p. 63.

10. E. Denoyer, R. Van Grieken, F. Adams, and D. F. S. Natusch, *Anal. Chem.*, **54**, 26A (1982).

11. B. Schueler, R. Nitsche, and F. Hillenkamp, *Scanning Electron Microsc.*, **II**, 597 (1980).

12. F. Hillenkamp, P. Feigl, and B. Schueler, in K. F. J. Heinrich, Ed., *Microbeam Analysis—1982*, San Francisco Press, San Francisco, 1982, p. 359.

13. P. Feigl, B. Schueler, and F. Hillenkamp, *Int. J. Mass Spectrom. Ion Phys.,* **47,** 15 (1983).

14. H. J. Heinen, S. Meier, H. Vogt, and R. Wechsung, *Int. J. Mass Spectrom. Ion Phys.,* **47,** 19 (1983).

15. P. K. D. Feigl, F. R. Krueger, and B. Schueler, *Mikrochim. Acta,* **II,** 85 (1984).

16. S. Nadahara, T. Kikuchi, K. Furuya, S. Furuya, and K. Hoshino, *Mikrochim. Acta,* **I,** 157 (1985).

17. T. Dingle, B. W. Griffiths, and J. C. Ruckman, *Vacuum,* **31,** 571 (1981).

18. T. Dingle, B. W. Griffiths, J. C. Ruckman, and C. A. Evans, Jr., in K. F. J. Heinrich, Ed., *Microbeam Analysis—1982*, San Francisco Press, San Francisco, 1982, p. 365.

19. C. A. Evans, Jr., B. W. Griffiths, T. Dingle, M. J. Southan, and A. J. Ninham, in R. Gooley, Ed., *Microbeam Analysis—1983*, San Francisco Press, San Francisco, 1983, p. 101.

20. T. Dingle and B. W. Griffiths, in A. D. Romig, Jr., and J. I. Goldstein, Eds., *Microbeam Analysis—1984*, San Francisco Press, San Francisco, 1984, p. 23.

21. N. S. Clarke, A. R. Davey, and J. C. Ruckman, in A. D. Romig, Jr., and J. I. Goldstein, Eds., *Microbeam Analysis—1984*, San Francisco Press, San Francisco, 1984, p. 31.

22. N. S. Clarke, A. R. Davey, and J. C. Ruckman in A. D. Romig, Jr., and J. I. Goldstein, Eds., *Microbeam Analysis—1984*, San Francisco Press, San Francisco, 1984, p. 40.

23. J. F. Eloy, *Microsc. Acta,* **Suppl. 2,** 307 (1978).

24. J. F. Eloy, *Rev. Méthod. Phys. Anal. (GAMS),* **5,** 157 (1969).

25. A. Chamel and J. F. Eloy, *Scanning Electron Microsc.,* **II,** 841 (1983).

26. A. Chamel and J. F. Eloy, *J. Phys.* **45,** C2-573 (1984).

27. E. Deloule and J. F. Eloy, *Chem. Geol.,* **37,** 191 (1982).

28. J. F. Eloy, *J. Phys.,* **45,** C2-265 (1984).

29. J. F. Eloy, *Scanning Electron Microsc.,* **II,** 563 (1985).

30. R. Stefani, *Trends Anal. Chem.,* **1,** 84 (1981).

31. R. J. Conzemius and H. J. Svec, *Anal. Chem.,* **50,** 1854 (1978).

32. R. J. Conzemius, F. A. Schmidt, and H. J. Svec, *Anal. Chem.,* **53,** 1899 (1981).

33. R. J. Conzemius, in K. F. J. Heinrich, Ed., *Microbeam Analysis—1982,* San Francisco Press, San Francisco, 1982, p. 369.

34. Zhao Shankai, R. J. Conzemius, and H. J. Svec, *Anal. Chem.,* **56,** 382 (1984).

35. J. A. J. Jansen and A. W. Witmer, *Spectrochim. Acta,* **37B,** 483 (1982).

36. H. J. Dietze and H. Zahn, *Exp. Tech. Phys.*, **20**, 389 (1972).

37. Y. A. Bykovskii, L. M. Babenkov, T. A. Basova, V. I. Belousov, V. M. Gladskoi, V. V. Gorshkov, V. G. Degtyarev, I. D. Laptev, and V. N. Novelin, *Prib. Tekh. Eksp.*, **2**, 163 (1977).

38. R. A. Bingham and P. L. Salter, *Anal. Chem.*, **48**, 1735 (1976).

39. I. D. Kovalev, G. A. Maksimov, A. I. Suchkov, and N. V. Larin, *Int. J. Mass Spectrom. Ion Phys.*, **27**, 101 (1978).

40. B. K. Furman and C. A. Evans, Jr., in R. H. Geiss, Ed., *Microbeam Analysis—1981*, San Francisco Press, San Francisco, 1981, p. 336.

41. B. K. Furman and C. A. Evans, Jr., in K. F. J. Heinrich, Ed., *Microbeam Analysis—1982*, San Francisco Press, San Francisco, 1982, p. 222.

42. R. J. Conzemius, D. S. Simons, Zhao Shankai, and G. D. Byrd, in R. Gooley, Ed., *Microbeam Analysis—1983*, San Francisco Press, San Francisco, 1983, p. 301.

43. A. H. Verbueken, F. Van de Vyver, M. De Broe, and R. Van Grieken, *CRC Crit. Rev. Clin. Lab. Sci..*, **24**, 263 (1987).

44. R. J. Cotter, *Anal. Chem.*, **56**, 485A (1984).

45. F. Brech and L. Cross, *Appl. Spectrosc.*, **16**, 59 (1962).

46. N. A. Peppers, E. J. Scribner, L. E. Alterton, R. C. Honey, E. S. Beatrice, I. Hardin-Barlow, R. C. Rosan, and D. Glick, *Anal. Chem.*, **40**, 1178 (1968).

47. W. Van Deyck, J. Balke, and F. J. M. J. Maessen, *Spectrochim. Acta*, **34B**, 359 (1979).

48. K. Laqua, in N. Omenetto, Ed., *Analytical Laser Spectroscopy*, Wiley, New York, 1979, Chap. 2.

49. D. Glick and K. W. Marich, *Clin. Chem.*, **21**, 1238 (1975).

50. L. J. Radziemski, T. R. Loree, D. A. Cremers, and N. M. Hoffman, *Anal. Chem.*, **55**, 1246 (1983).

51. H. S. Kwong and R. M. Measures, *Anal. Chem.*, **51**, 428 (1979).

52. S. Mayo, T. B. Lucatorto, and G. G. Luther, *Anal. Chem.*, **54**, 553 (1982).

53. M. W. Williams, D. W. Beekman, J. B. Swan, and E. T. Arakawa, *Anal. Chem.*, **56**, 1348 (1984).

54. R. Wechsung, F. Hillenkamp, R. Kaufmann, R. Nitsche, E. Unsöld, and H. Vogt, *Microsc. Acta*, **Suppl. 2**, 281 (1978).

55. H. Vogt, H. J. Heinen, S. Meier, and R. Wechsung, *Fresenius Z. Anal. Chem.*, **308**, 195 (1981).

56. M. J. Beesley, *Lasers and their Applications*, Taylor and Francis, London, 1972.

57. M. I. Cohen, in F. T. Arecchi and E. O. Schulz-Dubois, Eds., *Laser Handbook*, Vol. 2, North-Holland, Amsterdam, 1972, p. 1599.

58. R. Wurster, U. Haas, and P. Wieser, *Fresenius Z. Anal. Chem.*, **308**, 206 (1981).

59. C. A. McDowell, Ed., *Mass Spectrometry*, McGraw-Hill, New York, 1963.

60. E. Michiels, M. De Wolf, and R. Gijbels, *Scanning Electron Microsc.*, **III**, 947 (1985).

61. T. Mauney and F. Adams, in A. D. Romig, Jr., and J. I. Goldstein, Eds., *Microbeam Analysis—1984*, San Francisco Press, San Francisco, 1984, p. 19.

62. E. Michiels, "Instrumentele effecten in laser microprobe massa analyse en clusterionen distributies van oxiden en fluoriden," Ph.D. thesis, University of Antwerp (UIA), Antwerp, Belgium, 1985.

63. T. Mauney, "Instrumental effects in LAMMA, measurements of kinetic energy distributions and analysis of soot particles," Ph.D. thesis, Colorado State University, Fort Collins, Colorado, 1984.

64. A. H. Verbueken, "Application of laser microprobe mass analysis (LAMMA) to biology and medicine," Ph.D. thesis, University of Antwerp (UIA), Antwerp, Belgium, 1985.

65. T. Mauney and F. Adams, *Int. J. Mass Spectrom. Ion Proc.*, **63**, 201 (1985).

66. T. Mauney and F. Adams, *Int. J. Mass Spectrom. Ion Proc.*, **59**, 103 (1984).

67. E. Michiels, T. Mauney, F. Adams, and R. Gijbels, *Int. J. Mass Spectrom. Ion Proc.*, **61**, 231 (1984).

68. U. Fehn, *Int. J. Mass Spectrom. Ion Phys.*, **15**, 391 (1974).

69. U. Fehn, *Int. J. Mass Spectrom. Ion Phys.*, **21**, 1 (1976).

70. M. A. Rudat and G. H. Morrison, *Int. J. Mass Spectrom. Ion Phys.*, **27**, 249 (1978).

71. M. A. Rudat and G. H. Morrison, *Int. J. Mass Spectrom. Ion Phys.*, **29**, 1 (1979).

72. D. S. Simons, *Int. J. Mass Spectrom. Ion Proc.*, **55**, 15 (1983/1984).

73. T. Mauney and P. Van Espen, in A. D. Romig, Jr., and J. I. Goldstein, Eds., *Microbeam Analysis—1984*, San Francisco Press, San Francisco, 1984, p. 295.

74. N. Fürstenau, F. Hillenkamp, and R. Nitsche, *Int. J. Mass Spectrom. Ion Phys.*, **31**, 85 (1979).

75. N. Fürstenau, *Fresenius Z. Anal. Chem.*, **308**, 201 (1981).

76. N. Fürstenau and F. Hillenkamp, *Int. J. Mass Spectrom. Ion Phys.*, **37**, 135 (1981).

77. B. Jöst, B. Schueler, and F. R. Krueger, *Z. Naturforsch.*, **37A**, 18 (1982).

78. B. Schueler, P. K. D. Feigl, and F. R. Krueger, *Z. Naturforsch.*, **38A**, 1078 (1983).

79. B. Schueler, F. R. Krueger, and P. Feigl, *Int. J. Mass Spectrom. Ion Phys.*, **47**, 3 (1983).

80. F. R. Krueger, *Appl. Surf. Sci.*, **11/12**, 819 (1982).

81. F. R. Krueger, *Z. Naturforsch.*, **38A**, 385 (1983).

82. H. J. Heinen, *Int. J. Mass Spectrom. Ion Phys.*, **38**, 309 (1981).

83. L. Van Vaeck, J. Claereboudt, J. De Waele, E. Esmans, and R. Gijbels, *Anal. Chem.*, **57**, 2944 (1985).

84. D. M. Hercules, R. J. Day, K. Balasanmugam, T. A. Dang, and C. P. Li, *Anal. Chem.*, **54**, 280A (1982).

85. F. P. Novak, K. Balasanmugam, K. Viswanadham, C. D. Parker, Z. A. Wilk, D. Mattern, and D. M. Hercules, *Int. J. Mass Spectrom. Ion Phys.*, **53**, 135, (1983).

86. U. Seydel and B. Lindner, in A. Benninghoven, Ed., *Ion Formation from Organic Solids*, Springer-Verlag, Berlin, 1983, p. 240.

87. B. Lindner and U. Seydel, *Anal. Chem.*, **57**, 895 (1985).

88. F. Hillenkamp, *Int. J. Mass Spectrom. Ion Phys.*, **45**, 305 (1982).

89. F. Hillenkamp, in A. Benninghoven, Ed., *Ion Formation from Organic Solids*, Springer-Verlag, Berlin, 1983, p. 190.

90. P. Wieser and R. Wurster, in A. Benninghoven, Ed., *Ion Formation from Organic Solids*, Springer-Verlag, Berlin, 1983, p. 235.

91. P. Wieser, R. Wurster, and H. Seiler, *J. Phys.*, **45**, C2-261 (1984).

92. F. Bruynseels and R. Van Grieken, *Int. J. Mass Spectrom. Ion Proc.*, **74**, 161 (1986).

93. E. Michiels, A. Celis, and R. Gijbels, *Int. J. Mass Spectrom. Ion Phys.*, **47**, 23 (1983).

94. V. S. Antonov, V. S. Letokhov, and A. N. Shibanov, *Appl. Phys.*, **25**, 71 (1981).

95. G. S. Hurst, *Anal. Chem.*, **53**, 1488A (1981).

96. G. Krier, F. Verdun, and J. F. Muller, *Fresenius Z. Anal. Chem.*, **322**, 379 (1985).

97. R. Kaufmann, F. Hillenkamp, R. Wechsung, H. J. Heinen, and M. Schürmann, *Scanning Electron Microsc.*, **II**, 279 (1979).

98. Proceedings of the 1st LAMMA-Symposium, Düsseldorf, F.R.G., Oct. 8–10, 1980 (Organizers: F. Hillenkamp and R. Kaufmann), *Fresenius Z. Anal. Chem.*, **308**, 193 (1981).

99. Proceedings of the 2nd LAMMA-Symposium, Borstel, F.R.G., Sept. 1–2, 1983 (Organizers: U. Seydel and B. Lindner).

100. P. Moesta, U. Seydel, B. Lindner, and H. Grisebach, *Z. Naturforsch.*, **37C**, 748 (1982).

101. H. J. Heinen and R. Holm, *Scanning Electron Microsc.*, **III**, 1129 (1984).

102. R. Holm, G. Kämpf, D. Kirchner, H. J. Heinen, and S. Meier, *Anal. Chem.*, **56**, 690 (1984).

103. M. A. Hayat, *Principles and Techniques of Electron Microscopy. Biological Applications*, Vol. 1, Van Nostrand Reinhold, New York, 1970.

104. A. H. Verbueken, W. A. Jacob, P. M. Frederik, W. M. Busing, R. C. Hertsens, and R. E. Van Grieken, *J. Phys.*, **45**, C2-561 (1984).

105. S. Henstra, E. B. A. Bisdom, A. Jongerius, H. J. Heinen, and S. Meier, *Beitr. elektronenmikroskop. Direktabb. Oberfl.*, **13**, 63 (1980).

106. E. B. A. Bisdom, S. Henstra, A. Jongerius, H. J. Heinen, and S. Meier, *Neth. J. Agric. Sci.*, **29**, 23 (1981).

107. S. Henstra, E. B. A. Bisdom, A. Jongerius, H. J. Heinen, and S. Meier, *Fresenius Z. Anal. Chem.*, **308**, 280 (1981).

108. J. M. Beusen, L. Van't dack, and R. Gijbels, in *Extended Summaries (EUR 8853 EN) Third International Seminar "European Geothermal Update,"* Munich, F.R.G., Nov. 29–Dec. 1, 1983, p. 370.

109. P. Wieser, R. Wurster, and H. Seiler, *Atmos. Environ.*, **14**, 485 (1980).

110. R. Kaufmann, P. Wieser, and R. Wurster, *Scanning Electron Microsc.*, **II**, 607 (1980).

111. F. Bruynseels, H. Storms, and R. Van Grieken, *J. Phys.*, **45**, C2-785 (1984).

112. F. Bruynseels, H. Storms, T. Tavares, and R. Van Grieken, *Int. J. Environ. Anal. Chem.*, **23**, 1 (1985).

113. F. Freund, H. Kathrein, H. Wengeler, R. Knobel, and H. J. Heinen, *Geochim. Cosmochim. Acta*, **44**, 1319 (1980).

114. P. K. Dutta and Y. Talmi, *Fuel*, **61**, 1241 (1982).

115. N. E. Vanderborgh and C. E. R. Jones, *Anal. Chem.*, **55**, 527 (1983).

116. E. Michiels, A. Celis, and R. Gijbels, in K. F. J. Heinrich, Ed., *Microbeam Analysis—1982*, San Francisco Press, San Francisco, 1982, p. 383.

117. R. Kaufmann and P. Wieser, in K. F. J. Heinrich, Ed., *Characterization of Particles*, NBS Spec. Publ. 533, Washington, DC, 1980, p. 199.

118. R. Gijbels, P. Verlodt, and S. Tavernier, in K. F. J. Heinrich, Ed., *Microbeam Analysis—1982*, San Francisco Press, San Francisco, 1982, p. 378.

119. A. H. Verbueken, F. L. Van de Vyver, G. J. Paulus, W. J. Visser, G. A. Verpooten, M. E. De Broe, and R. E. Van Grieken, in P. Brätter and P. Schramel, Eds., *Trace Element Analytical Chemistry in Medicine and Biology*, Vol. 3, Walter de Gruyter, Berlin, 1984, p. 375.

120. W. Schröder, D. Frings, and H. Stieve, *Scanning Electron Microsc.*, **II**, 647 (1980).

121. A. H. Verbueken, F. L. Van de Vyver, R. E. Van Grieken, G. J. Paulus, W. J. Visser, P. D'Haese, and M. E. De Broe, *Clin. Chem.*, **30**, 763 (1984).

122. A. H. Verbueken, R. E. Van Grieken, G. J. Paulus, G. A. Verpooten, and M. E. De Broe, *Biomed. Mass Spectrom.*, **11**, 159 (1984).

123. R. Kaufmann, in K. F. J. Heinrich, Ed., *Microbeam Analysis—1982*, San Francisco Press, San Francisco, 1982, p. 341.

124. P. Surkyn and F. Adams, *J. Trace Microprobe Techniques*, **1**, 79 (1982).

125. P. Wieser, R. Wurster, and H. Seiler, *Scanning Electron Microsc.*, **III**, 1035 (1983).

126. P. Wieser, R. Wurster, and H. Seiler, *Scanning Electron Microsc.*, **IV**, 1435 (1982).

127. A. H. Verbueken, R. E. Van Grieken, G. J. Paulus, and W. C. de Bruijn, *Anal. Chem.*, **56**, 1362 (1984).

128. R. W. Linton, S. R. Bryan, P. F. Schmidt, and D. P. Griffis, *Anal. Chem.*, **57**, 440 (1985).

129. J. F. Muller, M. A. Gelot, P. Netter, and R. Kaufmann, *Biomedicine*, **36**, 380 (1982).

130. A. H. Verbueken, F. L. Van de Vyver, W. J. Visser, F. Roels, R. E. Van Grieken, and M. E. De Broe, *Biol. Trace Element Res.*, in press.

131. R. Kaufmann, F. Hillenkamp, and R. Wechsung, *Med. Prog. Technol.*, **6**, 109 (1979).

132. E. Michiels and R. Gijbels, *Spectrochim. Acta*, **38B**, 1347 (1983).

133. J. K. De Waele and F. C. Adams, *Scanning Electron Microsc.*, **III**, 935 (1985).

134. F. Bruynseels and R. Van Grieken, *Atmos. Environ.*, **19**, 1969 (1985).

135. H. J. Heinen, S. Meier, H. Vogt, and R. Wechsung, *Adv. Mass Spectrom.*, **8**, 942 (1980).

136. P. K. Dutta and Y. Talmi, *Anal. Chim. Acta*, **132**, 111 (1981).

137. L. Van Vaeck, J. De Waele, and R. Gijbels, *Mikrochim. Acta*, **III**, 237 (1984).

138. C. D. Parker and D. M. Hercules, *Anal. Chem.*, **57**, 698 (1985).

139. J. F. Muller, C. Berthé, and J. M. Magar, *Fresenius Z. Anal. Chem.*, **308**, 312 (1981).

140. J. A. Gardella, Jr., D. M. Hercules, and H. J. Heinen, *Spectrosc. Lett.*, **13**, 347 (1980).

141. J. A. Gardella, Jr., and D. M. Hercules, *Fresenius Z. Anal. Chem.*, **308**, 297 (1981).

142. F. J. Bruynseels and R. E. Van Grieken, *Spectrochim. Acta*, **38B**, 853 (1983).

143. F. J. Bruynseels and R. E. Van Grieken, *Anal. Chem.*, **56**, 871 (1984).

144. E. Michiels and R. Gijbels, *Anal. Chem.*, **56**, 1115 (1984).

145. E. Weber and F. Vögtle, *Chem. Ber.*, **109**, 1803 (1976).

146. W. H. Schröder, *Fresenius Z. Anal. Chem.*, **308**, 212 (1981).

147. L. Edelmann, *Fresenius Z. Anal. Chem.*, **308**, 218 (1981).

148. A. Verbueken, G. Paulus, F. Van de Vyver, G. Verpooten, M. De Broe, and R. Van Grieken, in *Proceedings of the 31st Annual Conference on Mass Spectrometry and Allied Topics*, Boston, MA, May 8–13, 1983, p. 67.

149. E. A. Jordan, R. D. Macfarlane, C. R. Martin, and C. J. McNeal, *Int. J. Mass Spectrom. Ion Phys.*, **53**, 345 (1983).

150. P. Wieser, R. Wurster, and U. Haas, *Fresenius Z. Anal. Chem.*, **308**, 260 (1981).

151. H. G. Fromme, M. Grote, J. Schaffstein, W. Holländer, and B. Bechtloff, *J. Phys.*, **45**, C2-389 (1984).

152. R. Niessner, D. Klockow, F. Bruynseels, and R. Van Grieken, *Int. J. Environ. Anal. Chem.*, **22**, 281 (1985).
153. E.-R. Krefting, G. Lissner, and H. J. Höhling, *Scanning Electron Microsc.*, **II**, 369 (1981).
154. J. M. Beusen, P. Surkyn, R. Gijbels, and F. Adams, *Spectrochim. Acta*, **38B**, 843 (1983).
155. U. Haas, P. Wieser, and R. Wurster, *Fresenius Z. Anal. Chem.*, **308**, 270 (1981).
156. D. S. Simons, in D. B. Wittry, Ed., *Microbeam Analysis—1980*, San Francisco Press, San Francisco, 1980, p. 178.
157. K. R. Spurny, J. Schörmann, and R. Kaufmann, *Fresenius Z. Anal. Chem.*, **308**, 274 (1981).
158. D. S. Simons, in K. F. J. Heinrich, Ed., *Microbeam Analysis—1982*, San Francisco Press, San Francisco, 1982, p. 390.
159. R. A. Fletcher and A. J. Fatiadi, in A. D. Romig, Jr., and J. I. Goldstein, Eds., *Microbeam Analysis—1984*, San Francisco Press, San Francisco, 1984, p. 14.
160. P. Wieser, R. Wurster, and L. Phillips, in V. A. Marple and B. Y. H. Liu, *Aerosols in the Mining and Industrial Work Environment*, Vol. 3, Ann Arbor Science, Michigan, 1983, p. 1169.
161. J. A. Small, J. A. Norris, and R. L. McKenzie, in R. Gooley, Ed., *Microbeam Analysis—1983*, San Francisco Press, San Francisco, 1983, p. 209.
162. R. Brace and E. Matijević, *J. Inorg. Nucl. Chem.*, **35**, 3691 (1973).
163. P. F. Schmidt, H. G. Fromme, and G. Pfefferkorn, *Scanning Electron Microsc.*, **II**, 623 (1980).
164. P. F. Schmidt and K. Ilsemann, *Scanning Electron Microsc.*, **I**, 77 (1984).
165. C. A. Andersen and J. R. Hinthorne, *Anal. Chem.*, **45**, 1421 (1973).
166. A. E. Morgan and H. W. Werner, *Mikrochim. Acta*, **II**, 31 (1978).
167. P. Englert and U. Herpers, *Inorg. Nucl. Chem. Lett.*, **16**, 37 (1980).
168. G. H. Morrison, *CRC Crit. Rev. Anal. Chem.*, **8**, 287 (1979).
169. R. L. Kaufmann, F. Hillenkamp, and E. Remy, *Microsc. Acta*, **73**, 1 (1972).
170. F. Hillenkamp, R. Kaufmann, R. Nitsche, E. Remy, and E. Unsöld, in T. Hall, P. Echlin, and R. Kaufmann, Eds., *Microprobe Analysis as Applied to Cells and Tissues*, Academic Press, London, 1974, p. 1.
171. R. Kaufmann, F. Hillenkamp, R. Nitsche, M. Schürmann, and E. Unsöld, *J. Microsc. Biol. Cell.*, **22**, 389 (1975).
172. P. Klein and J. Bauch, *Holzforschung*, **33**, 1 (1979).
173. P. Klein and J. Bauch, *Wood Fiber*, **13**, 226 (1981).
174. P. Klein and J. Bauch, *Fresenius Z. Anal. Chem.*, **308**, 283 (1981).
175. J. Bauch and W. Schröder, *Forstwissenschaft. Centralbl.*, **101**, 285 (1982).
176. J. Bauch, in B. Ulrich and J. Pankrath, Eds., *Effects of Accumulation of Air Pollutants in Forest Ecosystems*, D. Reidel, Dordrecht 1983, p. 377.

177. D. W. Lorch and H. Schäfer, *Z. Pflanzenphysiol.*, **101**, 183 (1981).

178. D. W. Lorch and H. Schäfer, *Fresenius Z. Anal. Chem.*, **308**, 246 (1981).

179. H.-P. Bochem and B. Sprey, *Z. Pflanzenphysiol.*, **95**, 179 (1979).

180. B. Sprey and H.-P. Bochem. *Fresenius Z. Anal. Chem.*, **308**, 239 (1981).

181. A. Mathey, *Nova Hedwigia*, **31**, 917 (1979).

182. A. Mathey, *Fresenius Z. Anal. Chem.*, **308**, 249 (1981).

183. A. Mathey, J. Gatzmann, and J. F. Muller, *Beitr. elektronenmikroskop. Direktabb. Oberfl.*, **14**, 631 (1981).

184. G. Heinrich, *Biochem. Physiol. Pflanzen*, **179**, 129 (1984).

185. N. Paris, A. Chamel, S. Chevalier, A. Fourcy, J.-P. Garrec, and A.-M. Lhoste, *Physiol. Veg.*, **16**, 17 (1978).

186. A. R. Chamel, A.-M. Andréani, and J.-F. Eloy, *Plant Physiol.*, **67**, 457 (1981).

187. A. R. Chamel, R. D. Marcelle, and J.-F. Eloy, *J. Amer. Soc. Hort. Sci.*, **107**, 804 (1982).

188. D. G. Strullu, A. Chamel, J.-F. Eloy, and J. P. Gourret, *New Phytol.*, **94**, 81 (1983).

189. A. Chamel and J.-F. Eloy, *J. Plant Nutr.*, **2**, 445 (1980).

190. S. De Nollin, W. Jacob, T. Garrevoet, A. Van Daele, and P. Dockx, *Sabouraudia*, **21**, 287 (1983).

191. M. Thibaut, M. Ansel, and H. Saëz, *Eur. J. Appl. Microbiol. Biotechnol.*, **11**, 55 (1980).

192. M. Thibaut and M. Ansel, in A. D. Romig, Jr., and J. I. Goldstein, Eds., *Microbeam Analysis—1984*, San Francisco Press, San Francisco, 1984, p. 264.

193. G. Siebert, E. Gabriel, R. Hannover, D. Henschler, E. J. Karle, H. Kasper, M. Mack, W. Romen, B. Schmauck, and K. Trautner, *Arch. Fisch Wiss.*, **32**, 43 (1982).

194. U.-R. Heinrich and H. O. Gutzeit, *Verh. Deut. Zool. Ges.*, 241 (1982).

195. U.-R. Heinrich, R. Kaufmann, and H. O. Gutzeit, *Differentiation*, **25**, 10 (1983).

196. U. Seydel and B. Lindner, *Fresenius Z. Anal. Chem.*, **308**, 253 (1981).

197. U. Seydel and B. Lindner, *Int. J. Quant. Chem.*, **20**, 505 (1981).

198. U. Seydel and H. J. Heinen, in A. Frigerio and M. McCamish, Eds., *Recent Developments in Mass Spectrometry in Biochemistry and Medicine*, Vol. 6, Elsevier, New York, 1980, p. 489.

199. R. Böhm, *Fresenius Z. Anal. Chem.*, **308**, 258 (1981).

200. B. Lindner and U. Seydel, *J. Gen. Microbiol.*, **129**, 51 (1983).

201. B. Lindner and U. Seydel, *Int. J. Mass Spectrom. Ion Phys.*, **48**, 265 (1983).

202. B. Lindner and U. Seydel, *J. Phys.*, **45**, C2-565 (1984).

203. U. Seydel, B. Lindner, and K. Brandenburg, in A. Frigerio, Ed., *Recent*

Developments in Mass Spectrometry in Biochemistry, Medicine and Environmental Research, Vol. 8, Elsevier, New York, 1983, p. 319.

204. U. Seydel, B. Lindner, J. K. Seydel, and K. Brandenbrug, *Int. J. Leprosy*, **50**, 90 (1982).

205. B. Lindner and U. Seydel, in R. Gooley, Ed., *Microbeam Analysis—1983*, San Francisco Press, San Francisco, 1983, p. 106.

206. U. Seydel, B. Lindner, U. Zähringer, E. Th. Rietschel, S. Kusomoto, and T. Shiba, *Biomed. Mass Spectrom.*, **11**, 132 (1984).

207. A. Meyer zum Gottesberge-Orsulakova and R. Kaufmann, *Scanning Electron Microsc.*, **I**, 393 (1985).

208. A. Orsulakova, R. Kaufmann, C. Morgenstern, and M. D'Haese, *Fresenius Z. Anal. Chem.*, **308**, 221 (1981).

209. A. Orsulakova and C. Morgenstern, *Arch. Otorhinolaryngol.*, **231**, 817 (1981).

210. A. Orsulakova, C. Morgenstern, R. Kaufmann, and M. D'Haese, *Scanning Electron Microsc.*, **IV**, 1763 (1982).

211. C. Morgenstern, H. Amano, and A. Orsulakova, *Am. J. Otolaryngol.*, **3**, 323 (1982).

212. H. Amano, A. Orsulakova, and C. Morgenstern, *Arch. Otorhinolaryngol.*, **237**, 273 (1983).

213. H. J. Heinen, F. Hillenkamp, R. Kaufmann, W. Schröder, and R. Wechsung, in A. Frigerio and M. McCamish, Eds., *Recent Developments in Mass Spectrometry in Biochemistry and Medicine*, Vol. 6, Elsevier, New York, 1980, p. 435.

214. H. J. Heinen and W. Schröder, *Biochem. Soc. Trans.*, **9**, 591 (1981).

215. W. H. Schröder, *Invest. Ophthalmol. Visual Sci.*, **22**, 276 (1982).

216. W. H. Schröder, *Fresenius Z. Anal. Chem.*, **324**, 244 (1983).

217. R. Kaufmann, *Scanning Electron Microsc.*, **II**, 641 (1980).

218. R. Hennekes, *Ber. Deut. Ophthalmol. Ges.*, **78**, 529 (1981).

219. J. L. Van Reempts, M. Borgers, S. R. De Nollin, T. C. Garrevoet, and W. A. Jacob, *J. Histochem. Cytochem.*, **32**, 788 (1984).

220. W. H. Schröder and G. L. Fain, *Biophys. J.*, **45**, 341a (1984).

221. W. H. Schröder and G. L. Fain, *Nature*, **309**, 268 (1984).

222. W. H. Schröder, in A. D. Romig, Jr., and J. I. Goldstein, Eds., *Microbeam Analysis—1984*, San Francisco Press, San Francisco, 1984, p. 30.

223. E. Gabriel, R. Wechsung, H. G. Klinger, and H. Gerhard, *Caries Res.*, **13**, 118 (1979).

224. E. Gabriel, S. Neumeyer, and K. J. Klinke, *Deut. Zahnärztl. Z.*, **34**, 716 (1979).

225. E. Gabriel, Y. Kato, and H. J. Heinen, *Caries Res.*, **14**, 175 (1980).

226. E. Gabriel, Y. Kato, and P. J. Rech, *Fresenius Z. Anal. Chem.*, **308**, 234 (1981).

227. R. Lehmann, H. Sluka, and P. F. Schmidt, *Die Quintessenz,* **3,** 613 (1982).

228. P. F. Schmidt, *Trace Elements Med.,* **1,** 13 (1984).

229. P. Chibon and J. F. Eloy, *J. Biol. Buccale,* **7,** 263 (1979).

230. N. H. Lithwick, M. K. Healy, and J. Cohen, *Surg. Forum,* **15,** 439 (1964).

231. H. M. Goldman, M. P. Ruben, and D. Sherman, *Oral Surg., Oral Med., Oral Pathol.,* **17,** 102 (1964).

232. D. B. Sherman, M. P. Ruben, H. M. Goldman, and F. Breck, *Ann. N.Y. Acad. Sci.,* **122,** 767 (1965).

233. N. Avram, Gh. Draganescu, and I. Marki, *Studia Biophys. (Berlin),* **51,** 69 (1975).

234. A. H. Verbueken, F. L. Van de Vyver, R. E. Van Grieken, and M. E. De Broe, *Clin. Nephrol.,* **24,** S58 (1985).

235. S. W. King, J. Savory, and M. R. Wills, *CRC Crit. Rev. Clin. Lab. Sci.,* **14,** 1 (1981).

236. A. Verbueken, G. Paulus, F. Van de Vyver, G. Verpooten, R. Van Grieken, and M. De Broe, *Anal. Proc.,* **20,** 287 (1983).

237. F. L. Van de Vyver, A. O. Vanheule, A. H. Verbueken, P. D'Haese, W. J. Visser, A. B. Bekaert, R. E. Van Grieken, N. Buyssens, W. De Keersmaecker, W. Van den Bogaert, and M. E. De Broe, *Contrib. Nephrol.,* **38,** 153 (1984).

238. M. E. De Broe, F. L. Van de Vyver, A. B. Bekaert, P. D'Haese, G. J. Paulus, W. J. Visser, R. Van Grieken, F. A. de Wolff, and A. H. Verbueken, *Contrib. Nephrol.,* **38,** 37 (1984).

239. W. J. Visser, F. L. Van de Vyver, A. H. Verbueken, M. H. F. Lentferink, R. E. Van Grieken, and M. E. De Broe, *Calcif. Tissue Int.,* **36,** S22 (1984).

240. F. L. Van de Vyver, A. H. Verbueken, W. J. Visser, R. E. Van Grieken, and M. E. De Broe, *J. Clin. Pathol.,* **37,** 837 (1984).

241. W. J. Visser, F. L. Van de Vyver, A. H. Verbueken, P. D'Haese, A. B. Bekaert, R. E. Van Grieken, S. A. Duursma, and M. E. De Broe, in N. D. Priest, Ed., *Metals in Bone* (Proc. EULEP Symp., Angers, France, Oct. 11–13, 1984), MTP Press, Lancaster, UK, 1985, p. 433.

242. H. Zumkley, P. F. Schmidt, H. P. Bertram, A. E. Lison, B. Winterberg, K. Spieker, and H. Losse, *Verhandl. Deut. Ges. Innere Med.,* **89,** 1227 (1983).

243. P. F. Schmidt, in H. Zumkley, Ed., *Spurenelemente,* Georg Thieme, New York, 1983, p. 12.

244. P. F. Schmidt and H. Zumkley, *J. Phys.,* **45,** C2-569 (1984).

245. H. Zumkley, P. F. Schmidt, H. P. Bertram, A. E. Lison, B. Winterberg, K. Spieker, H. Losse, and R. Barckhaus, *Trace Elements Med.,* **1,** 103 (1984).

246. J. F. Osborn, H. Newesely, C. Werhahn, H. Brückmann, and E. Gabriel,

in A. J. C. Lee, T. Albrektsson, and P.-I. Brånemark, Eds., *Clinical Applications of Biomaterials*, Wiley, New York, 1982, p. 219.

247. J. F. Osborn and E. Gabriel, *Deut. Zahnärztl. Z.*, **37**, 769 (1982).

248. P. F. Schmidt, *Beitr. Elektronenmikroskop. Direktabb. Oberfl.*, **13**, 51 (1980).

249. E. Unsöld, F. Hillenkamp, and R. Nitsche, *Analusis*, **4**, 115 (1976).

250. R. Kaufmann, F. Hillenkamp, R. Nitsche, M. Schürmann, and R. Wechsung, *Microsc. Acta,* **Suppl. 2**, 297 (1978).

251. B. Kaduk, K. Metze, P. F. Schmidt, and G. Brandt, *Virchows Arch. A. Path. Anat. Histol.*, **387**, 67 (1980).

252. Hj. Hirche, M. Schramm, and J. Heinrichs, *Pfluegers Arch. Eur. J. Physiol.*, **379**, R58 (1979).

253. Hj. Hirche, J. Heinrichs, H. E. Schaefer, and M. Schramm, *Fresenius Z. Anal. Chem.*, **308**, 224 (1981).

254. L. Edelmann, *Physiol. Chem. Phys.*, **12**, 509 (1980).

255. S. De Nollin, W. Jacob, and R. Hertsens, in T. Godfraind, A. G. Herman, and D. Wellens, Eds., *Calcium Entry Blockers in Cardiovascular and Cerebral Dysfunctions*, Martinus Nijhoff, Boston, 1984, p. 53.

256. S. De Nollin, M. Borgers, W. Jacob, and D. Wellens, *Angiology*, **36**, 297 (1985).

257. H. Wakasugi, T. Kimura, W. Haase, A. Kribben, R. Kaufmann, and I. Schulz, *J. Membrane Biol.*, **65**, 205 (1982).

258. H. J. Heinen, R. Wechsung, H. Vogt, F. Hillenkamp, and R. Kaufmann, *Biotech. Umschau*, **2**, 346 (1978).

259. R. Kaufmann, F. Hillenkamp, and R. Wechsung, *Eur. Spectrosc. News*, **20**, 41 (1978).

260. R. Kaufmann and F. Hillenkamp, *Ind. Res. Develop.*, 145 (April 1979).

261. R. W. Linton, J. D. Shelburne, D. S. Simons, and P. Ingram, in G. W. Bailey, Ed., *40th Annual Proceedings of the Electron Microscopy Society of America* (Washington, DC, 1982), p. 370.

262. A. H. Verbueken, J. Boelaert, G. J. Paulus, F. L. J. Van de Vyver, F. Roels, R. Van Grieken, and M. E. De Broe, *Néphrologie*, **4**, 95 (1983).

263. J. Bommer, R. Waldherr, P. H. Wieser, and E. Ritz, *Lancet*, **i**, 1390 (1983).

264. H. Zumkley, P. F. Schmidt, M. Elies, H. Vetter, W. Zidek, H. Losse, and H. G. Fromme, *J. Amer. Coll. Nutr.*, **3**, 303 (1984).

265. I. M. Arefyev, A. I. Boriskin, A. S. Bryukhanov, Yu. A. Bykovskiy, A. A. Komleva, I. D. Laptev, and R. I. Utyamyshev, *Sud. Med. Ekspert.* (*Moscow*), **25**, 35 (1982).

266. K. D. Kupka, W. W. Schropp, C. Schiller, and F. Hillenkamp, *Scanning Electron Microsc.*, **II**, 635 (1980).

267. K. D. Kupka, W. W. Schropp, C. Schiller, and F. Hillenkamp, *Fresenius Z. Anal. Chem.*, **308**, 229 (1981).

268. W. Jacob and P. Dockx, *J. Cutaneous Pathol.*, **10**, 418 (1983).

269. W. A. Jacob, S. De Nollin, R. C. Hertsens, and M. De Smet, *J. Trace Microprobe Techniques*, **2**, 161 (1984).

270. R. G. Wilson, L. Goldman, and F. Brech, *Arch. Derm.*, **95**, 490 (1967).

271. A. J. Chaplin, P. R. Millard, and P. F. Schmidt, *Histochemistry*, **75**, 259 (1982).

272. J. Peiffer, E. Danner, and P. F. Schmidt, *Clin. Neuropathol.*, **3**, 76 (1984).

273. H. Pau and R. Kaufmann, *Arch. Ophthalmol.*, **101**, 1935 (1983).

274. G. Wilhelm and K.-D. Kupka, *FEBS Lett.*, **123**, 141 (1981).

275. R. Kaufmann and P. Wieser, in I. Beddow and A. Vetter, Eds., *Modern Methods of Fine Particle Characterization*, Vol. II, CRC Press, Boca Raton, FL, 1982, p. 199.

276. M. J. Van Craen, E. A. Denoyer, D. F. S. Natusch, and F. Adams, *Environ. Sci. Technol.*, **17**, 435 (1983).

277. T. Mauney, F. Adams, and M. R. Sine, *Sci. Total Environ.*, **36**, 215 (1984).

278. E. Denoyer, T. Mauney, D. F. S. Natusch, and F. Adams, in K. F. J. Heinrich, Ed., *Microbeam Analysis—1982*, San Francisco Press, San Francisco, 1982, p. 191.

279. E. Denoyer, D. F. S. Natusch, P. Surkyn, and F. C. Adams, *Environ. Sci. Technol.*, **17**, 457 (1983).

280. A. J. Gay, A. P. von Rosenstiel, and P. J. van Duin, in R. H. Geiss, Ed., *Microbeam Analysis—1981*, San Francisco Press, San Francisco, 1981, p. 229.

281. P. K. Dutta, D. C. Rigano, R. A. Hofstader, E. Denoyer, D. F. S. Natusch, and F. Adams, *Anal. Chem.*, **56**, 302 (1984).

282. F. Adams, P. Bloch, D. F. S. Natusch, and P. Surkyn, in *Proceedings of the International Conference on Environmental Pollution*, Thessaloniki, Greece, Sept. 21–25, 1981, p. 122.

283. P. Wieser, R. Wurster, and H. Seiler, *Scanning Electron Microsc.*, **III**, 56 (1980).

284. H. Seiler, U. Haas, I. Rentschler, H. Schreiber, P. Wieser, and R. Wurster, *Optik*, **58**, 145 (1981).

285. P. Surkyn, J. De Waele, and F. Adams, *Int. J. Environ. Anal. Chem.*, **13**, 257 (1983).

286. F. Bruynseels, H. Storms, and R. Van Grieken, in T. D. Lekkas, Ed., *Proceedings of the International Conference on Heavy Metals in the Environment*, Vol. 1, Athens, Greece, Sept. 10–13, 1985, p. 189.

287. H. Kühme, *Meteorol. Rundsch.*, **36**, 119 (1983).

288. F. Verdun and J. F. Muller, *J. Phys.*, **45**, C2-819 (1984).

289. J. De Waele, P. Van Espen, E. Vansant, and F. Adams, in K. F. J. Heinrich, Ed., *Microbeam Analysis—1982*, San Francisco Press, San Francisco, 1982, p. 371.

290. J. K. De Waele, M. J. Luys, E. F. Vansant, and F. C. Adams, *J. Trace Microprobe Techniques*, **2**, 87 (1984).

291. J. De Waele, E. F. Vansant, P. Van Espen, and F. C. Adams, *Anal. Chem.*, **55**, 671 (1983).

292. J. K. De Waele, I. Verhaert, E. F. Vansant, and F. C. Adams, *Surface Interface Anal.*, **5**, 186 (1983).

293. J. K. De Waele, E. F. Vansant, and F. C. Adams, *Mikrochim. Acta*, **III**, 367 (1983).

294. J. K. De Waele, E. F. Vansant, and F. C. Adams, *Anal. Chim. Acta*, **161**, 37 (1984).

295. J. K. De Waele, J. J. Gijbels, E. F. Vansant, and F. C. Adams, *Anal. Chem.*, **55**, 2255 (1983).

296. J. K. De Waele, I. M. Swenters, and F. C. Adams, *Spectrochim. Acta*, **40B**, 795 (1985).

297. K. Spurny, *Sci. Total Environ.*, **23**, 239 (1982).

298. K. Spurny, *Environ. Health Perspectives*, **51**, 343 (1983).

299. K. Spurny, F. Pott, W. Stöber, H. Opiela, J. Schörmann, and G. Weiss, *Am. Ind. Hyg. Assoc. J.*, **44**, 833 (1983).

300. K. Spurny, *Sci. Total Environ.*, **30**, 147 (1983).

301. H. H. Weinke, E. Michiels, and R. Gijbels, *Int. J. Mass Spectrom. Ion Phys.*, **47**, 43 (1983).

302. F. Freund, H. Wengeler, H. Kathrein, and H. J. Heinen, *Mat. Res. Bull.*, **15**, 1019 (1980).

303. E. B. Steel, D. S. Simons, J. A. Small, and D. E. Newbury, in A. D. Romig, Jr., and J. I. Goldstein, Eds., *Microbeam Analysis—1984*, San Francisco Press, San Francisco, 1984, p. 27.

CHAPTER

6

INDUCTIVELY COUPLED PLASMA SOURCE MASS SPECTROMETRY

A. L. GRAY

University of Surrey, Guilford, Surrey, United Kingdom

1. INTRODUCTION

In recent years techniques of multielement trace analysis have developed very rapidly following the introduction of plasma emission sources for atomic emission spectrometry. Quite apart from their considerable powers of detection, which provide solution detection limits at levels of between 1 ng/mL and 1 μg/mL for most elements (1), they also offer the analyst very simple sample preparation and introduction methods and high sample throughput rates, typically of 1 sample per minute.

The most common plasma used for emission spectrometry has been the inductively coupled plasma (ICP), which offers a relatively long sample dwell time at high temperature and thus gives very efficient sample volatilization, dissociation, and excitation. This will be discussed in more

257

detail later, but a major advantage arising from these characteristics is a high degree of immunity to interelement and matrix interferences, which, coupled with the high sensitivity referred to, has enabled major advances to be made in multielement trace analysis of a wide range of samples (2).

In the face of this rapid growth of plasma emission spectrometry since the introduction of the first commercial instruments in 1974, it is not surprising that multielement inorganic mass spectrometry has, during that time, suffered a relative eclipse. Although offering higher ultimate sensitivities, it has usually proved to be slower and more costly and to require more skillful operation and interpretation. Its use has continued therefore only in applications where its performance has been unattainable by other means, principally for multielement concentration measurements in solids at very low levels of 1 μg/g and below and where the isotope ratio determination inherent in mass spectrometry is of importance.

The high temperatures available at atmospheric pressure in electric plasmas make them very efficient excitation sources. The analyte species may spend as long as several milliseconds in a region at a temperature between 5000 and 10,000 K, and in these conditions most elements will also be ionized to some extent, many of them approaching 100% ionization to the singly charged state and some ionized to higher states. Such plasmas, therefore, offer some very attractive characteristics as ion sources. All the advantages of rapid and easy sample introduction are available-when they are used for emission sources, and their high freedom from interferences should also be available if ions could be easily extracted from them into a mass spectrometer. A number of other advantages also arise from the process of ionization at atmospheric pressure in a region of high gas density. The principal advantages of ICP mass spectrometry (ICP-MS) may be summarized thus:

1. Sample introduction at atmospheric pressure.
2. High gas temperature—complete sample vaporization and dissociation.
3. High proportion of sample atoms ionized.
4. Mainly singly charged ions.
5. Small ion energy spread.
6. External ion source—no "vacuum" ions.
7. Source at low potential, compatible with simple mass analyzers.

To accompany these advantages, two drawbacks must be considered:

1. High gas temperature—5000 K.
2. High gas pressure—1000 mbar.

Both these disadvantages are inherent in the good atomization, dissociation, and ionization characteristics that are desired. A reduction of either temperature or pressure in the excitation region would remove most of the advantages.

During the process of surveying the currently available techniques used for inorganic ion sources in 1970 it was realized that many of the problems with the available sources originated in the high-vacuum environment in which the processes of sample volatilization, dissociation, and ionization occurred. At around that time the process of flame reaction diagnostics by mass spectrometry had reached a considerable degree of maturity (3, 4), ions being readily extracted from atmospheric pressure flames at 3000 K into a quadrupole mass analyzer. A brief review of this technique had been given in 1964 (5), the same year as the first publications on the ICP as an emission source (6, 7), which described the whole process of flame mass spectrometry from the introduction of nebulized solutions to ion analysis and detection, although it was proposed only as a flame diagnostic technique and not for solution analysis. It was therefore a natural next step to consider the application of this technique to the higher temperature, cleaner atmosphere of an electric plasma (8). The feasibility of so doing was demonstrated first using a small dc plasma (9), and this showed that very low detection limits could be obtained from solution samples. The plasma used, however, did not provide a sufficiently high temperature environment to avoid volatilization interferences and to fully ionize elements of ionization energy above about 8 eV. This work suggested that if the plasma could be replaced by an ICP most of these difficulties could be overcome. There remained two major problems associated with the ICP to solve: the higher gas temperature and the presence of the intense rf fields associated with the plasma excitation that might be expected to couple with the ion-extraction system and to inject rf currents into the ion detection circuits. These problems were successfully overcome (10) to yield an ion source–mass analyzer combination that offers many of the advantages initially envisaged, and this development in several laboratories has now produced two commercial instruments of high performance.

2. ELECTRICAL PLASMAS AND THE ICP

Although the plasma state was originally defined as being matter in completely ionized form, the term has become more generally used to denote any highly ionized gas. Such a state of ionization can be produced by a variety of means, of which the simplest is the application of heat, and even a simple chemical flame is a plasma in this sense.

For the analysis of atomic species, a chemical flame is unattractive as a source of ions because of the high population of molecular and chemically reactive species present in it and its temperature, which is limited by the available reaction processes to a maximum of about 3000 K. This is too low to produce a high level of ionization in all but the most easily ionized atoms such as the alkaline elements, for which substantial ionization in flames has been reported (11).

The use of an electric discharge, such as a spark, arc, or glow discharge as an ion source is as old as mass spectroscopy itself. However, under the conditions of low pressure that are compatible with the mass analyzer, these operate at temperatures too low to thermally dissociate many inorganic molecules unless high energies are applied as in arc or spark sources. Major problems arise in producing steady-state controlled environments in vacuum capable of dissociating many molecules thermally, since the temperatures required are higher than the melting points of the materials of construction. Such temperatures are readily achieved in an electric discharge in a gas at atmospheric pressure, and such discharges have been explored for use as excitation sources for emission spectrometry for many years (12). These discharges, usually maintained in an inert gas such as argon, produce a central filament of heated gas by passing a high current thorugh it between two electrodes, often of tungsten or carbon. Although very high temperatures of 10,000 K or more may be achieved, it is extremely difficult to introduce an externally supplied sample, usually supported in an inert gas flow, into the hottest region of the arc. The strong thermal and density gradients and the force of radiation pressure itself ensure that the major part of the sample passes through the annular region surrounding the hot filament where it is only partially dissociated, excited, and ionized. This problem was experienced with the small dc plasma used to demonstrate the feasibility of using an atmospheric pressure plasma as an ion source (9). These plasmas tend to show strong interference effects and have therefore found only limited application in emission spectrometry.

A further drawback of dc plasmas lies in the presence of electrodes in contact with the plasma itself that erode and contribute contamination to the plasma. This may be avoided by the use of an electrodeless plasma, of which the simplest is the microwave-induced plasma (MIP). In this system microwave energy is supplied to the plasma gas from an excitation cavity around the discharge tube; this permits the gas stream to remain uncontaminated and enables very high power densities to be achieved with power levels of only 100–200 W. However, although very high excitation temperatures can be obtained, these exist only along a central filament as with the dc plasma, and the same problems of sample introduction exist. In addition, the mean gas temperature is low, rarely exceeding 2500 K, and thus poor volatilization and dissociation of many species result in strong matrix effects. Nevertheless, such a plasma was used as an ion source for mass spectroscopy by Douglas and French (13), and high sensitivity was obtained for low ionization energy elements with detection limits in the 0.1–1 μg/L range. Matrix effects were severe, and the gas temperature was too low to dissociate NO, which formed from oxygen and nitrogen present in the sample solution. NO has an ionization energy of 9.4 eV and was present in high enough concentration to prevent effective ionization of elements with ionization energies above this. These plasmas have found a place as emission sources for the analysis of gas streams such as chromatograph eluents, but they have not been used extensively for the analysis of solutions.

Since it was first described in 1964, the ICP has gradually come to dominate the field as a multielement emission source for solution analysis because of its unique combination of properties, which are produced by the structure of the plasma (14). The ICP typically used as an ion source in the research programs and the commercial instruments that have resulted from them is similar to that used for emission analysis, the most obvious change being horizontal mounting of the torch to provide direct access to the tail flame for the ion extraction system. The use of the standard system has obvious advantages in making the step from emission to ion source, but as better understanding of the ion formation and extraction process develops it is probable that ICP designs will evolve specifically for this application. The form of the plasma is shown in Fig. 6.1*a* and *b*. The plasma torch in Fig. 6.1*a* is a quartz tube 18 mm o.d. and

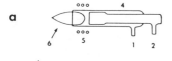

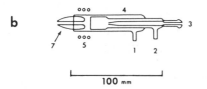

Figure 6.1. Inductively coupled plasma torch and flame. (*a*) Unpunched plasma. (*b*) Plasma punched by injector gas flow. 1, Coolant flow inlet; 2, auxiliary gas flow; 3, injector gas flow inlet; 4, torch body; 5, load coil; 6, unpunched fireball or flame; 7, annular punched flame.

about 100 mm long. The tip of the torch is surrounded by a coaxial water-cooled copper excitation (load) coil of two to four turns, which is coupled to the rf power supply. A coaxial inner tube, 14-mm o.d., is mounted inside the outer tube, leaving an annular gap of about 1 mm, and terminates just before the load coil. Argon is introduced tangentially to both inner tube and annular region so that a spiral flow is produced. The highest flow, 10–15 L/min, is fed to the annulus, producing a high-velocity flow to cool the inner surface of the outer tube at the position of the load coil. A small flow, called the plasma or auxiliary flow, of up to 1 L/min is used in the inner tube. The load coil is supplied with rf power, usually at 27.12 MHz, through an automatic matching network at power levels between 1 and 2.5 kW. When power is applied to the coil, the argon is initially cold and nonconducting. A spark from a Tesla coil is used to provide an initial stream of electrons that rapidly pick up energy from the load coil and form an intense plasma fireball in the mouth of the torch, kept off the walls by the high-velocity coolant flow. In this form, although intensely hot at the center and containing many ions of the plasma gas, the fireball is of little practical value, because any sample atoms introduced to the gas flow would be forced to skirt the plasma by the strong thermal gradient. In the practical form shown in Fig. 6.1*b*, the torch contains a further central injector tube that terminates in a capillary, usually of 1.5 mm i.d., through which the sample carrier gas flow of 0.5–1 L/min is supplied. This produces a jet of high velocity that punctures the base of the fireball and forms a cooler central channel through it, surrounded by the remains of the original fireball in the form of an annulus of hot gas at about 10,000 K. This annulus forms a single turn, which couples to the load coil and so maintains the plasma. The central channel is heated by conduction and radiation from the annulus to a temperature approaching 8000 K in the center, which reduces as the gas leaves the torch mouth and interdiffuses with the gas from the annulus. A typical temperature profile along the torch axis is shown in Fig. 6.2.

The hot annulus emits intense continuum radiation in the visible and ultraviolet from argon recombination processes, but with dry argon in the injector flow the tail flame beyond the torch mouth is almost invisible. The addition of water to the argon flow produces oxygen and hydrogen band emission in the visible region, and the flame may then be seen as a white cone extending some 25 mm from the end of the torch. Beyond this region the flame is normally colorless unless an analyte with a strong visible emission is introduced. The region from about 10 to 30 mm from the torch mouth is normally used for emission spectrometry as the signal-to-background ratio is highest here.

This method of introducing the sample gas flow is unique to the ICP

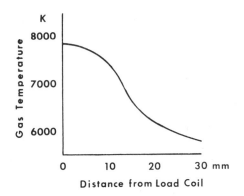

Figure 6.2. Axial profile of gas temperature of central chanel in annular plasma flame shown in Fig. 6.1.

and is possible only because it complements the method of energy addition. It produces an ideal environment for successive volatilization, dissociation, excitation, and ionization as the sample travels up the temperature gradient approaching the load coil and passes along the axis through the high-temperature region and into the tail flame. During this passage the sample is confined by the higher-temperature annular region, and, apart from a small part that fails to penetrate the fireball, the majority of the sample has to follow this route.

The most common method of introducing sample is as a gas-supported aerosol from a pneumatic nebulizer. In these devices the sample carrier gas flow is blown across the mouth of a capillary tube, along which the sample solution is either drawn by the gas flow or pumped by a small peristaltic pump, typically at a rate of about 1 mL/min. The form of the nebulizer may be crossflow, in which the gas tube blows a jet across the mouth of the sample tube, or concentric, in which the gas flow is blown across the sample tube from a concentric outer tube. Many forms of pneumatic nebulizers are available, but all have a common characteristic—they produce a wide range of droplet sizes in which the major fraction of the solution is carried in droplets of about 20 μm diameter. These are too large to be carried more than a few centimeters in the gas stream before they are lost on the walls of the cloud chamber into which the nebulizer discharges. Droplets below about 4 μm diameter, which constitute about 5% of the population, are small enough to remain supported in the gas and are therefore carried into the plasma. The usual efficiency of sample delivery to the plasma by these nebulizers is only 1–2%. This is of little importance in many situations, since the detection power of the technique is determined by the signal-to-noise ratio but requires considerable volumes of solution, most of which runs to the drain. Pneumatic nebulizers that recirculate the waste sample have been developed to deal with this problem (15). Other methods of sample introduction may also be used; they will be discussed in Section 8.

As the gasborne sample aerosol droplets approach the plasma, the water evaporates, leaving the sample as small solid particles of 0.1 μm diameter or less, which are rapidly vaporized as they enter the plasma. However, in the presence of a high matrix concentration where the dissolved solids content of the solution may be 1% or more, the particles may be much larger than this. It is important that these larger particles be completely volatilized or some of the analyte will be carried right through the plasma and be lost. Plasmas of the size and power normally used are able to vaporize solid particles of refractory materials of up to 1 μm diameter.

The operation of the plasma as an ion source may be described in terms of the equilibria established along the central channel, which may in turn be defined by the temperature. However, the temperature itself may differ according to the parameter used to define it, and in an argon plasma four temperatures are usually defined:

T_g	Gas temperature defined by the kinetic energy of the neutral atom population
T_e	Electron temperature defined by the kinetic energy of the electrons
T_{exc}	Excitation temperature defined by the populations of energy levels of excited states
T_i	Ionization temperature defined by the population of the various ionization states

The plasma is said to be in thermal equilibrium if all these temperatures agree. Full thermal equilibrium with the surroundings is clearly impracticable, but local thermal equilibrium may be achieved if conditions are reasonably uniform over a few mean free paths of the phenomenon being considered. It is generally agreed that the ICP is not in local thermal equilibrium (LTE) at power levels of 1–2 kW, but measurements of plasma temperature by optical means generally show gas temperatures of about 6000 K in the central channel, while excitation temperatures are found to be 7000–7500 K, and the ionization temperature about 8000 K at 10 mm beyond the load coil (16). This suprathermal characteristic of the ICP is no disadvantage for use as an ion source as long as the gas temperature T_g is sufficiently high to fully volatilize and dissociate the most difficult samples expected and the ionization temperature T_i is sufficient to produce adequate ionization of the elements of interest but not so high that multiply charged ions are produced to an unacceptable extent.

The degree of ionization α of a species is defined by the ratio of the ion population n_i to the total population

$$\alpha = \frac{n_i}{n_t} \tag{1}$$

The equilibrium among neutral atoms, ions, and electrons is defined as a function of temperature by the Saha equation, which in convenient logarithmic form is

$$\log S_n = \frac{3}{2} \log T_i - \frac{5040}{T_i} E_i + \log \left(\frac{Z_i}{Z_a}\right) + 15.684 \tag{2}$$

where S_n is the ionization constant for the species

$$S_n = \frac{\alpha}{1 - \alpha} n_e \tag{3}$$

T_i is the ionization temperature, E_i is the ionization energy for the species, Z_i, Z_a are partition coefficients for ions and atoms, and n_e is the electron population.

At 10 mm beyond the load coil the value of n_e is about 3×10^{15} cm^{-3} (16). If this value is assumed, the value of T_i may be calculated from the Saha equation if measurements of the degree of ionization of two elements with different ionization energies are made. Values determined in this way for a 1200-W ICP ion source are about 8000 K (17), in reasonable agreement with optical measurements.

The degree of ionization as a function of species ionization energy for $T_i = 8000$ K is summarized in Table 6.1. Elements of ionization energy below 7 eV are fully ionized. Up to 10.5 eV the degree of ionization exceeds 20%, and this includes all elements except 12—the five noble gases, the halogens fluorine, chlorine, bromine; and hydrogen, carbon,

Table 6.1. Degree of Ionization in Plasma for $T_i = 8000$ K[a]

E_i (eV)	<7	8	9	10	11	12	13
α	>0.98	0.91	0.71	0.36	0.12	0.03	0.01

[a] Value of n_e assumed 3×10^{15} cm^{-3}.

nitrogen, and oxygen. Experimentally better ionization than shown in Table 6.1 is obtained above 9 eV, which may be due to a lower value of n_e than that assumed. In practice, good sensitivity is obtained for elements up to 10.5 eV, thus including most of the analytically important elements. Even those of higher ionization energy such as carbon (11.26), oxygen (13.62), chlorine (13.02), and bromine (11.85) can be detected, although at lower sensitivity, and measurements of isotope ratios are possible, for example, on carbon and oxygen.

Some doubly charged ions do occur, but, even for elements with the lowest second ionization energies, their intensity may be kept below a few percent by suitable choice of operating conditions. This will be discussed further in Section 6.

3. ION EXTRACTION

The basic processes of ion extraction from chemical flames have been described in the literature (3, 4, 5, 18); these have been adapted and developed for the ICP along the lines used for molecular beam formation (19, 20). In order to prevent the formation of a boundary layer of cooler gas over the sampling aperture (10, 21), which would interpose a lower-temperature reaction zone between the plasma and the mass analyzer, thus distorting sampling, it is necessary to use large extraction apertures whose diameter exceeds 100 λ, where λ is the mean free ion path in the plasma. Since λ is about 1.5 μm, the minimum acceptable diameter is 150 μm, but in practice aperture diameters from 0.5 to 1.0 mm are used. The amount of gas admitted by these large apertures would require a very large pump in the first stage if it were to operate at the usual pressure of about 10^{-3} torr required for a supersonic molecular beam stage. This is avoided by using a much higher pressure in this stage of about 1 torr, as suggested by Greene et al. (20) and used by Douglas and French when sampling an MIP (13). At this pressure the large gas flow entering the aperture can be removed by a modest mechanical pump with a capacity of about 300 L/min.

The arrangement of a typical ion extraction interface is shown in Fig. 6.3. The plasma tail flame is allowed to impinge on the water-cooled front plate of the first stage, in the center of which is mounted the sampling aperture, drilled in the tip of a removable metal cone. Some provision is usually made for lateral adjustment of the plasma torch so that slight asymmetry in torch construction can be accommodated. The complete ICP head assembly containing the torch can also usually be adjusted in position along the system axis so that the extraction point in the tail flame

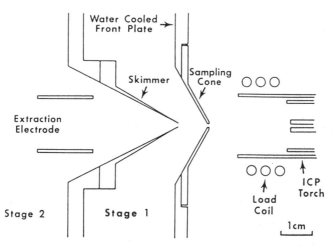

Figure 6.3. Arrangement of typical ion extraction interface. Typical operating pressures: stage 1, 1–3 torr; stage 2, 10^{-3}–10^{-4} torr.

can be varied, and it can also be retracted by a fixed amount to provide access to the front plate for changing apertures and skimmers. These adjustments are normally necessary only after refitting a plasma torch following cleaning or replacement.

The 0.5–1-mm wide extraction or sampling apertures are drilled in the tip of a shallow metal cone, which usually tapers in thickness at the tip so that the diameter/length ratio of the aperture is approximately unity and a sharp edge is presented to the plasma at the lip. The most satisfactory material for this demanding role is pure nickel, as it has a sufficiently high thermal conductivity to maintain a temperature of about 500°C at the tip when in the tail flame. In the plasma flame, where there is normally a high concentration of atomic oxygen, nickel forms a coating of oxide, which produces a rough surface. This is not easily removable because it is insoluble in acids, but it may be removed by light abrasive cleaning. Unfortunately, more-corrosion-resistant alloys are usually poorer conductors so that they offer no practical advantages. Nickel apertures are very durable and will usually operate for hundreds of hours provided that samples containing high acid concentrations are not introduced into the plasma. Sample acid concentrations above 5% are undesirable for other reasons as well, in particular because they can contaminate the first-stage pump.

The gas pressure in the first stage is maintained at 1–3 torr so that there is a steep pressure gradient through the aperture. This drives a jet of gas from the central channel of the plasma into the first stage, where it is intercepted by a second aperture of 0.5–1.5-mm diameter in the tip of a second sharper cone, the skimmer. The skimmer, usually also made of nickel, extracts the core of the expanding jet from the aperture and passes it into the second stage. The edges of the skimmer are sharp to minimize the shock effects produced in the jet, which is traveling at supersonic velocity. Spacing between the two apertures varies, depending on the systems, from 2 to 10 mm. Although the extracted gas enters the aperture at a temperature of about 6000 K, the very rapid expansion and resultant decrease in the collision frequency of the plasma gas atoms sharply reduce the temperature of the gas and effectively freeze the composition. Since the transition through the aperture takes place in about 1 μs, only reactions that are faster than this can occur while the gas is cooling but not fully expanded. Possible fast reactions include ion–molecule and attachment processes so that some molecules can be formed from the atoms present in the plasma. Ionic recombination is relatively slow, so there is little change in the ionic composition of the sample.

The mean free path in front of the skimmer is comparable to the skimmer aperture diameter of about 1 mm, so the flow through the skimmer occurs in the transition region and there are many collisions with the skimmer walls. Once past the skimmer, the pressure is considerably lower, usually below 5×10^{-4} torr, and the flow becomes fully molecular. In this second stage it is therefore necessary to use an electrostatic lens system to separate ions from the neutral species, which are pumped away, and focus them into the mass analyzer. This lens usually contains a small disk on the axis to obstruct the direct path for light from the plasma, which otherwise would create a high background count in the ion detector.

The ion lens systems used are simple arrangements of electrically biased cylinders or meshes whose purpose is to collect the ions emerging from the skimmer in an expanding cloud and focus them into a beam of circular cross section at the entrance aperture of the mass analyzer. Where the beam has to pass through a differential pumping aperture an intermediate focal point is required so that as many ions as possible are transmitted. Although intermediate potentials of up to a few hundred volts may be used, the ions enter the mass analyzer at low energy in order to obtain the longest dwell time in the analyzing region.

The pumping arrangements in the second stage vary between the system developed by Douglas and French at Sciex Ltd. (22) and that developed by Gray (23, 24), which forms the basis of the instrument built by Vacuum Generators Ltd. (25). In the former, only one vacuum stage is used behind the skimmer, and this is pumped by a large cryopump of about 10^5 L/s capacity. The operating pressure achieved in this stage is reported as being below 10^{-5} torr. In Gray's system the second stage is

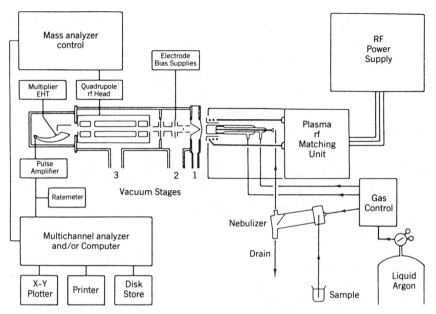

Figure 6.4. Typical ICP source mass spectrometer system, showing use of three vacuum stages as in Ref. 24.

pumped by a diffusion pump of at least 1200 L/s to about 5×10^{-4} torr, and the ions are focused by the lens through a differential aperture into a third stage containing the mass analyzer, which operates at about 2×10^{-6} torr. This stage is pumped by a smaller diffusion pump of about 200 L/s capacity. The overall arrangement may be seen in Fig. 6.4.

The ion extraction interface in the form currently used has proved simple and reliable. Although samples of up to 1% dissolved solids content may be analyzed, very little contamination occurs even in the first stage, and none appears to penetrate beyond the skimmer. After running samples with a heavy or volatile matrix it may be necessary to clean the skimmer, but in normal daily use this is needed only at weekly or monthly intervals. It is usual to provide a gate valve behind the skimmer so that skimmer and aperture may be removed readily without interfering with the high-vacuum stages. The ion lens and mass analyzer operate in high-vaccum stages and remain clean, so no attention is normally required.

4. MASS ANALYSIS AND ION DETECTION

The ion beam produced after the ion lens system is cylindrical in cross section and contains ions of a mean energy of 0–30 eV with an energy spread of about 5 eV (FWHM). Such an ion beam is well suited for analysis by a quadrupole mass analyzer. The type of quadrupole used by Gray and in the V.G. Plasma Quad is the V.G. Mass Lab 12-12S, based on a type developed for GC-MS use. This uses 12-mm diameter analyzing rods preceded and followed by short rods of the same diameter to which only rf supplies are fed. The function of these small rods is to reduce the fringing fields at the entrance to the main rods so that low-energy ions are not prevented from entering. The dc pole bias, the mean dc potential of the rod system above ground, may thus be set slightly below the mean ion energy of the entering ions so that they traverse the rods at low velocity and spend as long as possible in the analyzing field. This maintains the ion transmission up to much higher masses than is possible without any form of electrode system reducing the fringing field, and enables useful resolution to be obtained over a mass range of 0–300 daltons, which is adequate for elemental analysis.

In the type of analyzer used, the resolution over the whole mass range may be chosen by the user so that $M/\Delta M$ is between 1 and 3 on the 5% valley definition, where M is the mass transmitted and ΔM is the peak width at a peak height of 5%. Although the peaks are roughly triangular, they are not perfectly shaped and have small tails and precursors that contribute to the signal at the adjacent mass numbers. In a good instrument these are usually kept to below 10^{-5} of the peak height at $M - 1$ and below 10^{-6} at $M + 1$. This measure of abundance sensitivity is as significant as the mass resolution in reducing spectral interference. Since the ion mass transmitted by a quadrupole analyzer is determined by a single dc potential, the analyzer is readily controlled either manually or by computer program. The sweep speed is limited by the time constants of the rf/dc circuits to about 10,000 daltons/s, so that a sweep of the full elemental mass range may be completed in about 30 ms. This enables transient signals lasting only a few seconds to be adequately analyzed provided that the data system recording the ion detector output can handle information at an adequate rate. A full sweep across the whole mass range is of value when dealing with an unknown sample, but it may make poor use of the available integrating time if large regions of the spectrum are of no interest. In such circumstances a restricted mass range is of greater value, and under computer control this may be arranged to cover just those regions of the spectrum of interest or even to concentrate on only a few elements of importance in the peak-switching mode.

Ions leaving the exit aperture of the mass analyzer may be detected by any conventional ion detector. To obtain the highest possible sensitivity, it is desirable to detect individual ions, and a pulse-counting detector is normally used. This is usually a channel electron multiplier operated with a high negative potential at its mouth. The multiplier is placed just off axis to avoid being struck by residual photons. The electron multiplier pulse is then passed to a suitable amplifier and discriminator and then to the data-handling system. These detectors are capable of high detection efficiencies and can count individual ions up to arrival rates of over 10^6 s^{-1}. Above about $3 \times 10^6 \ s^{-1}$, however, the detector response becomes too slow, and counting losses increase to an unacceptable extent. However, a dynamic range of over 10^5 is obtained, which may be extended by operating either the main detector or a second one in an integrated current mode.

Although the lifetime of a channel multiplier detector is long and they are tolerant of occasional exposure to air, their gain does eventually fall; their life can be prolonged by avoidance of exposure to the excessive ion rates ($\sim 10^9 \ s^{-1}$) resulting from very large peaks. Some automatic protection is often arranged to reduce ion lens transmission when the analyzer is scanned over major peaks.

5. OVERALL SYSTEM AND DATA HANDLING

A typical arrangement for mass analysis from an ICP source is shown in Fig. 6.4. The plasma system is shown on the right of the figure with the torch mounted horizontally on the side of the rf matching unit. Although it is basically similar to an optical emission plasma, some precaution is usually taken to minimize the dc offset produced when the plasma contacts the grounded sampling cone, which otherwise may lead to high plasma potentials above ground. This may consist of a special coil geometry to minimize capacitative coupling to the plasma or, in the case of the Sciex (22) instrument, the use of a load coil center-tapped for the ground connection.

The interface stage between the sampling cone and skimmer at a pressure of 1–3 torr is followed by the ion lens in the second stage, and the mass analyzer and ion detector in the third stage. The ion lens elements in these two stages are supplied with dc potentials from regulated, adjustable supplies. High-voltage supplies to the quadrupole rods and to the electron multiplier are fed through a bulkhead at the end of the analyzer vacuum housing.

The signal pulses from the ion-detector pulse amplifier may be handled in a variety of ways. In the research instruments at the University of Surrey and the British Geological Survey, the pulses are fed to a multichannel analyzer (MCA) used in the multiscaler mode (MCS). When the mass analyzer is used in a scanning mode, the multichannel scaler memory address is scanned in synchrony with the transmitted mass so that signals from each mass region of the scan are stored repeatedly in the same group of channels. During successive scans a spectrum is thus built up in the memory. Usually memory groups of 1024 or 2048 channels are employed for a scan width of up to 250 daltons so that each peak occupies at least 8 channels. Channel dwell times of as little as 10 μs may be used, giving scan lengths down to 20 ms for the full width of 250 daltons. However, this is about the useful maximum scan rate of the mass analyzer for the full mass range, and it is more usual to use dwell times between 50 μs and 1 ms. The rapid scan rate of 1–20 scans/s is of great value when analyzing samples from transient sources such as flow injection, electrothermal volatilization, or laser ablation, which may be present in the plasma for only a few seconds.

It is usual to use a 1-min integrating period for an analysis in which between 60 and 600 individual scans may be accumulated in the memory for display and data processing. Such MCA systems offer full data readout options and some limited processing functions but do not provide integrated control of the whole system. This may be obtained with the addition of a computer, which is then used to control the scan function of the mass analyzer. The MCA is then used as an intermediate fast store, and at the end of the integrating period it rapidly transfers the memory contents to the computer for subsequent processing and readout.

A simpler alternative without the use of the intermediate MCA is adopted in the Sciex instrument (22): Signals are directly accumulated as they are produced in the computer memory, the permissible rate of scan then being determined by the computer.

The use of computer control of the mass analyzer in the commercial instruments provides much greater flexibility of operation. Although a conventional scan over a chosen mass range may still be used, operation is not limited to this. A peak-switching mode may be chosen, with the mass set on a series of peaks in succession or performing short scans across them, so that maximum use is made of the integrating time. A more sophisticated mode still may be used in which the integrating time in each region can be varied in accordance with the signal level present. In this way, longer counting times may be given for weaker peaks, thus maintaining more uniform counting statistics for all elements of interest. This can be particularly valuable in isotope ratio measurements.

The software developed for the commercial instruments was initially based on ICP-AES experience with provision for storage of standard spectra, calibration curves, and instrument standardization and provides direct readout of selected element concentration and isotope ratios where required. Because of the inherent isotope ratio information available from mass spectrometry, the programs also provide information on potential interferences from adjacent elements with overlapping isobars.

With computer control of the system, additional accessories may be included in the overall program. The most important of these are associated with sample introduction, and the most common is a multiposition sample changer, which can accommodate up to 100 samples with interlaced standards and blanks to provide unattended operation. Such changers usually provide a wash cycle to flush out the nebulizer system after each integration.

The services required to operate such a system are relatively easy to provide. A laboratory environment at a controlled temperature of $\pm 5°C$ or better is preferable, with humidity of 60% or less. Three-phase electric power at about 10 kV-A is needed, and a cooling water flow of about 300 L/h at a temperature of less than 20°C is usually required. Exhaust ducting from the plasma and vacuum pumps is necessary. The argon supply must be capable of delivering up to about 15 L/min of 99.95% pure argon at a pressure of about 4 bar and may be derived from gas cylinders or, more conveniently, from liquid argon tanks.

6. PERFORMANCE

The normal mode of introducing sample to the plasma is by pneumatic nebulization, and the majority of applications reported so far have used this method. Conventional crossflow or concentric nebulizers at a sample uptake rate of 1 mL/min produce a concentration of sample ions of about 10^7 cm^{-3} in the plasma for a fully ionized element at a concentration in solution of 1 ng/mL. Such a concentration results in a count rate of between 10^2 and 10^3 count/s for a monoisotopic element. The background count rate is usually between 10 and 20 counts/s, or a concentration equivalent to between 0.2 and 0.01 ng/mL.

The sample is normally introduced to the nebulizer at a controlled rate by a peristaltic pump. Between samples, deionized water or a blank solution is introduced. When the uptake tube is transferred to a sample, the response appears a few seconds after the sample reaches the nebulizer and reaches a steady level in about 10 s. The response decays to between 10^{-3} and 10^{-5} of the peak value in 2–3 min for most elements. Most of the memory occurs in the nebulizer system and is similar to that seen also in ICP-AES, but some elements can show memory from deposits on the sampling cone or, in extreme cases of a heavy matrix, on the skimmer. These are easily removed by cleaning, which requires a brief shutdown.

The nature of the response to a cerium nitrate solution at a concen-

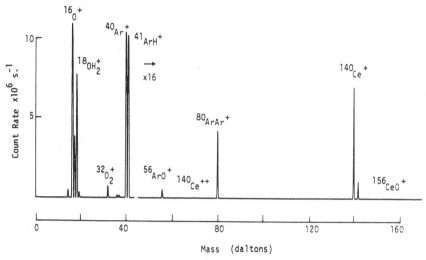

Figure 6.5. Spectrum obtained from 1 μg/mL solution of cerium in 1% HNO$_3$. Vertical scale above 42 daltons expanded × 16. Low-intensity peaks detailed in Table 6.2.

tration of 1 μg/mL in 1% HNO$_3$ is shown in Fig. 6.5. The spectrum covers the mass range from 4 to 160 daltons and shows the major background peaks and smaller peaks for $^{140}Ce^+$ and $^{142}Ce^+$, the former with an intensity of 5 × 10^5 counts/s. The major background peaks are always present when solutions are introduced; their response is saturated at about 10^7 counts/s. There are two main groups, the lighter based on oxygen and the heavier on argon, and both include molecules containing hydrogen as well. Hydrogen and oxygen constitute about 30% of the atoms and ions in the plasma; most of the remainder are due to argon. The response to hydrogen is not shown in this spectrum, but since both hydrogen and oxygen have ionization energies (13.60, 13.62 eV) considerably below that of argon (15.76 eV) they actually provide a higher ion population than argon. The largest peaks of the lighter group are $^{16}O^+$, $^{17}OH^+$, and $^{18}OH_2^+$, and smaller peaks are seen at $^{14}N^+$ and $^{19}OH_3^+$. It is believed that the molecular ions are not present in the plasma itself but are formed during the extraction process. The heavier peaks consist of $^{40}Ar^+$ and $^{41}ArH^+$, usually of similar height. Smaller dimer peaks of $^{16}O_2^+$ and $^{40}Ar_2^+$ are also found at 32 and 80 daltons. The $^1H_2^+$ dimer peak is again not displayed. Smaller peaks are also found for the atomic and molecular ions corresponding to the minor isotopes of H, O, and Ar such as $^{36}Ar^+$ and $^{37}ArH^+$, which may be seen clearly in the spectrum.

Other atomic and molecular ions also occur at much lower levels, some of which may be seen in the expanded portion of Fig. 6.5. Among the most serious are $^{56}ArO^+$ (and to a much lesser extent $^{54}ArN^+$) and $^{80}ArAr^+$. The molecular ions are the product of ion–molecule and condensation reactions that occur between the major components of the plasma gas during the expansion of the extracted gas into the first stage. Since the heaviest atom present at high concentration is ^{40}Ar and, apart from the attachment of hydrogen to a binary molecule, molecular ions normally involve only two components, such peaks are found only between 17 and 80 daltons, and many of them are very small, equivalent to sample concentrations of less than 1 ng/mL. Typical equivalent concentration values of background peaks and the ions with which they interfere are shown in Table 6.2.

A number of other ion peaks, not related to the analyte species, may also be seen at trace level in spectra. Both copper and zinc may be seen at a level of a few nanograms per milliliter. These ions are commonly present as contaminants in solvent acids and often in deionized water. For ultratrace analysis, the purest water and solvents are necessary. Nickel peaks may be seen at a level equivalent to 2 ng/mL; they arise from the sampling aperture and skimmer from sputtering by reactive oxygen ions. The extent of this effect depends on the plasma potential, and it may be reduced or avoided by measures that reduce the plasma potential below the sputtering threshold. This nickel background has little effect on the detection limit for nickel at this level.

The response to cerium at 1 μg/mL in Fig. 6.5 consists of a single ion peak for each isotope. Despite the low second ionization energy (10.85 eV), doubly charged ions have a low intensity, and the doubly charged response is only 0.35% of the total. Cerium is a very refractory element with a high bond dissociation energy for CeO of 8.2 eV. Although probably fully dissociated in the plasma, the atomic ion is present in an excess population of oxygen, and as the plasma gas is cooled some recombination is possible. The gas extracted from the bulk plasma is quenched very rapidly in the expansion, but gas entering the region of the boundary layer, which forms around the sampling aperture over the cone surface, is delayed in a region of intermediate temperature for a period that may be as long as several microseconds. This provides opportunities for re-formation of oxides of strongly bound species, and some of these may then be entrained in the extracted gas at the periphery of the aperture. A small oxide response is thus obtained, in this case about 0.35% of the total response. A very small hydroxide peak of 0.06% is also found.

Some other elements have lower second ionization energies than cerium (e.g., barium, 10.00 eV) or slightly higher oxide bond strengths, but in general both doubly charged ions and oxides are less than 1% of the

Table 6.2. Molecular Ion Interferences from 1% HNO_3 Solution

m/z	Probable Ion	Equivalent Conc.[a] (ng/mL)	Ion Affected
28	$^{14}N_2^+$	31	$^{28}Si^+$
29	$^{14}N_2{}^1H^+$	6	$^{29}Si^+$
30	$^{14}N^{16}O^+$	23	$^{30}Si^+$
31	$^{14}N^{16}O^1H^+$	0.7	$^{31}P^+$
32	$^{16}O_2^+$	1492	$^{32}S^+$
33	$^{16}O_2{}^1H^+$	45	$^{33}S^+$
34	$^{16}O^{18}O^+$	8	$^{34}S^+$
35	$^{16}O^{18}O^1H^+$	2	$^{35}Cl^+$
36	$^{36}Ar^+$	365	$^{36}S^+$
37	$^{36}Ar^1H^+$	315	$^{37}Cl^+$
38	$^{38}Ar^+$	81	—
39	$^{38}Ar^1H^+$	76	$^{39}K^+$
40	$^{40}Ar^+$	107×10^3	$^{40}Ca^+$
41	$^{40}Ar^1H^+$	105×10^3	$^{41}K^+$
42	$^{40}Ar^2H^+$	21	$^{42}Ca^+$
52	$^{40}Ar^{12}C^+$	2	$^{52}Cr^+$
54	$^{40}Ar^{14}N^+$	5	$^{54}Fe^+$
56	$^{40}Ar^{16}O^+$	66	$^{56}Fe^+$
76	$^{36}Ar^{40}Ar^+$	4	$^{76}Se^+$
78	$^{38}Ar^{40}Ar^+$	1	$^{78}Se^+$
79	$^{38}Ar^{40}Ar^1H^+$	0.7	$^{79}Br^+$
80	$^{40}Ar^{40}Ar^+$	532	$^{80}Se^+$
81	$^{40}Ar^{40}Ar^1H^+$	1	$^{81}Br^+$

[a] Equivalent concentration calculated from response to 1 μg/mL Ce in 1% HNO_3 shown in Fig. 6.5.

total response. This was not the case in some of the earlier reported work (24, 26, 27), where considerable populations of doubly charged species were found for elements with low second ionization energies, especially at high injector flow rates. These high levels were found to be a function of excitation coil design, which influenced the field gradients in the plasma, and of the sample injector flow rate (17) and were reduced by modification of coil geometry (10) or the use of a balanced load coil (28).

The spectrum in Fig. 6.5 includes a number of molecular peaks at low level containing nitrogen produced by ion–molecule reactions with nitric acid in the sample solution. Some nitrogen is still present in the plasma even for a deionized water sample, since dissolved air is present; air may enter the gas lines by diffusion, and some may be entrained at the sampling aperture. In addition, some residual air is normally present in the argon. However, because of the high ionization potential of nitrogen (14.55 eV), the nitrogen-containing molecular peaks are relatively small and difficult to show in an expanded spectrum, but their extent may be seen in Table 6.2. They cause few interference problems. Other acids such as hydrochloric, sulfuric, or phosphoric acids cause more serious interferences. The relatively low ionization potential of sulfur and phosphorus in particular (10.36, 10.48 eV) and chlorine (13.02 eV) results in larger molecular peaks, which occur at higher masses than those including nitrogen. The most important of these interferences are listed in Table 6.3. Similar interferences are found if these elements are present in high concentrations in the analyte matrix, for example, chlorine in brine samples.

The size of these interfering peaks from 1% acids is in most cases small, only ^{48}SO approaching an equivalent concentration of 1 µg/mL. Their extent is a measure of the rapidity of the expansion process behind the sampling aperture in the interface and of the disturbance of the jet formed after the aperture by the presence of the skimmer. It has been found possible to reduce their size by paying attention to the aerodynamics in this stage, and further reduction is likely as design improves.

Many of these potential interferences may be avoided by the use of an alternative isotope of the analyte, although this may cause some loss of sensitivity; in some cases, such as the interferences by ^{51}ClO and $^{75}ClAr$

Table 6.3. Principal Interferences from Molecular Ions Containing Acid or Matrix Elements

Element	Molecular Ion	m/z	Interference
N	N_2^+	28	^{28}Si
N	N^5O^+	30	^{28}Si
N	ArN^+	54	^{54}Fe
P	PO^+	47	^{47}Ti
P	P_2^+	62	^{62}Ni
P	Ar^5P^+	71	^{71}Ga
S	SO^+	48	^{48}Ti
S	S_2^+	64	^{64}Zn
S	SO_2^+	64	^{64}Zn
S	ArS^+	72	^{72}Ge
Cl	ClO^+	51	^{51}V
		53	^{53}Cr
Cl	$ArCl^+$	75	^{75}As
		77	^{77}Se

with ^{51}V and ^{75}As, no satisfactory alternative exists. The use of HCl or the presence of Cl in the matrix presents some problems with these elements.

Avoidance of these interferences requires particular care in sample preparation. Insofar as possible, high acid concentrations should be avoided, and whenever possible nitric acid should be used. When an interference cannot be avoided, the analysis may still be completed, although with lower precision, by subtracting the response from a blank solution of the same acid concentration.

Apart from these instances, which cause problems in relatively few applications, the elemental response is in the form of singly charged ion peaks of the element or metal, and the mass resolution available from a quadrupole analyzer enables adjacent mass peaks to be separated. The resolution needed to distinguish nominally coincident molecular peaks, however, would be at least 4000 across the mass range, which is beyond the present capability of quadrupole analyzers.

The transmission provided by these analyzers provides a roughly uniform sensitivity [count rate/(μg/mL)] over most of the mass range. For light elements below argon the response falls as shown in Fig. 6.6, which shows the spectrum obtained from a multielement standard solution containing 26 elements ranging from lithium to bismuth each at a concentration of 1 μg/mL.

This standard was prepared in 4% HCl, and the spectrum shows the ^{51}ClO, ^{53}ClO, ^{75}ArCl, and ^{77}ArCl peaks already referred to. Allowing for the multiple isotopes of many elements, the peak height across the mass range is reasonably constant above argon. The region around the oxygen and argon peaks is displayed again in expanded format in Fig. 6.7 after subtraction of the blank. This has removed all but minor traces of the blank peaks, including the ClO peaks, and some response in the blank to residual lithium and sodium in the nebulizer. The decrease in response to light elements may now be clearly seen.

With a relatively uniform response across the periodic table and with a uniform random background level of 10–20 counts/s, it is expected that detection limits are similar for most elements within the limitations of isotope abundance and high ionization energy for some elements. At low mass there is additional background from contamination, which is mainly from the common transition elements and from the molecular peaks already discussed. As can be seen in Table 6.4, therefore, detection limits below 80 daltons are poorer than above.

The ability to make successive rapid integrations on blank solutions enables detection limits to be defined in the manner commonly used in atomic spectrometry by specifying the smallest concentration that pro-

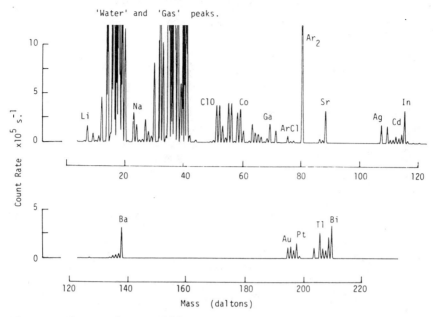

Figure 6.6. Spectrum from a multielement standard solution containing 26 elements, each at 1 μg/mL in 4% HCl. Upper portion shows mass range below 120 daltons; lower portion shows range above 120 daltons.

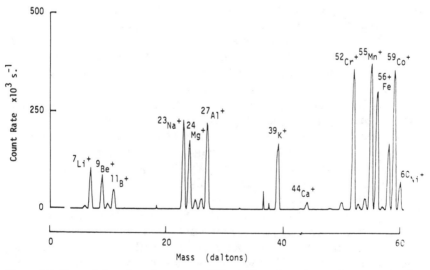

Figure 6.7. Expanded portion of spectrum of Fig. 6.6 in region of gas and water peaks after blank stripping in data system.

279

Table 6.4. Elemental Detection Limits[a] (3σ)

Element	Limit (ng/mL)	Element	Limit (ng/mL)	Element	Limit (ng/mL)
Li	0.09	Zn	0.21	I	0.02*
Be	0.03	Ga	0.03*	Te	0.08
B	0.37	Ge	0.02	Cs	0.03*
Na	0.12*	As	0.04	Ba	0.03*
Mg	0.13*	Se	0.79	La	0.03*
Al	0.16	Br	2.4*	Ce	0.05
Ca	6.2*	Rb	0.08*	W	0.05
Sc	0.07*	Sr	0.03*	Pt	0.06*
Ti	0.07*	Zr	0.05*	Au	0.06
V	0.03*	Mo	0.04	Hg	0.02
Cr	0.06	Ag	0.03	Pb	0.05
Mn	0.10	Cd	0.06	Bi	0.05*
Ni	0.10	In	0.07	Th	0.02
Co	0.03*	Sn	0.06	U	0.01*
Cu	0.32*	Sb	0.05*		

[a] Determined from 10 integrations of 10 s each on 1% HNO_3 blank and a single 10-s integration on the element solution in 1% HNO_3 at 100 ng/mL or 1 μg/mL. Asterisk (*) denotes values obtained on multielement scan, corrected to 10 s integrating time as for single-ion values.

duces a statistically significant response above background, using the concentration equivalent to three times the standard deviation of the blank response (3σ level). The standard deviation is usually determined from successive integrations on the blank solution. When this is done on one element at a time (single-ion monitoring), integration periods of 10 s are commonly used for blank and signal. On a computer-controlled instrument this may be done in the peak-switching mode. In the scanning mode, the integration time for each element is usually much shorter, between 0.1 and 1 s. The standard deviation of the blank appears to vary as the square root of the integrating time, and thus the detection limits observed when scanning are up to a factor of 10 worse than for single-ion monitoring. Table 6.4 shows experimentally obtained detection limits.

The spectra in Figs. 6.5–6.7 show a wide range of peak heights. Identifiable peaks that are just above the background level, which is equivalent to an analyte concentration of about 0.03 ng/mL, may be found on expanding the scale. At the upper end of the dynamic range, the detector and counting system begin to show nonlinearity above about 5×10^6 counts/s. If background correction is applied, the dynamic range thus approaches 6 decades. The plasma operates linearly over a wider range than this. Using the detector in a dc mode, a linear response has been obtained up to about 0.3% for transition metals in solution; this overall range of 10^8 cannot be realized with one detecting system.

Depending on the nature of the matrix, solutions with dissolved solids contents of up to 1% can be analyzed with little interelement interference from the matrix except for that caused by elements of low ionization energy such as the alkalis, alkaline earths, and lanthanides. These cause no problems at a concentration of up to 1000 μg/mL but do produce ionization suppression above this limit because of the high electron population they contribute to the plasma, which suppresses the ionization of other elements [see Eq. (3), Section 2]. This suppression is about 10% at 2000 μg/mL (0.2%). Suppression may also be caused by processes that do not depend on ionization; for example, a high solids content increases the problem of transporting and volatilizing the droplets from the nebulizer. If there is any resulting loss of transport efficiency or if the microparticulates from desolvated droplets do not fully vaporize in their passage through the plasma, the response to the analyte may drop. Care is needed in calibrating the system for analysis in these conditions, for example, by using matrix matching, standard addition, or isotope dilution methods. However, selection of the best operating conditions to minimize the effects is also possible and can contribute considerably to successful analysis of high matrix samples.

7. APPLICATIONS

Once the basic development of the continuum flow interface had been completed at Surrey (23, 24), a second research instrument was built and then installed at the British Geological Survey, London. Using this instrument from mid-1983 for the analysis of mineralogical samples, Date has pioneered the use of the technique in a geological laboratory, and a number of the applications described below are from that work. Commercial instruments, although introduced to the market in 1983, only began to come into use in analytical laboratories in 1984.

7.1. Analysis of Mineralogical Samples

Multielement analysis of geological samples by ICP-AES has become a routine technique, and an extensive methodology has developed for the necessary dissolution and matrix treatment of the samples involved. Much of this methodology is directly applicable to ICP-MS, although some modifications may be required to reduce the final acid concentration in the samples as suggested in Section 6.

In many cases ICP-AES analysis is adequate, but, particularly for heavier elements, mass spectrometry has considerable advantages over emission spectrometry because of the greatly reduced interference problems associated with the simpler spectra. This is especially true for the rare earth group of elements (REEs). These may be analyzed by emission spectrometry using a high-resolution instrument, but the method is time-consuming whereas ICP-MS analysis is straightforward. Every rare earth element has at least one interference-free isotope. The less abundant heavy members of the group are monoisotopic or nearly so (e.g., ^{175}Lu abundance is 97.4%) and thus have lower detection limits. Most of the more common REEs have two or three isotopes free from overlap and are therefore suitable for isotope dilution analysis. Experimentally determined detection limits for REEs are shown in Table 6.5. A spectrum of the Canadian Geological Survey standard silicate rock syenite SY-3 in solution at 1 g in 100 mL is shown in Fig. 6.8, and calibration curves are given for a number of REEs in Fig. 6.9; these are linear over more than 4 decades. Results obtained on two standard reference silicate rocks by Date are shown in Table 6.6. The data were derived from a single 1-min integration in each case for a blank solution in 1% HNO_3, a reference standard solution containing each REE at 1 μg/mL in 1% HNO_3, and the sample solutions. Sample solutions were prepared at 0.5 g in 100 mL. Except for silica, which was removed during dissolution with HF, the matrix elements are present in solution. The agreement obtained with the accepted values for these standards is excellent for such a rapid analysis.

The same technique may be used for multielement determination of a wide range of elements, and the values obtained by Date on three further

Table 6.5. Multielement Detection Limits on Rare Earth Elements^a (3σ)

Element	Limit (ng/mL)	Element	Limit (ng/mL)
^{139}La	0.08	^{159}Tb	0.08
^{140}Ce	0.24	^{164}Dy	0.28
^{141}Pr	0.15	^{165}Ho	0.07
^{142}Nd	0.20	^{166}Er	0.27
^{152}Sm	0.33	^{169}Tm	0.10
^{153}Eu	0.13	^{174}Yb	0.21
^{158}Gd	0.46	^{175}Lu	0.04

^{a} Determined on isotopes shown on a single 1-min integration on solution at 100 ng/mL. Blank σ values from 10 integrations of 1 min. Scan width 70 daltons. Estimated single-ion values may be obtained by multiplying these values by 0.3.

Table 6.6. Rare Earth Element Data for Two International Standard Reference Silicate Rocks

		GSP-1		NIM-G	
Element	Ion Used	ICP-MS	Ref. 29	ICP-MS	Ref. 29
Lanthanum	$^{139}La^+$	160	195	110	105[c]
Cerium	$^{140}Ce^+$	370	360	200	200
Praseodymium	$^{141}Pr^+$	46	50[c]	21	—
Neodymium	$^{146}Nd^+$	170	190[c]	69	68[c]
Samarium	$^{147}Sm^+$	20	25[c]	12	16[c]
Europium	$^{153}Eu^+$	1.7	2.4	0.3	0.4
Gadolinium	$^{157}Gd^+$	12	15	15	11[c]
Terbium	$^{159}Tb^+$	1.0	1.4	2.3	3
Dysprosium	$^{163}Dy^+$	4.2	5.7	17	15
Holmium	$^{165}Ho^+$	0.7	—	3.6	3
Erbium	$^{166}Er^+$	1.8	3	11	10[c]
Thulium	$^{169}Tm^+$	0.2	—	1.8	2
Ytterbium	$^{172}Yb^+$	1.0	1.9	11	14
Lutetium	$^{175}Lu^+$	0.2	0.2	1.7	2[c]

[a] Concentrations in μg/g.
[b] Values from multilaboratory analyses.
[c] Values suggested by Abbey (29) as "usable."

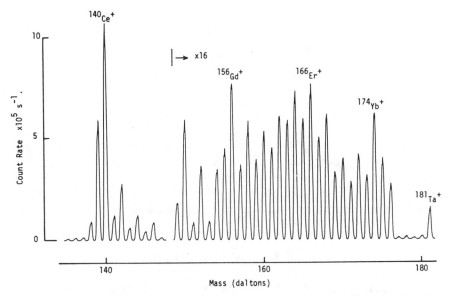

Figure 6.8. Rare earth element spectrum for international standard rock syenite SY-3 in solution at 1.0 g in 100 mL (29).

283

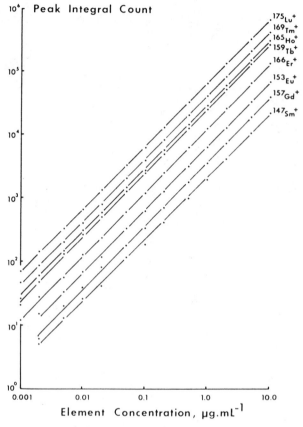

Figure 6.9. Calibration curves for rare earth elements (29).

standard rock solutions over the full mass range are shown in Table 6.7. Only a single run was made for each of the solutions, which were prepared at a concentration of 0.1 g in 100 mL. In spite of the short integrating time for each isotope during the 1 min scan from 4 to 250 daltons, the agreement with certified values is excellent.

The method is being extended by Date to gold and the platinum group metals, and preliminary results using a fire assay collection procedure are very promising.

7.2. Analysis of Water

The analysis of groundwater samples is of considerable importance in geological survey work, in the control of pollution sources, and in the

Table 6.7. Multielement Determination in International Rock Standards[a]

Element	Ion	Standard G-2		Standard SY-1		Standard MRG-1	
		ICP-MS	Cert.	ICP-MS	Cert.	ICP-MS	Cert.
Lithium	$^{7}Li^{+}$	21	35	120	121	—	4
Beryllium	$^{9}Be^{+}$	—	2.4	17	24	—	—
Mn (as MnO)	$^{55}Mn^{+}$	0.03%	0.03%	0.40%	0.40%	0.16%	0.17%
Cobalt	$^{59}Co^{+}$	5.1	5	17	18	75	86
Zinc	$^{66}Zn^{+}$	89	84	230	228	150	190
Rubidium	$^{85}Rb^{+}$	160	170	130	170	7.5	8
Strontium	$^{88}Sr^{+}$	500	480	180	258	239	260
Molybdenum	$^{95}Mo^{+}$	0.7	0.9	5.8	3.8	2.1	—
Barium	$^{138}Ba^{+}$	2270	1900	230	300	39	50?
Thallium	$^{205}Tl^{+}$	0.5	1.2?	1.3	1.4	0.2	—
Lead	$^{206-208}Pb^{+}$	31	30	470	475	3.1	10
Thorium	$^{232}Th^{+}$	27	25	1340	1338	1.2	1?
Uranium	$^{238}U^{+}$	2.7	2.1	2510	2500	0.5	0.3

[a] Concentration in µg/g, unless otherwise stated.

Table 6.8. Trace Elements in NBS Water SRM 1643a[a]

Element	Ion	NBS	ICP-MS Mean	RSD (%)[b]
Beryllium	$^9Be^+$	19	21	20
Vanadium	$^{51}V^+$	54	52	6
Chromium	$^{52}Cr^+$	17	18	12
Manganese	$^{55}Mn^+$	32	34	5
Cobalt	$^{59}Co^+$	19	21	7
Zinc	$^{66}Zn^+$	69	57	11
Arsenic	$^{75}As^+$	77	76	5
Strontium	$^{88}Sr^+$	243	297	7
Molybdenum	$^{98}Mo^+$	97	134	9
Silver	$^{107}Ag^+$	2.8	3.5	16
Cadmium	$^{114}Cd^+$	10	13	22
Barium	$^{138}Ba^+$	47	74	17
Lead	$^{208}Pb^+$	27	31	8

[a] Concentration in ng/mL.
[b] Based on 10 determinations.

monitoring of water supplies, and the lowest possible detection limits are usually required on a multielement basis. The ability to detect unexpected elements is important in most of these applications, and scanning of the full mass range is usually necessary. Thus during a 1-min integration the actual integration period for each isotope is only about 0.25 s, and at low concentrations, even for monoisotopic elements, the total count integral may be quite small and statistical precision relatively poor. An example of such an analysis on the NBS reference water standard SRM1643a, reported by Date and Gray (26), is shown in Table 6.8. This sample was run for 10 successive 1-min integrating periods, using 2048 channels to store each spectrum. A deionized water blank and a multielement standard solution containing 33 elements, each at 1 μg/mL, were also run for a single 1-min integration.

The total integrating time on each isotope of the SRM for the 10 runs was thus 2.5 s, and on the laboratory standard only 0.25 s. Total time on the instrument was thus only about 15 min. Table 6.8 shows the results for 13 elements over the mass range from beryllium to lead. The multi-isotopic elements were determined by integrating on the largest isotope except for zinc, where ^{66}Zn was used.

The mean values determined in this way compare very well with the certified concentrations, even though the individual results show considerable variation due to the small integrated count for some of the isotopes.

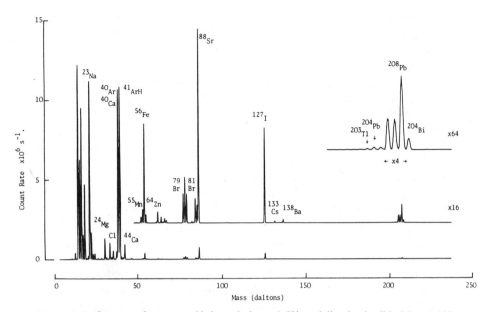

Figure 6.10. Spectrum from natural brine solution at 0.5% total dissolved solids (Na at 0.2%) showing majors and traces. Peak count rate (saturated) of ^{23}Na, 11×10^6 counts/s. Approximate trace concentrations (ng/mL): Fe, 2600; Zn, 500; Cs, 70; Pb, 330; Tl, 10. Peak count rate ^{208}Pb, 60,433 s^{-1}.

Many water samples contain high salt concentrations. The spectrum, again from a 1-min integration across the whole mass range, shown in Fig. 6.10 is typical for a brine containing 25% dissolved solids. It was diluted to 0.5% total solids, of which sodium was at 2000 µg/mL. The spectrum shows the major elements, many of whose peaks were saturated, ^{23}Na for example being at 11×10^6 counts/s. Magnesium, chlorine, and calcium, including the low-abundance ^{44}Ca isotope, are clearly seen. Trace elements are detectable in the two partial spectra, which are at successively increased expansions of the original data. Thallium is clearly seen at 10 ng/mL, the ^{203}Tl isotope, of equivalent concentration 3 ng/mL, giving a peak count rate of 1367 s^{-1}, well above the blank level of about 20 s^{-1}. Determinations down to below 1 ng/mL are thus clearly possible at this high solids content.

7.3. Analysis of Samples in the Life Sciences

There are increasing requirements for trace analysis in the life sciences, particularly in animal and human studies of metabolism and environmental pollution uptake. Much of the routine work in the medical services is carried out by AAS, but increasingly frequently there are situations where the need to monitor several elements at levels at the bottom of the flame AAS range arises and where other methods such as neutron activation analysis are inconvenient or slow. Soils, plant materials, foodstuffs, and body fluids are typical analytical samples, and in many cases they contain complex matrices.

Blood is one of the more difficult sample types. Because it contains suspended solid matter, simple dilution to reduce the viscosity is not satisfactory for a normal crossflow nebulizer, which is liable to block under these conditions. High-solids nebulizers may be used, or alternatively the sample may be ashed and redissolved. The typical calibration plots in Fig. 6.11 for cadmium, selenium, and lead were obtained as follows. Samples were ashed in HNO_3 and HNO_3/H_2O_2 and redissolved in 5% HNO_3 at a dilution of 1:1 compared to the whole blood. Samples with a range of standard additions were prepared so that calibration curves in the correct matrix could be plotted. The analysis was carried out for each element in turn, using 25-s integrations with the mass analyzer set to the wanted ion, which provided better statistics at low levels than multielement scanning. Excellent linearity is obtained, enabling the natural levels in the original blood to be readily determined—below 1 ng/mL in the case of cadmium. Some problems arose in these samples with the determination of chromium and manganese owing to the presence of small molecular peaks at 52 and 55 daltons, which were due to ^{52}ArC from incompletely removed carbon and to $^{55}ArNH$ due to the 5% HNO_3 solution used. More appropriate sample preparation was needed to enable these elements to be determined.

An example from the analysis of food is shown in Fig. 6.12, where the detection of lead is shown at a concentration of 2.8 ng/mL in ashed rice flour that had been redissolved in a mixture of 1% HNO_3 and 3% HCl. The spectrum in Fig. 6.12a is a small part of a scan over a range 100–220 daltons stored in 1024 channels. Although the total integration time was 1 min, the portion shown occupied only 6 s. The individual channel points are clearly seen, each peak occupying about 6 memory channels. The spectrum in Fig. 6.12b was obtained on the same sample by integrating over the narrower range 200–210 daltons for 1 min so that each peak occupies about 120 channels. The height of the ^{208}Pb peak is about the same in the two spectra because the dwell time in each channel is unchanged, but the total count in each peak is about 20 times greater for the narrower scan. This increased count gives better statistics, and it is possible to distinguish the ^{204}Pb peak from the background if peak integrals are taken for this spectrum, whereas it is quite indistinguishable with

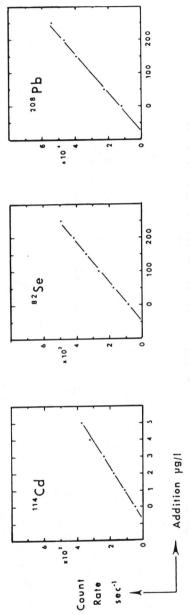

Figure 6.11. Calibration curves for cadmium, selenium, and lead in ashed blood samples obtained by standard addition.

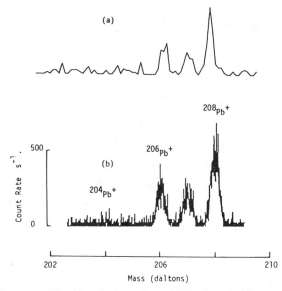

Figure 6.12. Spectrum of lead in ashed rice flour. (*a*) Portion of wide-range spectrum over lead isotopes, effective integration time 6 s. (*b*) Narrow scan over lead isotopes, 1 min integration. Sample dissolved in 1% HNO_3 + 3% HCl. Lead concentration 2.8 ng/mL; concentration of ^{204}Pb is 0.04 ng/mL.

the wider scan. The concentration of ^{204}Pb in this sample is approximately 40 pg/mL.

Organic materials are frequently analyzed by other atomic spectrometric methods such as ICP-AES and AAS, and many routine methods of sample preparation, usually involving preliminary ashing, have been developed. Many of these methods are directly suitable for analysis using the ICP-MS, but whenever possible it is preferable that the final solution be made in nitric acid at the 1% level for reasons explained in Section 6.

7.4. Measurement of Isotope Ratios

In the case of a quadrupole mass analyzer, using a channel electron multiplier as an ion detector, it is in principle only necessary to measure peak areas for the isotopes concerned to determine the isotope ratios. This can be done in two principal ways. A mass range can be set to cover the necessary span and the spectrometer scanned continuously across this or, alternatively, using the peak-switching mode, isotope integrals can be accumulated for each in turn, the cycle being repeated regularly to minimize the effect of plasma noise. The ICP is a relatively noisy ion source, the noise spectrum containing a number of components in the range of 0.1–10 Hz so that in order to avoid the influence of such fluctuations in signal it is necessary either to scan across the peaks with a repetition rate higher than the main noise frequency or to use a relatively long dwell time on each peak in turn. Both of these methods are used in practice; instruments based on the work at Surrey are able to scan at rates up to 100 s^{-1}, and such rates are used in these instruments for isotope ratios (24, 30). Douglas and his group, using an instrument with a slower data transfer rate, have made use of the longer peak integration times possible with peak switching (28, 31), typically using 2 s per isotope. Results, which are given below, are similar for either method.

The obtainable precision ranges between 0.1 and 1% on a series of repeated runs. At low sample concentrations, precision is limited by counting statistics, an integral of more than 10^6 counts being required for 0.1% precision. This can usually be accumulated for major isotopes from elemental concentrations of about 1 μg/mL in a few minutes. In the pulse-counting mode, improved statistics cannot be increasingly obtained by using higher analyte concentration, since at rates of above $5 \times 10^5 \text{ s}^{-1}$ counting loss in the largest peak becomes significant and accuracy becomes impaired. A larger integrated count may, of course, be obtained by using longer periods, but since statistical precision increases only with the square root of integrating time the practical upper limit is ~5 min. It would also seem advisable to scan over only the isotopes of interest to ensure that the maximum use is made of counting time, but again it is found that precision does not improve above the range 0.1–1%. Hence, the loss of precision obtained by using lower concentrations and wider scans is less than would be expected on purely statistical grounds. It is thought that fatigue or hysteresis effects occur in the channel multiplier if it is required to operate at high gain and high counting rates for a large proportion of the available integrating time, which occurs with narrow scans.

As with other types of mass analyzers, some bias exists in the measured intensity ratios, and it is usual to run certified isotopic materials to correct for this. Most of the work reported so far has been on lead, for which NBS standard SRM 981 can be used as it is close in isotopic composition to natural lead. A set of measurements on the Surrey instrument on this standard gave the results shown in Table 6.9 on a solution at 1 μg/mL. A total of 12 runs of 1 min each were made, and the total measurement time including direct printout of the ratios was approximately 20 min. The mean values for the three ratios show some bias from the certificate values, the absolute values of bias being approximately the same as the

**Table 6.9. Lead Isotope Ratio Measurements
on NBS SRM 981**

	204/206	207/206	208/206
Certificate values	0.05904	0.9146	2.168
Mean measured value	0.05965	0.9179	2.159
Bias (%)	+1.03	+0.36	−0.40
Experimental RSD (%)	1.29	0.35	0.32
Count rate RSD (%)	1.10	0.38	0.31

experimental relative standard deviations. These in turn are very similar to the relative standard deviation expected from counting statistics.

The value of bias determined in this way was used by Date in the determination of lead isotope ratios in a series of galena concentrates provided by the Canadian Geological Survey, for which the isotope ratios had previously been determined by thermal ionization mass spectrometry. For the ICP determination, the samples were first dissolved in dilute HCl and then evaporated to dryness, the final solution being prepared in 1% HNO_3 at ~5 μg Pb per milliliter.

A scan range of 203–210 daltons was set using 1024 channels with a scan rate of 10 s^{-1}. A total integration time of 5 min was used for each sample, allowing a few minutes equilibration time for each new sample. The NBS standard 981 was run in the same way at the same concentration, six similar integrations being used. The total integrated count was sufficient to provide a counting precision of about 0.5% for the ^{204}Pb isotope, which largely controls the overall precision.

The values obtained for five of these samples are shown in Table 6.10, together with the values determined by thermal ionization (TIMS) and the difference δ (in percent) between the results from the two methods. There is good agreement between the two methods.

Similar results have been reported by Smith et al. (31), who used an Elan (Sciex) instrument for a similar series of Canadian samples that included several of those analyzed by Date. These authors used an integration time of 10 min per sample at a concentration of about 40 μg/mL, consuming between 1 and 2 mg for each sample measurement, more than 10 times that used for Table 6.10.

Work on the simultaneous determination of multielement concentrations and isotope ratios has also been reported by Date (32). In spite of the wide scan required for this, giving an integration time on each of the lead isotopes at the top end of the spectrum of only about 0.5 s, agreement with thermal ionization determination is within 0.7% in the worst case for an integration time of 1 min on a solution at 10 μg/mL. A suite of 30

Table 6.10. Lead Isotope Ratios in Canadian Lead Mineral Samples[a]

Sample	206/204			207/204			208/204		
	ICP-MS	TIMS	$\bar{\delta}$ (%)	ICP-MS	TIMS	$\bar{\delta}$ (%)	ICP-MS	TIMS	$\bar{\delta}$ (%)
SP 1631	13.40	13.35	0.37	14.55	14.50	0.34	33.17	33.19	0.06
SYA 78 324	14.71	14.63	0.55	15.04	15.00	0.27	34.27	34.10	0.50
SP 3991	18.38	18.28	0.55	15.70	15.66	0.26	38.30	38.16	0.37
TQ 65-14	20.37	20.35	0.10	15.85	15.85	0.00	39.59	39.74	0.38
TQ 70-150	25.74	25.82	0.31	17.25	17.32	0.40	40.38	40.31	0.17
NBS SRM 981 (certificate)		16.94			15.49			36.72	

[a] $\bar{\delta}$ = absolute value of difference between ICP-MS and TIMS values, as percentage of the latter. TIMS values from R. I. Thorpe, Geological Survey of Canada (personal communication) except for NBS certificate value on SRM 981.

sample solutions were run in this way to determine isotope ratios in a total time of less than 1 h. The use of pathfinder elements in the location of ore deposits is a potentially important application of this (31).

The ability to determine isotope ratios to reasonable precision of better than 1% as rapidly as this opens up the possibility of using isotope ratios in survey analysis where extensive series of samples must be analyzed. It makes practicable both the use of enriched isotopic mixtures for isotope dilution analysis, a method of high accuracy for ultratrace concentration determination (33), and the use of artificially enriched stable isotopes in tracer studies (34). The latter technique offers considerable advantages to the research worker in, for example, metabolism studies in animals and humans, since the use of radioactive materials with the attendant risk and half-life problems is avoided.

Hitherto the measurement of isotope ratios has been a relatively slow process, involving sophisticated sample preparation routines, although within these limitations extremely high accuracy could be achieved. This new technique offers moderate accuracy together with multielement concentration determination in an analysis time of a few minutes per sample and is expected to lead to a considerable extension of the use of isotope ratio measurements in the future.

8.   ALTERNATIVE METHODS OF SAMPLE INTRODUCTION

The preceding discussion has been entirely concerned with the analysis of solution samples introduced by pneumatic nebulizer. In spite of their low efficiencies, these nebulizers are almost universally used with ICPs because of their reliability and convenience. A number of workers have substituted ultrasonic nebulizers, which, because of the better droplet size distribution, allow an increased loading of sample into the plasma. This can give up to a tenfold increase of response to the analyte but only at a similar increase in matrix effects. Since the blank response is often determined by the matrix, little advantage may arise in practice.

Continuous nebulization of sample solutions at the rate of 1–2 mL/min for analyses that may require several integrations of 1 min or more in length is quite acceptable in many circumstances where there is adequate sample available, such as in many mineralogical analyses. In some situations, however, this is not acceptable, and a method of introducing small samples is needed. Not all samples are soluble without unacceptable contamination, however, and a number of alternative methods have been investigated for the direct introduction of solid samples.

A review of methods of sample introduction to the ICP for atomic emission analysis has been given by Fassel (35), and many of these methods are directly transferable to ICP-MS. Three have already been explored in conjunction with the ICP ion source and are briefly described in Sections 8.1–8.3.

8.1. Electrothermal Volatilization

The ICP requires sample to be introduced in a gas stream as a vapor or as a finely divided aerosol of droplets or microparticulates. A small volume of liquid or solid sample may be introduced to the gas stream as vapor by flash evaporation from a filament or boat heated by a current pulse. A system of the type used for graphite furnace atomic absorption analysis was used by the author (24) for solution samples of 5–10 μL, and microsample introduction has also been reported from a rhenium filament. Detection limits of a few picograms were obtained for several elements. The sample vapor is produced in a period of a few seconds and is transferred to the plasma as a plug of gas. If a fast scanning system is available, a complete spectrum may be obtained consisting of 10 or more full scans taken within this time. A fast scan also enables isotope ratios to be measured on such minute samples. Alternatively, for highest sensitivity only a single ion peak could be monitored. This method may be particularly convenient when, in solution nebulization, interference is experienced from one of the background peaks, such as the interference of $^{32}O_2$ with ^{32}S (24).

8.2. Flow Injection

For small samples of more than the few microliters conveniently handled by electrothermal volatilization, an alternative method using the flow injection technique was first applied to mass spectrometry by Houk and Thompson (36). They were extracting ions from the ICP in the boundary-layer mode using small apertures, the first ion extraction method used with the ICP (37). A continuous-flow ultrasonic nebulizer was used to produce an aerosol of reference blank solution. The sample solution volume between 50 and 200 μL was injected into this flow so that a plug of sample was transported to the nebulizer. A burst of signal lasting about 20 s was obtained for each injection. Successive injections provided relative standard deviations between 6 and 10%. A mass analyzer and multichannel scaler capable of high scan rates were used, making it possible to determine isotope ratios with precisions of 1–3% for magnesium and nickel.

More recently, using the continuum-flow ion extraction technique,

McLeod and Date reported the use of flow injection for the analysis of blood serum (32) using 200-μL samples. A synthetic serum standard was used that contained 0.1 μg/mL cadmium and 1500 μg/mL sodium, providing a precision on seven successive shots of 0.45% RSD. The duration of the response to each injection was about 20 s, thus affording ample time for multielement analysis or the determination of isotope ratios.

Sample introduction by flow injection is thought to be a particularly useful method for the analysis of samples containing a high matrix concentration. The upper limit of dissolved solids concentration tolerable for continuous nebulization is determined largely by solids condensation on the tip of the sampling aperture. Introducing the sample into the normal nebulizer uptake by flow injection ensures that the high matrix is present only during the period of signal accumulation, thus reducing solids deposition to the minimum.

8.3. Laser Ablation of Solid Samples

One of the most attractive methods of introducing solid samples to the ICP is laser ablation into the injector gas flow. The sample may be nonconducting and remains at atmospheric pressure, so minerals and metals may be analyzed with equal ease (38). A ruby laser operating in the normal mode (694 nm) was used at pulse energies of up to 1.5 J and repetition rates up to 1 Hz. Samples were in the form of disks 32 mm in diameter by 3 mm thick. Mineral sample compacts of this size were prepared by pelletting the material, after grinding to 300 mesh and mixing with 20% of a binder, into a standard aluminum cup of the type used for X-ray fluorescence analysis. Samples were mounted on a small turntable in a borosilicate glass housing with their surface at the focus of the laser beam. The plasma injector flow was passed over the sample surface and then through a plastic tube 0.8 m long to the ICP torch. A single laser shot of 0.5 J on the sample produced a pit about 0.5 mm in diameter and 0.5 mm deep, removing about 0.2 mg of material.

If the mass analyzer is set to a single ion mass, each laser pulse produces a transient response very similar to that obtained from electrothermal volatilization or flow injection. With the length of tube used, the response to each pulse persisted for about 15 s and successive shots could be made at intervals of about 30 s. The precision obtained on 11 successive shots on a mineral sample, using a fresh surface for each shot, was 6.3% RSD when monitoring ^{24}Mg at a concentration of 2.85%.

For multielement analysis or isotope ratio determination, the duration of the response peak is sufficient for the completion of many scans at the usual rate of 10 s^{-1}. The spectra shown in Figs. 6.13 and 6.14 illustrate

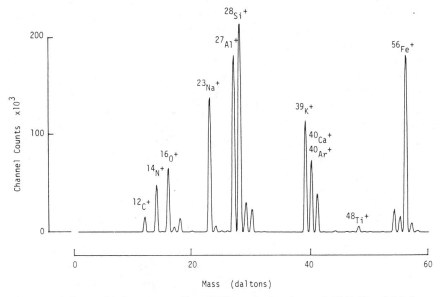

Figure 6.13. Laser ablation spectrum from BGS standard rock sample PN1. Two 0.5-J shots integrated together. Major atomic component concentrations (%): Na, 5.2; Mg, 0.21; Al, 6.1; Si, 26.3; K, 5.3; Ca, 1.4; Ti, 0.18; Mn, 0.15; Fe, 4.6.

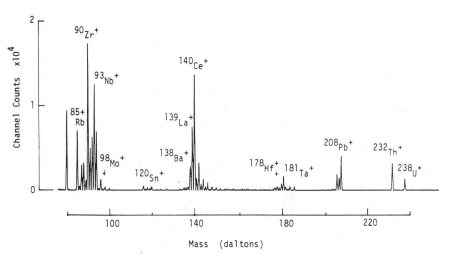

Figure 6.14. Trace elements in BGS standard PN1 sample from 10 laser shots of 0.5-J integrated together. Trace concentrations (μg/g): Zn, 150; Ga, 21; Cu, 20; Rb, 220; Sr, 200; Zr, 700; Nb, 150; Mo, 12; Sn, 8; Ba, 120; La, 150; Ce, 130; Pb, 20; Th, 20; U, 3.

the possibilities of this method of sample introduction on a British Geological Survey standard rock sample, phonolite 1 (PN1). In Fig. 6.13 the spectrum over the range 0–60 daltons obtained from two 0.5-J laser shots is shown. Major constituents are clearly seen in the trace up to ^{56}Fe. Trace elements at higher masses may be seen in more detail in Fig. 6.14, where the spectrum obtained from 10 shots is shown on an expanded vertical scale. Sensitivity may be appreciated from the concentrations given; tin, for example, clearly displayed in the spectrum, is present at 8 μg/g, with lead and thorium at 20 μg/g. On narrow scans the detection limit using 10 shots is found for many elements to be below 10 ng/g. On this basis the method compares well in sensitivity with alternatives and merits further investigation.

9. CONCLUSIONS

The use of an atmospheric-pressure ICP as an ion source for elemental analysis by quadrupole mass spectrometry is as yet a very young technique, only dating from 1982 in its present form even in the research laboratory. Commercial instruments are now coming into service in increasing numbers, but as yet little independent assessment of its strengths and weaknesses has been made. Nevertheless the attractive features of simple, rapid, and flexible sample introduction, wide element coverage at high sensitivity, and simple data readout and interpretation have attracted a lot of attention, especially among analysts who have hitherto had little experience with mass spectrometry. Even at this early stage, however, the first users report that it is capable of analyses that were previously impracticable.

In its present form the technique does not represent any final stage of development; its current performance has been achieved largely by adopting plasma and mass analysis methods developed for other purposes. There is no doubt that as its use expands new methodologies will be developed in place of the modified ICP-AES techniques used at present and that as user feedback influences instrument development improvements in performance, stability, and convenience will be obtained. The most novel area is the interface between plasma and vacuum system, and the evolution of this since the first design has already resulted in a considerable decrease in molecular interferences (10). However, for the immediate future the most significant developments will undoubtedly be in applications methodology.

A significant impetus to the acceptance of the technique is likely to come as new manufacturers introduce instruments. At least two more are likely to reach the marketplace in the next year or two. It seems very likely that the hybrid nature of the method will bring the advantages of mass spectrometry to the notice of an increasingly larger number of analysts who have regarded it in the past as essentially a research technique. It is hoped that this will contribute significantly to a revival of interest in inorganic mass spectrometry as a hitherto neglected but very powerful tool in the hands of the analytical chemist.

ACKNOWLEDGMENTS

The work at the University of Surrey and at the British Geological Survey is supported by the British Geological Survey (NERC) and the Metals and Mineral Substances Programme of the European Community under contract number MSM 104 UK(H).

This chapter is published by permission of the Director, British Geological Survey (NERC).

REFERENCES

1. V. A. Fassel, *Pure Appl. Chem.,* **49,** 1533–1545 (1977).
2. S. Greenfield, H. McD. McGeachin, and P. B. Smith, *Talanta,* **23,** 1–14 (1976).
3. R. M. Fristrom, *Int. J. Mass Spectrom. Ion Phys.,* **16,** 15–32 (1975).
4. A. N. Hayhurst, *I.E.E.E. Trans. Plasma Sci.,* **PS-2,** 115–121 (1974).
5. T. M. Sugden, in R. I. Reed, Ed., *Mass Spectrometry,* Academic, New York, 1965, pp. 347–358.
6. S. Greenfield, I. L. Jones, and C. T. Berry, *Analyst,* **89,** 713–720 (1964).
7. R. H. Wendt and V. A. Fassel, *Anal. Chem.,* **37,** 920–922 (1965).
8. British Patent 1,371,104 (1971). Gray assigned ARL Ltd.
9. A. L. Gray, *Analyst,* **100,** 289–299 (1975).
10. A. L. Gray, *Spectrochim. Acta,* **40B,** 1525–1537 (1985).
11. A. F. Ashton and A. N. Hayhurst, *Combustion Flame,* **21,** 69–75 (1973).
12. C. D. Keirs and T. J. Vickers, *Appl. Spectrosc.,* **31,** 273–283 (1977).
13. D. J. Douglas and J. B. French, *Anal. Chem.,* **53,** 37–41 (1981).
14. V. A. Fassel, *Science,* **202,** 183–191 (1978).
15. P. Hulmston, *Analyst,* **108,** 166 (1983).
16. J. F. Alder, R. M. Bombelke, and G. F. Kirkbright, *Spectrochim. Acta,* **35B,** 165 (1980).
17. A. L. Gray, *Spectrochim. Acta,* **41B,** 1/2, 151, (1986).

18. A. N. Hayhurst, D. B. Kittelson, and N. R. Telford, *Combustion Flame,* **28,** 67–80, 123–135, 137–143 (1977).

19. C. A. Stearns, F. J. Kohl, G. C. Fryburg, and R. A. Miller, Nat. Bur. Stand. Spec. Publ. 561, Vol. 1, 1979, pp. 303–355.

20. F. T. Greene, J. E. Beachey, and T. A. Milne, Nat. Bur. Stand. Spec. Pub. 561, Vol. 1, 1979, pp. 431–442.

21. A. R. Date and A. L. Gray, *Spectrochim. Acta,* **38B,** 29–37 (1983).

22. D. J. Douglas, *Can. Res.,* **16,** 55–60 (1983).

23. A. R. Date and A. L. Gray, *Analyst,* **108,** 159–165 (1983).

24. A. L. Gray and A. R. Date, *Analyst,* **108,** 1033–1050 (1983).

25. Plasma Quad, V. G. Isotopes Ltd., Winsford, Cheshire, CW7 3BX, England.

26. A. R. Date and A. L. Gray, *Spectrochim. Acta,* **40B,** 115–122 (1985).

27. A. L. Gray, *ICP Inf. Newslett.,* **10,** 200–202 (1984).

28. D. J. Douglas and R. S. Houk, *Prog. Anal. Atom. Spectrosc.,* **8,** 1–18 (1985).

29. S. Abbey, Geological Survey of Canada, Papers 80-14 and 83-15.

30. A. R. Date and A. L. Gray, *Int. J. Mass Spectrom. Ion Phys.,* **48,** 357–361 (1983).

31. R. G. Smith, E. J. Brooker, D. J. Douglas, E. S. K. Quan, and G. Rosenblatt, *J. Geochem. Explor.,* **21,** 385–395 (1984).

32. A. R. Date, *ICP Inf. Newslett.,* **10,** 202–206 (1984).

33. K. G. Heumann, *Trends Anal. Chem.,* **1,** 357 (1982).

34. E. R. Klein and P. D. Klein *Stable Isotopes,* Academic, New York, 1979.

35. V. A. Fassel, *Anal. Chem.,* **51,** 1290A, 1979.

36. R. S. Houk and J. J. Thompson, *Biomed. Mass Spectrom.,* **10,** 107 (1983).

37. R. S. Houk, V. A. Fassel, G. D. Flesch, H. J. Svec, A. L. Gray, and C. E. Taylor, *Anal. Chem.,* **52,** 2283–2289 (1980).

38. A. L. Gray, *Analyst,* **110,** 551–556 (1985).

CHAPTER

7

ISOTOPE DILUTION MASS SPECTROMETRY

KLAUS G. HEUMANN

University of Regensburg,
Regensburg, Federal Republic of Germany

1. INTRODUCTION

The production of accurate results is a common problem for all trace and microanalyses. Normally, the variation in the analytical results of different laboratories or of different methods is much greater with lower concentrations of the element to be determined. Normally, the coefficient of variation is also higher for multielement methods than for mono- or oli-

301

Table 7.1. Hierarchy of Analytical Methods with Respect to Accuracy

Analytical Data	Analytical Method
True value	No method known
Definitive value	Definitive method, e.g., IDMS
Reference method value	Reference method
Assigned value	Routine method

goelement methods. There are basically two ways to obtain accurate results in micro and trace analyses. One can either use standard reference materials for calibrating analytical methods, in which case, the restricted number of reference materials that are available for the different analytical problems limits the calibration, or else one can use highly accurate methods. In connection with accuracy, a hierarchy of analytical methods can be established as shown in Table 7.1 (1, 2). There is no known analytical method that is able to produce the "true value" in all cases. Therefore, only a more or less accurate approximation of the true value is the best that can be achieved.

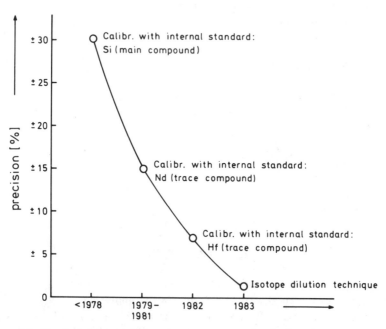

Figure 7.1. Uranium determination at nanogram per gram and microgram per gram levels in geochemical samples by spark source mass spectrometry (10–12).

Definitive values can be obtained by definitive methods. One of the best-known definitive methods is isotope dilution mass spectrometry (IDMS). Today, IDMS is used for element determination (3–7) as well as for molecular analyses, especially in clinical chemistry (8, 9).

IDMS can be applied with most of the ionization methods used in mass spectrometry. Mass spectrometry is a sensitive method, but highly accurate and precise results can be obtained only with the isotope dilution technique. Jochum and coworkers (10–12) compared the precision by various calibration methods for spark source mass spectrometric determinations of uranium in geochemical samples and obtained the results shown in Fig. 7.1. Before 1979, the determination of uranium traces was calibrated by relative sensitivity factors using the silicon content of the geological sample as an internal standard; the precision was approximately ±30%. The precision was increased to approximately 15% and 7% by calibrating with the trace elements neodymium and hafnium, respectively. However, the best results with relative standard deviations of 1–2% were achieved when uranium was determined by IDMS.

2. PRINCIPLES OF IDMS

2.1. Isotope Dilution Technique

The principles of IDMS are illustrated by the schematic mass spectrum of a lead analysis shown in Fig. 7.2. A known quantity of a spike isotope, normally a stable isotope with minor natural abundance, is added to the sample. In some cases, a long-lived radioisotope is applied. Enriched

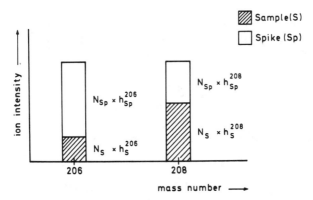

Figure 7.2. The principle of IDMS illustrated by the schematic mass spectrum of lead (2).

^{206}Pb is used in the example shown in Fig. 7.2. After the spike isotope is added, the sample and the spike isotopes must be mixed completely. This implies that a solid sample must be chemically decomposed. With homogeneous solutions, one has to make sure that a total isotopic exchange between the sample and the spike isotopes takes places. In solutions, only a few metal ions have very slow exchange rates between the free cation and complex-forming agents (13, 14); hence, most of the metal ions to be determined are quickly diluted by the spike isotope if there is no stable organometallic compound present. In the case of nonmetals, often no isotope exchange reaction exists between different species of the same element. Therefore, either one must change the conditions in the solution in order to carry out an isotope exchange for the total element determination or one can use such a system for the species analysis of elements (see Section 5.5). When a complete isotopic exchange has taken place between sample and spike, the loss of substance during the isolation procedure normally has no effect on the analytical result.

After the isotope-diluted element is isolated, the isotope ratio R is measured in the mass spectrometer. The two isotopes selected for the measurement of R are the spike isotope and, normally, the isotope with the highest abundance in the sample (reference isotope) if there is no interference from an isobaric nuclide of another element (see Section 2.4). The ratio $R = {}^{208}$Pb$/{}^{206}$Pb is measured in the example of lead analysis. As can be seen in Fig. 7.2, the ion intensities of the lighter (index 1) and heavier (index 2) isotopes in the mass spectrum are identical to the sum of the sample portion and the spike portion. Hence, Eq. (1) follows:

$$R = \frac{N_S h_S^2 + N_{Sp} h_{Sp}^2}{N_S h_S^1 + N_{Sp} h_{Sp}^1} \qquad (1)$$

where N is the number of atoms, h the isotope abundance (%), and subscripts S and Sp indicate sample and spike, respectively.

If one solves Eq. (1) for N_S, it follows that

$$N_S = \frac{N_{Sp}(h_{Sp}^2 - R h_{Sp}^1)}{R h_S^1 - h_S^2} \qquad (2)$$

After transformation of Eq. (2) for the content G_S of the sample, one obtains

$$G_S = 1.66 \times 10^{-18} \frac{M}{W_S} N_{Sp} \left(\frac{h_{Sp}^2 - R h_{Sp}^1}{R h_S^1 - h_S^2} \right) \qquad [\mu g/g] \qquad (3)$$

where M is the atomic weight of the element to be determined and W_S the sample weight (g).

For monoisotopic elements, h_S^1 or h_S^2 becomes zero. In this case an artificial long-lived radioactive isotope can be used as a spike. The determination of iodine by IDMS is an example. In this analysis an ^{129}I-enriched spike is applied, whereas natural iodine consists fully of ^{127}I ($h_S^1 = 100\%$; $h_S^2 = 0\%$). Then Eq. (3) can be reduced to

$$G_S = 1.66 \times 10^{-20} \frac{M}{W_S} N_{Sp} \left(\frac{h_{Sp}^2}{R} - h_{Sp}^1 \right) \qquad (4)$$

Similar derivations are possible for all other combinations of isotopes, such as when the sample element consists of two or more isotopes and the spike consists of one isotope. An example is the analysis of natural uranium with a pure ^{235}U or ^{233}U spike.

Equations (3) and (4) do not take the blank correction into consideration. Therefore, in all analyses the blank has to be determined in the same way as the sample. One of the advantages of IDMS is that small amounts of a blank as well as its variation can easily be determined because the analytical procedure can be followed by the spike isotope. The detection of a blank is limited by the precision of the isotope ratio measurement of the spike (see Section 2.3).

In most analyses, the sample isotope abundances h_S are identical with the tabulated natural abundances of the elements (15). If there is a sample with unknown isotopic composition of the elements to be determined, the isotope abundances h_S must be assessed separately. This has to be taken into consideration especially in the analysis of samples that have been irradiated, for example, in a nuclear reactor. Possible isotope variations in nature are very small for most of the elements, so they do not usually influence the result obtained by IDMS using the tabulated isotope abundances of the elements. H, B, C, N, O, S, Ar, Ca, Sr, and Pb are important exceptions in nature (16). For argon and calcium this is caused by the decay of the natural radionuclide ^{40}K, for strontium by the decay of ^{87}Rb, and for lead by the decay of uranium and thorium. Therefore, this effect has to be taken into consideration for geological samples that are enriched in potassium, rubidium, uranium, and thorium. The influence of natural isotope variations on IDMS results for hydrogen, carbon, nitrogen, and oxygen is small because the heavier isotope of the elements always has a very low abundance h_S^2. One can see from Eq. (3) that the denominator of the term in parentheses can be influenced by an isotope variation only if the R value is very small. Under conditions normally used in IDMS, this is not the case, because optimum R values are near 1 (see Section

**Table 7.2. Relative Analytical Error with Respect to Possible
$^{13}C/^{12}C$ Isotope Variations in Nature[a]**

$^{13}C/^{12}C$ Variation Relative to Tabulated Values (%)	Relative Analytical Error (%)				
	$R = 0.1$	$R = 0.5$	$R = 1$	$R = 2$	$R = 5$
0.5	0.05	0.01	0.01	0.01	0.01
1	0.14	0.03	0.03	0.02	0.01
2	0.27	0.07	0.05	0.03	0.03
3	0.41	0.10	0.07	0.05	0.04
4	0.55	0.14	0.09	0.07	0.05

[a] Calculation on the basis of tabulated values: $h_S^{13} = 1.1\%$, $h_S^{12} = 98.9\%$ (15).

2.2). In determining small element amounts in the sample or for blank analyses, an excess of the spike with the enriched heavier isotope is used, so that the R value increases compared to the normal situation. In Table 7.2, relative analytical errors (rounded values) are listed for different R values depending on possible $^{13}C/^{12}C$ variations, assuming that the real $^{13}C/^{12}C$ ratio in the sample deviates by the percentage given in the first column from the tabulated isotopic composition of carbon (15), which is used for the calculation of IDMS results. In nature, extreme variations of up to 4% in the isotopic composition of carbon are possible (16). In this case, the relative error is about 0.5% for $R = 0.1$, whereas for $R \geq 1$ the error is always less than 0.1%. This means that the error in IDMS caused by isotope variations in nature is negligible for double-isotope elements when one isotope has a high abundance and the other a very low abundance. In contrast to this, the isotopic composition of elements of comparable isotope abundances and showing isotope variations in nature must be determined.

The determination of lead, which is becoming more and more important in IDMS (see Section 5.4), is an example of an analysis where inaccurate results are obtained if the real isotope abundances of the sample are not taken into consideration. Table 7.3 gives lead isotope ratios of a selection of samples (17–21). Common lead consists of the isotopes ^{204}Pb, ^{206}Pb, ^{207}Pb, and ^{208}Pb. The isotopes ^{206}Pb, ^{207}Pb, and ^{208}Pb are formed as end products of the radioactive decay of uranium and thorium. Therefore, ores that contain significant amounts of uranium show higher $^{206}Pb/^{204}Pb$ ratios than those with low uranium levels. A tremendous isotope shift can be seen in the case of a phosphate ore, and a significant shift in common lead ores. The values listed in Table 7.3 are extreme values found in the

Table 7.3. Lead Isotope Ratios in Selected Samples

Sample	Location or Origin	^{206}Pb/^{204}Pb	^{207}Pb/^{206}Pb	^{208}Pb/^{206}Pb	Ref.
Ore	Phosphate ore, Morocco (SRM,* BCR 32)	39.24	0.429	0.977	17
	Lead ores				
	Missouri	21.78	0.772	1.872	18
	Broken Hill, Australia	17.04	0.884	2.179	18
	Idaho	16.45	0.951	2.210	18
Coal mine	Hopkins County, KY	19.72	0.799	1.964	18
	Niederberg, Germany	18.25	0.838		19
	Foster Creek, MT	17.64	0.888	2.128	18
Gasoline additives	Houston, TX (1970)	19.43	0.820	2.012	18
	Hong Kong (1968)	17.00	0.917	2.168	18
Aerosol	San Diego, CA (1974)	19.02	0.826	2.028	18
	Helsinki, Finland (1979)	17.39	0.891	2.132	20
Soil	Chester, SC (1968)	19.43	0.809	2.008	18
	Würzburg, Germany (1985)	18.61	0.865	2.107	17
	San Francisco, CA (1968)	17.49	0.888	2.138	18
Sewage sludge	SRM, BCR 144, and BCR 145	18.46	0.865	2.092	17
Plants	Water plant (SRM, BCR 60)	18.19	0.856	2.089	21
	Olive leaves (SRM, BCR 62)	17.73	0.865	2.095	21
Chemicals	Lead standard solution (EGA, 1983)	19.29	0.812	2.019	21
	Lead standard solution (Merck, 1983)	17.34	0.893	2.146	21

[a] SRM = Standard reference material

Table 7.4. Advantages and Disadvantages of IDMS

Advantages	Disadvantages
Precise and accurate analysis	Destructive
Nonquantitative isolation of the substance to be analyzed	Chemical preparation of sample necessary
Ideal internal standardization	Time-consuming
Multielement as well as oligo- and monoelement analyses possible	Relatively expensive
High sensitivity with low detection limits	

literature. The isotope ratio of lead products thus depends on the origin of the ore from which they have been synthesized (see additives of gasoline and chemicals in Table 7.3). The lead isotope ratio of environmental samples is necessarily influenced by the emission of exhaust gases from motor vehicles and from the combustion of coal.

The advantages and disadvantages of IDMS are summarized in Table 7.4. The high precision and accuracy can be attributed to the fact that no quantitative isolation of the elements to be determined has to be carried out. A quantitative isolation of trace or micro amounts of an element or its compound is very often a problem in separation processes. On the other hand, a stable isotope used as a spike is the best internal standard to apply. Multielement determinations can be carried out by IDMS with mass spectrometric multielement methods such as spark source mass spectrometry and ICP-MS (see Section 3), whereas oligo- and monoelement analyses are possible with thermal ionization, electron impact, and field desorption mass spectrometry. High sensitivities can be obtained for all elements if one selects the optimum ionization method. This yields detection limits for trace elements in all matrices below the microgram per gram level.

The two major disadvantages of IDMS are its destructive nature and the fact that if the matrix is not a homogeneous solution the sample must be chemically treated in order to carry out the spiking process. The method is also more time-consuming than a number of other analytical methods, for example, those that need no chemical preparation. IDMS is relatively expensive with respect to instrumentation and mass spectrometric know-how. However, the costs of spike isotopes are, in most cases, no longer a limiting factor.

2.2. Optimization of Spike Addition

Eqs. (1) and (2) make clear that the uncertainty of an IDMS analysis is mainly caused by:

Errors in the determination of the spike concentration.

Isotope fractionation effects in the ion source of the mass spectrometer during the measurement of R and spike isotope abundances h_{Sp}.

A reverse isotope dilution technique with standard solutions of natural isotope abundances is often used for the determination of the spike concentration. This can be performed on minute amounts or very small spike concentrations. If a pure spike solution is applied, interferences caused by other elements are negligible for the reverse-IDMS technique as well as for other analytical methods. The precision and accuracy of these spike analyses are generally better than 0.3% (22–24). The isotope fractionation in an ion source normally ranges within a few parts per thousand, as long as a high molar fraction of the sample amount has not evaporated (25–27). Such a fractionation is small compared to other uncertainties in the analysis. Still, analytical results influenced by this effect can be corrected by means of either synthetic isotope mixtures or established isotopic reference materials (28). As an alternative, one can also measure R, h_S, and h_{Sp} under the same mass spectrometric conditions, so that systematic errors caused by isotopic fractionation neutralize each other [see Eq. (2)]. Statistical errors in spike addition can be reduced by weighing the spike solution and avoiding transfer of the solution with a pipet. The statistical error in the determination of the isotopic abundances of the spike can be decreased by multiple measurements.

Under the assumption that the isotopic abundance errors of the sample and the spike do not contribute significantly to the standard deviation of the analytical result, one obtains the approximation equation for the standard deviation of N_S by the laws of propagation of errors (22, 29):

$$s^2(N_S) \approx s^2(N_{Sp}) + f^2(R)s^2(R) \tag{5}$$

where s is the standard deviation and $f(R)$ is the error multiplication factor for the R value.

Whereas the standard deviation of N_{Sp} for the amount of spike added can be ignored under the above-mentioned conditions, any statistical error in the isotope ratio measurement R is increased by the error multiplication factor $f(R)$. Hence, an essential precondition for a precise analytical result to be obtained by IDMS is the precise determination of R. It follows (29) that

$$f(R) = \frac{[(h^2/h^1)_S - (h^2/h^1)_{Sp}]R}{[R - (h^2/h^1)_S][(h^2/h^1)_{Sp} - R]} \tag{6}$$

If the amount $f(R)$ becomes minimum, the optimum value of R (R_{opt}) is reached; that is, taking the differential of Eq. (6),

$$R_{opt} = [(h^2/h^1)_S(h^2/h^1)_{Sp}]^{1/2} \tag{7}$$

Using Eqs. (1) and (6), one can calculate the error multiplication factor using the atomic ratio N_S/N_{Sp} for different isotopic enrichments of the spike. The diagrams for a chloride analysis using a ^{37}Cl-enriched spike and for an iron analysis using a ^{57}Fe-enriched spike are shown in Figs. 7.3 and 7.4, respectively. Curves similar to those shown in Figs. 7.3 and 7.4 are obtained for all analyses where bi-isotopic and polyisotopic elements are determined using a bi-isotopic or polyisotopic spike. The diagrams clearly demonstrate that with a higher isotopic enrichment of the spike the error multiplication factor decreases and differs much less from the minimum over a large range of the atomic ratio N_S/N_{Sp}.

When the element to be determined or the spike solution is monoisotopic, for example, iodine, a linear relation is obtained between the error multiplication factor and the ratio N_S/N_{Sp}. Nevertheless, using IDMS as an analytical method, one should calculate the range for the optimum N_S/N_{Sp} ratio and try to realize this range in the analysis to minimize the

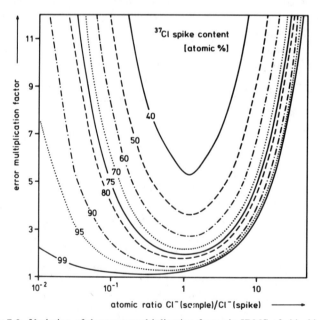

Figure 7.3. Variation of the error multiplication factor in IDMS of chloride (32).

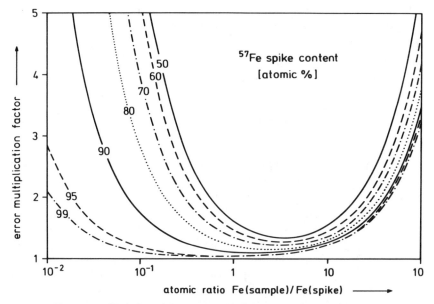

Figure 7.4. Variation of the error multiplication factor in IDMS of iron.

statistical error of the result. Additional calculations of the optimum spike addition for different elements and compounds are given in the literature (4, 22, 30–33). These do not take into account the fact that the highest precision in isotope ratio measurement can usually be achieved for ratios of approximately unity.

2.3. Limitations of IDMS

The detection of traces of elements by IDMS is limited mainly by blank problems, by the precision of the isotope ratio measurement, and by the minimum amount of the element to be isolated for a successful mass spectrometric measurement. The detection limit in IDMS is usually caused by the variation in the blank. This is especially valid for elements for which contamination is a critical problem during separation, as for lead and thallium determinations in biological samples (34) and for boron in metals (23) in the nanogram per gram level. Usually, the detection limit is defined as

$$\text{Detection limit} = 3s(\text{blank}) \tag{8}$$

where $s(\text{blank})$ is the standard deviation of the blank.

When the blank does not influence the detection limit, a minimum amount of the element to be determined has to be present in the sample. This amount must be high enough to change the isotope ratio of the spike significantly, which means that the isotope ratio R of the isotope-diluted sample must fulfill the following condition if h^1 is the spike isotope:

$$R \geq \left(\frac{h^2}{h^1}\right)_{Sp} + 3s(\text{spike}) \tag{9}$$

where s(spike) is the standard deviation of the isotope ratio measurement of the spike. A relative precision better than 1% can be achieved in the isotope ratio determination for most of the ionization methods in mass spectrometry (see Section 3). The deviation of the R value from the isotope ratio of the spike increases when smaller amounts of spike are used under the same conditions, but the amount of spike cannot be reduced infinitely because every isolation process of the isotope-diluted element requires a substantial amount of substance. It ranges between a few nanograms and a few micrograms, depending on the element and the separation process used. Because the isolation yields for different separation processes are often considerably below 100%, the total amount of the isotope-diluted element is not identical with the minimum amount of substance necessary for a mass spectrometric measurement. However, the analytical result becomes less precise with an increasing ratio of amount of spike to amount of sample. This is due to the higher error multiplication factor (see Section 2.2).

With due optimization of the factors discussed above that restrict the detection limit of IDMS, one can normally determine nanogram amounts even in unfavorable systems. With optimum conditions the detection limit can be less than 1 pg.

2.4. Preparation of Spike Solutions

As mentioned in Section 2.1, the best form in which to apply a spike is a solution. Weighing the added amount normally results in a higher accuracy than adding the spike solution to the sample with a pipet (see Section 2.2). The concentration of the spike solution depends on the analytical problem to be solved. The addition of a solution amount ranging between 0.5 and 2 g is an established procedure because of the small relative weighing errors that can occur under these conditions. The stability of the spike solution with regard to concentration and isotopic composition has to be guaranteed. Spike compounds that are chemically unstable must be stablized by the addition of chemicals; for example, a nitrate spike is stabilized by addition of traces of phenylmercuric acetate. Metal spike solutions, which are usually stored in ultracleaned plastic vessels, can be acidified to prevent adsorption effects. Nevertheless, when a spike solution of very low concentration must be applied at or below the nanogram per gram level, a fresh preparation from a more concentrated stock solution is preferable.

There are 272 stable isotopes and a number of long-lived radioactive ones that, in principle, can be used as enriched spikes. Today, most of these are commercially available at a reasonable price (e.g., see Refs. 35–39). As a few micrograms or less of an enriched isotope is necessary for one analysis, the cost of the spike added to the sample does not normally restrict the use of the method. In a few cases where the element consists only of two natural isotopes, one with very high natural abundance and the other with an abundance of 0.1% or less (^{50}V, ^{138}La, ^{180}Ta), the spike costs can contribute significantly to the overall expense of the analysis.

In the selection of the spike isotope the following must be considered:

There is an advantage to using one of the isotopes with minor abundance in the sample as a spike (see Section 2.1).

If possible, the spike should not interfere with any isobaric nuclide of another element or with the mass of a compound ion.

In cases where the element to be determined contains more than one isotope in minor abundance, the spike isotope can be selected on the basis of financial criteria; for example, highly enriched ^{42}Ca or ^{44}Ca is less expensive than ^{46}Ca or ^{48}Ca.

In the analysis of monoisotopic elements, one should check to determine whether a long-lived radioactive isotope is applicable.

Table 7.5 provides a list of recommended spike isotopes taking into consideration the selection criteria discussed. The isotope with the highest abundance in the sample is normally selected as a reference isotope for the isotope ratio measurement, except when there is an interference of this isotope caused by an isobaric nuclide of another element. Possible interferences of the reference or the spike isotope with isobaric nuclides of other elements are also listed in the table. Interfering elements are shown in parentheses when a discrimination of the interfering nuclide can be achieved by the ionization process usually applied for the element to be determined (see Section 3). Half-lifes of long-lived radioisotopes of artificial or natural origin that are used as reference or spike isotopes are also listed.

Table 7.5. Recommended Spike and Reference Isotopes of Elements for IDMS

Element	Reference Isotope				Spike Isotope			
	Mass Number	Natural Abundance (%)	Half-Life (yr)	Interfering Element [b]	Mass Number	Natural Abundance (%)	Half-Life (yr)	Interfering Element [b]
H	1	99.985			2	0.015		
He	4	~100			3	10^{-4}		
Li	7	92.5			6	7.5		
Be	9	100			10	—	1.6×10^6	(B)
B	11	80.1			10	19.9		
C	12	98.90			13	1.10		
N	14	99.634			15	0.366		
O	16	99.762			18[a]	0.200		
F	19	100			—	—		
Ne	20	90.51			22[a]	9.22		
Na	23	100			—	—		
Mg	24	78.99			25	10.00		
Al	27	100			26	—	7.2×10^5	Mg
Si	28	92.23			30	3.10		
P	31	100			—	—		
S	32	95.02			34[a]	4.21		
Cl	35	75.77			37	24.23		
Ar	40	99.600		(K, Ca)	38	0.063		
K	39	93.2581			41[a]	6.7302		
Ca	40	96.941		K, (Ar)	42[a]	0.647		
Sc	45	100			—	—		
Ti	48	73.8		Ca	49[a]	5.5		
V	51	99.750			50	0.250		
Cr	52	83.789			53[a]	9.501		Ti, Cr

Element	A	%	$t_{1/2}$ (yr)	Product	A	%	$t_{1/2}$ (yr)	Product
Mn	55	100			53	—	3.7×10^6	Cr
Fe	56	91.72			57[a]	2.2		
Co	59	100			—	—		
Ni	60[a]	26.10			62[a]	3.59		
Cu	63	69.17			65	30.83		
Zn	66[a]	27.9			67[a]	4.1		
Ga	69	60.1			71	39.9		
Ge	72[a]	27.4			73	7.8		
As	75	100			—	—		
Se	80	49.7		(Kr)	82[a]	9.2	1×10^{19}	(Kr)
Br	79	50.69			81	49.31		
Kr	84	57.0		(Sr)	83[a]	11.5		
Rb	85	72.165			87	27.835	4.8×10^{10}	Sr
Sr	88	82.58			86[a]	9.86		(Kr)
Y	89	100			—	—		
Zr	90	51.45			91[a]	11.22		
Nb	93	100			92	—	3.6×10^7	Zr, Mo
Mo	95[a]	15.92			97[a]	9.55		
Tc	99	—	2.1×10^5	(Ru)	97	—	2.6×10^6	(Mo)
Ru	101[a]	17.0			99[a]	12.7		
Rh	103	100			—	—		
Pd	105[a]	22.33			102	1.02		Ru
Ag	107	51.839			109	48.161		
Cd	114	28.73	4×10^{14}	Sn	111[a]	12.80		
In	115	95.7		Sn	113	4.3		Cd
Sn	118[a]	24.22			117[a]	7.68		
Sb	121	57.3			123	42.7		Te
Te	126[a]	18.95		(Xe)	125[a]	7.14		
I	127	100			129	—	1.6×10^7	(Xe)

Table 7.5. Recommended Spike and Reference Isotopes of Elements for IDMS (*continued*)

Element	Reference Isotope				Spike Isotope			
	Mass Number	Natural Abundance (%)	Half-Life (yr)	Interfering Element[b]	Mass Number	Natural Abundance (%)	Half-Life (yr)	Interfering Element[b]
Xe	132	26.9		(Ba)	134[a]	10.4		(Ba)
Cs	133	100			135[a]	—	2×10^6	Ba
Ba	138	71.7		La	135[a]	6.592		
La	139	99.91			138	0.09	1.4×10^{11}	Ba, Ce
Ce	140	88.48			142[a]	11.08		Nd
Pr	141	100			—	—		
Nd	146[a]	17.19			145[a]	8.30		
Pm	—	—			—	—		
Sm	152	26.7		Gd	149[a]	13.8		
Eu	153	52.5			151	47.8		
Gd	157[a]	15.65			155[a]	14.8		
Tb	159	100			—	—		
Dy	163[a]	24.9			161[a]	18.9		
Ho	165	100			—	—		
Er	166	33.6			167[a]	22.95		
Tm	169	100			—	—		
Yb	174	31.8		Hf	173[a]	16.12		
Lu	175	97.41			176	2.59	3.6×10^{10}	Yb, Hf
Hf	178[a]	27.297			177[a]	18.606		
Ta	181	99.988			180	0.012	$>10^{13}$	Hf, W
W	182[a]	26.3			183[a]	14.3		
Re	187	62.60	5×10^{10}	(Os)	185	37.40		
Os	189[a]	16.1			188[a]	13.3		
Ir	193	62.7			191	37.3		
Pt	195	33.8			194[a]	32.9		

Element	Isotope	Abundance	Half-life (y)	Isotope	Abundance	Half-life (y)	
Au	197	100	—	—	—	—	
Hg	202	29.80	—	201[a]	13.22	—	
Tl	205	70.476	—	203	29.524	—	
Pb	208	52.4	—	207[a]	22.1	—	
Bi	209	100	—	210	—	3×10^6	
Po	—	—	—	—	—	—	
At	—	—	—	—	—	—	
Rn	—	—	—	—	—	—	
Fr	—	—	—	—	—	—	
Ra	—	—	—	—	—	—	
Ac	—	—	—	—	—	—	
Th	232	~100	1.4×10^{10}	230	10^{-4}	7.5×10^4	
Pa	—	—	—	—	—	—	
U	238	99.2745	4.5×10^9	235	0.7200	7×10^8	
Np	237	—	2.1×10^6	236	—	1.2×10^5	
Pu	239	—	2.4×10^4	242	—	3.8×10^5	
Am	243	—	7.4×10^3	241	—	4.3×10^2	
Cm	247	—	1.6×10^7 Bk	248	—	3.4×10^5	U

[a] Spike/reference isotope not with the lowest/highest natural abundance.
[b] Parentheses indicate that interfering isobaric nuclide can be discriminated by ionization process.

Only a small number of elements cannot be analyzed by IDMS. Thus the great analytical power of this technique is obvious.

3. IONIZATION METHODS AND THEIR PREPARATION TECHNIQUES

3.1. Survey of Methods

Precise and accurate isotope ratio measurements are a major requirement for IDMS. In principle, all ionization methods of mass spectrometry can be used to determine isotope ratios with a more or less high accuracy. The most important ionization methods with regard to element analyses are listed in Table 7.6, which is divided into two groups of ionization methods. With the methods of the first group it is possible to obtain mono- or oligoelement determinations, whereas the methods of the second group allow multielement determinations. When using thermal ionization, electron impact, and field desorption mass spectrometry, the quantitative de-

Table 7.6. Ionization Methods for Element Analyses in Mass Spectrometry

Ionization Method	Preferred Compounds
Mono- or Oligoelement Determination	
Thermal ionization	
Positive ions	Metals with low first ionization potential
Negative ions	Nonmetals with high electron affinity
Electron impact	Noble gases, H_2, CO_2, N_2, SO_2; metals after chelation and GC separation
Field desorption	Alkalis and alkaline earths
Multielement Determination	
Spark source	All elements
Plasma source (ICP, MIP, glow discharge)	All metals and semimetals
Laser	All elements (microanalysis)
SIMS	All elements (surface and microanalysis)

termination will almost invariably be carried out by the isotope dilution technique. Quantitative determinations with spark source mass spectrometry, with the different plasma sources (ICP = inductively coupled plasma; MIP = microwave-induced plasma), with laser excitation, and with secondary ion mass spectrometry (SIMS) are usually obtained using relative sensitivity coefficients for the various elements. Nevertheless, a number of analyses were also done with spark source IDMS. Up to now, the other methods applied the isotope dilution technique only in rare instances.

The principles of spark source, glow discharge, SIMS, laser, and ICP mass spectrometry are described in Chapters 2–6, and a fundamental description of thermal ionization and field desorption mass spectrometry with respect to element analysis is given in Sections 3.2 and 3.5 of this chapter. Special features of sample preparation techniques used in the isotope dilution method are discussed in later sections.

3.2. Thermal Ionization

The production of atomic or molecular ions at the hot surface of a metal filament is called thermal ionization. An ion source with a single-, double-, or triple-filament arrangement is used for the evaporation and for the ionization process. A schematic diagram of a double-filament ion source is given in Fig. 7.5. Using a single-filament ion source, the evaporation and ionization processes of the substance to be determined are carried out on the same filament surface. In the case of a double- or triple-filament ion source, one or two filaments are used for the evaporation of the compound and the gaseous sample molecule (atom) is adsorbed on the surface of the other filament; then, after an electron transfer from the molecule to the filament (production of positive ions) or by the inverse effect (negative ions), the ion can desorb from the filament surface. This process is the same with double- and triple-filament ion sources, but with the triple-filament arrangement a direct comparison of two different samples can be obtained under the same ion source conditions.

A double-filament is used instead of a single-filament ion source if it is useful to separate the evaporation from the ionization process. This is valid for the determination of all elements with which the sample compound is evaporated at low temperatures, but a sufficient ion yield is reached only at high filament temperatures, for example, for the determination of volatile selenium compounds by negative thermal ionization mass spectrometry (40). Furthermore, the isotope fractionation effect from evaporation of the sample is higher within the same period of mea-

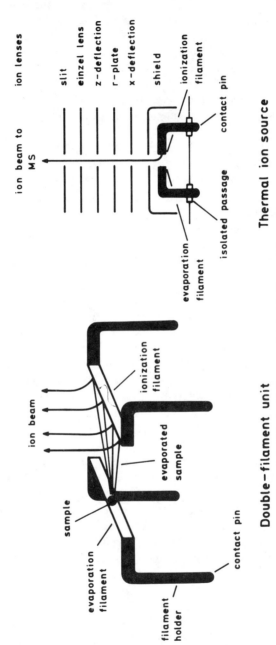

Double-filament unit

Thermal ion source

Figure 7.5. Schematic diagram of a double-filament thermal ion source.

surement for a single-filament than for a double-filament ion source as was found, for example for alkali and alkaline earth elements (25, 41).

Positive as well as negative ions can be obtained by the thermal ionization process (PTI = positive thermal ionization; NTI = negative thermal ionization). The quantitative correlations between the ion production and the physical parameters responsible for this ionization process were first considered by Langmuir in 1925 (42). Since then, many other publications have described this ionization process (e.g., 43, 44). The ion yield β, the ratio of the number of ions to the number of ions and neutral atoms, is usually represented by the following equations for positive and negative ion formation:

$$\beta^+ = \frac{N^+}{N^0 + N^+} = \left[1 + \frac{g_0}{g_+} \exp\left(\frac{I - W}{kT}\right)\right]^{-1} \tag{10}$$

$$\beta^- = \frac{N^-}{N^0 + N^-} = \left[1 + \frac{g_0}{g_-} \exp\left(\frac{W - EA}{kT}\right)\right]^{-1} \tag{11}$$

where N is the number of ions or atoms; g is a statistical factor; I is the first ionization potential of the analyte; W is the work function of electrons for the filament material; EA is the electron affinity of the analyte; and the indices 0, $+$, and $-$ indicate a neutral atom, a positive ion, and a negative ion, respectively.

As can be seen from Eqs. (10) and (11), high yields of positive and negative ions can be obtained for atoms or molecules with low ionization potential and with high electron affinity, respectively. A high electron work function of the filament material is advantageous in increasing the positive ion beam, whereas for the production of negative ions a material with a low W value should be selected.

First ionization potentials and electron affinities of a few important elements and compounds, as well as electron work functions and melting points of some filament materials that are used in thermal ionization mass spectrometry are listed in Tables 7.7–7.9. From the data it appears that the formation of positive or negative thermal ions is usually an endothermic process when a pure filament material is used. This means that high ionization filament temperatures will increase the ion yield according to Eqs. (10) and (11). Therefore, filament materials with melting points above 2500°C are necessary, which favors the use of tantalum, rhenium, or tungsten. Previous experience has shown that sufficiently high positive-ion beams can be produced for all elements with first ionization potentials less than 7 eV, such as the alkalis, the alkaline earths, the lanthanides, and the actinides. Special ionization techniques must usually be applied

Table 7.7. First Ionization Potentials of Selected Elements Analyzed by PTI-IDMS

Element	I (eV)	Element	I (eV)
K	4.34	Sn	7.34
Li	5.39	Pb	7.42
Ce	5.47	Ag	7.58
Nd	5.49	Ni	7.64
Sr	5.70	Cu	7.73
Ca	6.11	Pd	8.34
Tl	6.11	Cd	8.99
Ti	6.82	Te	9.01
Mo	7.10	Zn	9.39

Source: Ref. 45.

to elements with a first ionization potential of more than 7 eV, such as the ones listed in Table 7.7. The most common methods are the silica gel and the resin bead techniques. In the formation of negative thermal ions, the ion yield can be increased by reducing the filament work function of electrons, for example, by the use of thoriated filament materials or the addition of a lanthanum compound to the surface of the metal filament (48).

Either the sample to be analyzed can be fixed to the filament by depositing a few drops of the sample solution on the filament surface followed by evaporation of the solvent to dryness, or it can be electrodeposited on the filament by using microelectrodeposition methods (49). The silica gel technique was suggested by Akishin and coworkers (50) and

Table 7.8. Electron Affinities of Selected Atoms and Molecules Detected by NTI

Atom or Molecule	EA (eV)
NO_2	3.91
Cl	3.61
BO_2	3.56
F	3.45
Br	3.36
CN	3.17
I	3.06
Se	2.12
S	2.07
Te	1.96

Source: Refs. 45–47.

Table 7.9. Electron Work Functions and Melting Points of Filament Materials Used in Thermal Ionization IDMS

Filament Material	Melting Point (°C)	W (eV)
Pt	1772	5.13
Ta	2996	4.30
Re	3180	4.98
W	3410	4.58
W(Th)[a]		2.7
W(LaB$_6$)[b]		2.7

Source: Refs. 44 and 45.

[a] W(Th) = thoriated tungsten.

[b] W(LaB$_6$) = lanthanum hexaboride covered tungsten.

Cameron and coworkers (51) and has been applied in modified forms in a number of investigations (e.g., 24, 52). In this technique a colloidal suspension of silica gel, a nitric acid solution of the sample, and phosphoric acid are normally used. For example, it is possible to determine lead (19, 24, 51, 52), cadmium (24, 53–55), copper (53, 55), silver (56, 57), palladium (57, 58), tellurium (59), zinc (60), and tin (61) using the silica gel procedure. If boric acid is substituted for phosphoric acid, higher ion yields can be obtained for some metals, for example, for chromium and nickel (62).

A V-shaped filament is used for the resin bead technique as is shown in Fig. 7.6 (63). The element to be determined is absorbed at a resin bead, using a strongly basic anion exchange resin. Afterwards, the resin bead is fixed to the filament, with collodium (63), for example. This technique was initially developed for the determination of small amounts of actinides (64). When the resin bead is heated on the metal surface, the resin is partly decomposed, and the ionization occurs mainly at the interface between filament and resin bead, as was shown by investigations of Smith and coworkers with uranium (65). The resin bead technique is also useful to determine transition metals that form volatile oxides in a high oxidation state such as molybdenum, ruthenium, and technetium (66, 67). The elements are retained on the filament by the resin bead, possibly by a reduction of the element. In addition to these special techniques, carbonizing the filament material (68) or adding chemical compounds, such as Ta$_2$O$_5$, as powders to the filament (69) also increase ionization efficiency and produce more stable ion beams. The formerly preferred boric acid and borax technique (70)—fusion of a small bead of borax to the filament—for ionization of heavy metals with high ionization potentials is

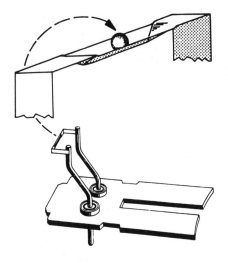

Figure 7.6. Filament construction for the application of the resin bead technique (63).

applied only rarely today. But boron is still analyzed using borax-producing $Na_2BO_2^+$ ions in the PTI mode (70, 71).

Figure 7.7 shows the elements for which negative thermal ions have been detected (44) and for which either negative thermal IDMS techniques have been developed or are potentially applicable (48). Table 7.10 lists the elements and their compounds that have been analyzed by NTI-MS. For nonmetals, with their typically high ionization potentials, a successful

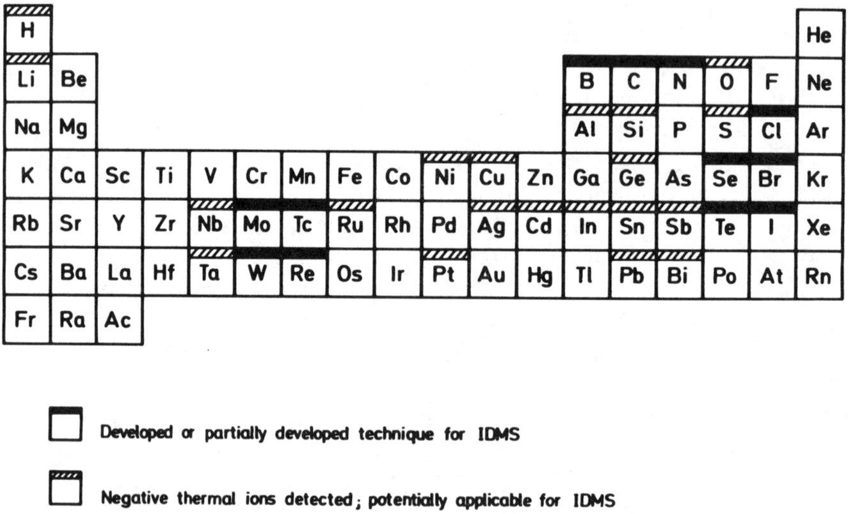

H																	He
Li	Be											B	C	N	O	F	Ne
Na	Mg											Al	Si	P	S	Cl	Ar
K	Ca	Sc	Ti	V	Cr	Mn	Fe	Co	Ni	Cu	Zn	Ga	Ge	As	Se	Br	Kr
Rb	Sr	Y	Zr	Nb	Mo	Tc	Ru	Rh	Pd	Ag	Cd	In	Sn	Sb	Te	I	Xe
Cs	Ba	La	Hf	Ta	W	Re	Os	Ir	Pt	Au	Hg	Tl	Pb	Bi	Po	At	Rn
Fr	Ra	Ac															

☐ Developed or partially developed technique for IDMS

▨ Negative thermal ions detected; potentially applicable for IDMS

Figure 7.7. Elements that can be measured by NTI and its applicability for IDMS (44, 48).

Table 7.10. Elements Analyzed by NTI-MS

| Element | Compound Preferred for Measurement | Measured Ions | | Refs. |
		Most Abundant	Other	
B	H_3BO_3, $Na_2B_4O_7$	BO_2^-	BO^-	23,46,72,73
C	AgCN	CN^-		74
N	Nitron nitrate	NO_2^-		75–77
Cl	AgCl	Cl^-		32,78–82
Se	H_2SeO_3	Se^-	SeO^-, SeO_2^-	40
Br	AgBr	Br^-		83–85
Mo	$(NH_4)_6Mo_7O_{24}$	MoO_3^-	MoO_2^-	66,86
Tc	NH_4TcO_4	TcO_4^-	TcO_3^-, TcO_2^-	66,86
Te	TeO_2	Te^-		40
I	AgI	I^-		87–91
W	Na_2WO_4	WO_3^-	WO_2^-	86,92
Re	NH_4ReO_4	ReO_4^-	ReO_3^-, ReO_2^-	86,92

application of thermal ionization mass spectrometry was possible only after the development of NTI techniques. The formation of negative thermal ions instead of positive ions in transition metals is especially useful if isobaric interferences of other elements can occur. For example, the most abundant negative thermal ion of technetium is TcO_4^-, whereas molybdenum forms MoO_3^- as the most oxygen-rich ion. This means that no interference can occur in the NTI mode between the different technetium isotopes and the isobaric molybdenum nuclides, thus solving a great problem in positive thermal ionization mass spectrometry (93). The major advantage of using the NTI technique is the selective ionization of inorganic anions. Therefore, there are usually no problems with interferences or with background peaks.

As mentioned before, the ion yield of negative thermal ions can be increased by using a material with an added compound that reduces the electron work function, instead of a pure rhenium, tungsten, or tantalum filament. An intense negative thermal ion beam was found for a number of species when a thoriated filament material was used or when a lanthanum, alkali, and alkaline earth salt was added to the filament surface (48, 72). The major filament preparation techniques applied today are summarized in Table 7.11 and correspond to the data and references listed in Table 7.10. In the case of thoriated tungsten [W(Th)], an activation has to be carried out by carbonizing the filament (87, 94). In the lanthanum and barium additions, about 50 ng and 5 µg of the elements is deposited on the filament as lanthanum nitrate and barium hydroxide solution, respectively. Afterwards, the solutions are heated to dryness.

Although the formation of negative thermal ions is usually an endothermic reaction, the ion intensity does not increase continuously with increasing filament temperature, as is shown in Fig. 7.8 for the Se^- ions produced from a H_2SeO_3 and a $BaSeO_3$ sample (40). This contradicts the theoretically expected effect with respect to Eq. (11) and can be attributed to secondary effects, for example, to nonequilibrium states, electron emission from the filament, adsorption effects at the metal surface, and evaporation effects of the sample. For most elemental analyses carried out with NTI, one will find a temperature where the ion intensity reaches a maximum (23, 40, 75, 80, 85, 87). This maximum filament temperature must be selected for measurements with high sensitivity.

In Fig. 7.9, all elements are marked that have been determined with IDMS either as the element itself or in a compound (2). Horizontal lines in the element box symbolize the use of positive thermal ions, vertical lines the use of negative ones. A long-lived radioactive spike must be used with the elements marked by a black bar at the top of the element box due to the lack of stable isotopes (see Table 7.5 and Section 2.4). The

Table 7.11. Techniques Used in Measurement of Negative Thermal Ions

Filament			
Double/Single	Material[a]	Preparation	Measured Ions
Double	Re (eva + ion)	None	CN^-, NO_2^-, Cl^-
	Re (eva + ion)	Lanthanum (ion)	BO_2^-, Br^-, I^-
	Re (eva + ion)	Barium	Se^-, Te^-
	Re (eva), W(Th) (ion)	Carbonizing (ion)	Br^-, I^-
Single	Re	Lanthanum	BO_2^-, Cl^-, Br^-, I^-, MoO_3^-, TcO_4^-, WO_3^-, ReO_4^-

[a] eva = evaporation filament; ion = ionization filament.

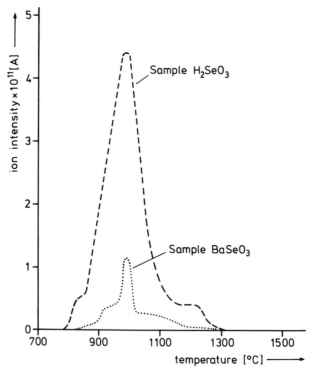

Figure 7.8. Dependence of the Se^- intensity on the temperature of the ionization filament (40).

elements beryllium, aluminium, manganese, niobium, and bismuth, which are marked in this way, are the only monoisotopic elements for which no isotope dilution technique has yet been developed, although the theoretical conditions are fulfilled in these cases. The tremendous power of thermal ionization mass spectrometry for trace analysis is shown by the fact that most of the elements can be determined with this ionization technique. Therefore, thermal ionization IDMS is now preferred in mass spectrometry if only one or a few elements within a sample have to be determined. Since the precision of the isotope ratio measurement by thermal ionization is usually better than 0.1% for magnetic sector field instruments (6) and better than 0.5% for quadrupole filter instruments (48), the accuracy of IDMS for trace analyses with this technique will be in the same range as long as no other errors (see Section 5.3) influence the result.

In the PTI mode the most abundant ions are usually the singly charged

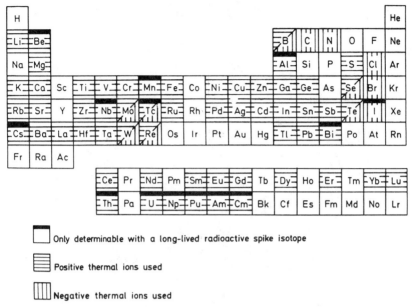

Figure 7.9. Elements determinable by thermal ionization IDMS (2).

atomic ions. Due to the high second ionization potential of all elements (>10 eV), no doubly charged thermal ions can be observed under normal ionization conditions. Cluster ions occur very seldom; for example, the cluster ion K_6^+ has been observed (95). In contrast to that, some metal compounds produce high metal oxide ion intensities in PTI-MS. This is true especially with the lanthanides and actinides. In addition to atomic ions, the lanthanides (96, 97), thorium (98), neptunium (99), and plutonium (99) form monoxide ions MeO^+. Also, the oxide ions UO^+ and UO_2^+ could be observed when uranium is present (98, 100), and MeO^+ thermal ions are emitted from a filament with a titanium (69, 101), zirconium (101), or vanadium sample (101). Whereas TiO^+, ZrO^+, and VO^+ are the most abundant ions in single-filament techniques, Me^+ becomes the major ion in double-filament techniques. However, the intensity ratio of atomic to oxide ions (Me^+/MeO^+) strongly depends on the filament temperature, as shown in Fig. 7.10 for the light lanthanides lanthanum, cerium, praseodymium and neodymium (97). The results shown in the diagram were obtained with lanthanide nitrate samples in a double-filament ion source using those temperatures of the evaporation filament with which a stable ion beam can be maintained for several hours. It is obvious that the Me^+/MeO^+ ratio increases with the atomic number of the lan-

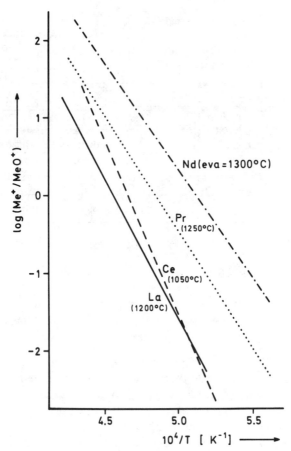

Figure 7.10. Me$^+$/MeO$^+$ ratio of the light lanthanides as a function of the temperature of the ionization filament (eva = evaporation filament) (97).

thanides, which is due to the decreasing dissociation energy of the MeO molecules in this elemental order (102). A linear correlation between the logarithm of Me$^+$/MeO$^+$ and $1/T$, as shown in Fig. 7.10, was also found in a single-filament ion source for the lanthanides, but with a significant shift of the Me$^+$/MeO$^+$ ratio to lower values for the corresponding temperatures (96).

One consequence of the temperature dependence of the Me$^+$/MeO$^+$ ratio is that lanthanum and cerium are preferably measured as monoxide ions in PTI, whereas the heavier lanthanides, such as neodymium, are determined via the atomic ions. The different atomic and oxide ion intensities also allow an effective discrimination in the interference of isobaric nuclides, for example, between ^{142}Ce and ^{142}Nd. On the other hand, metal oxide ions interfering with atomic ions have to be taken into account, especially in the determination of the less abundant isotopes. For example, $^{160}GdO^+$ ions interfere with $^{176}Lu^+$. It was also found that MeF^+ ions can sometimes interfere with isobaric masses (69). These types of interferences have not been considered in the data given in Table 7.5.

Although thermal ionization IDMS is generally viewed as a monoelement method, oligoelement analyses are also possible under certain conditions. If the type of thermal ionization technique is the same for a number of elements, and if successive formation of the different element ions can be obtained by increasing the temperature of the ionization and/or evaporation filament, then a determination of these elements is possible within the same sample measurement. This type of simultaneous determination of elements using thermal ionization mass spectrometry was carried out, for example, for thallium, cadmium, and lead (24, 34, 54, 103) and for copper, nickel, chromium, and zinc (62, 103) with the silica gel technique; for vanadium, titanium, zirconium, and hafnium with the double-filament PTI mode (101); for several lanthanides with the single- and double-filament technique (96, 97); for uranium and plutonium with the resin bead technique (104); and for chlorine, bromine, and iodine with NTI (89, 105). As an example the ion intensities of Tl^+, Cd^+, and Pb^+ ions obtained with the silica gel technique are shown in Fig. 7.11 as a function of the filament temperature (106). From the diagram it follows that thallium should be measured at a temperature of approximately 800°C, cadmium at 1200°C, and lead at 1300°C. The major advantage of an oligo- compared to a monoelement determination by thermal ionization

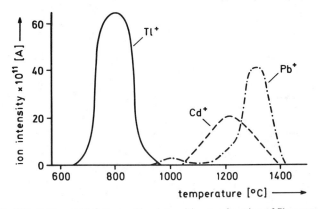

Figure 7.11. Tl^+, Cd^+, and Pb^+ thermal ion intensities as a function of filament temperature using the silica gel technique (0.5 nmol Tl, 1 nmol Pb, 4 nmol Cd; intensities are isotope corrected) (106).

mass spectrometry is the time savings for the analysis. Previous experience has shown that similar precision can be obtained for single- and oligoelement determinations.

3.3. Spark Source Mass Spectrometry

The principles of spark source mass spectrometry are described in Chapter 2, and therefore only the special features of the sample preparation for isotope dilution using this type of mass spectrometric instrumentation will be discussed here. In accordance with the principle of ion formation in a spark source, the sample has to be prepared as an electrode. This means that the sample first has to be dissolved for the spiking process, and then the solution must be converted into a conducting electrode. The fundamental studies in this field were carried out by Paulsen and co-workers before 1970 (107, 108). Others have also used this method during the last 15 years (e.g., 109–112).

The procedure for an analysis by spark source isotope dilution mass spectrometry (SS-IDMS) is shown in Fig. 7.12 (2). After spiking, the sample must be dissolved, for example, by acid. In cases where interferences in the spark source mass spectrum can influence the analysis result, it is useful to separate the elements that are to be determined, by ion chromatography or other means. Then the solution is heated to dryness. The preparation of electrodes can be carried out by mixing the dried sample with either graphite powder or a suitable metal powder such as gold or

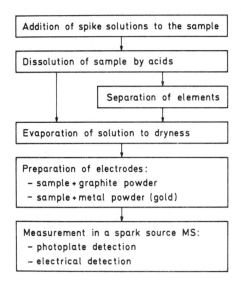

Figure 7.12. Sample preparation for spark source IDMS (2).

silver powder. Afterwards, this mixture is pressed into an electrode as in the analysis of nonconducting materials by SSMS without the isotope dilution technique. In principle, all elements for which spike isotopes are available (see Section 2.4) can be determined in a multielement analysis by SS-IDMS. Nevertheless, there are practical limitations to the number of elements that can be analyzed simultaneously, for example, interferences. Also SS-IDMS analyses are usually somewhat less precise than analyses carried out by thermal ionization IDMS. Therefore, the major application of SSMS in the isotope dilution technique is for multielement analysis.

3.4. Electron Impact Ionization

Electron impact ionization is the best known ionization technique in mass spectrometry. A detailed description of this ionization process is given in all fundamental books on mass spectrometry (e.g., 113, 114). All elements that are preferably determined by electron impact ionization IDMS are marked in Fig. 7.13 (2). This ionization method is especially useful for trace analyses of noble gases; it is used, for example, to analyze argon in geochronology (115). Traces of hydrogen and nitrogen are analyzed by measuring the isotope ratio in the diatomic molecules H_2 and N_2, whereas carbon, oxygen, and sulfur can be determined by the dioxides CO_2 and SO_2. If other compounds of these elements must be determined by electron impact ionization IDMS, they should be converted into the molecules mentioned prior to the mass spectrometric measurement. Major contributions to the mass spectrometric isotope ratio measurements of these elements have been made by Friedmann (116), Craig (117), and Dansgaard (118).

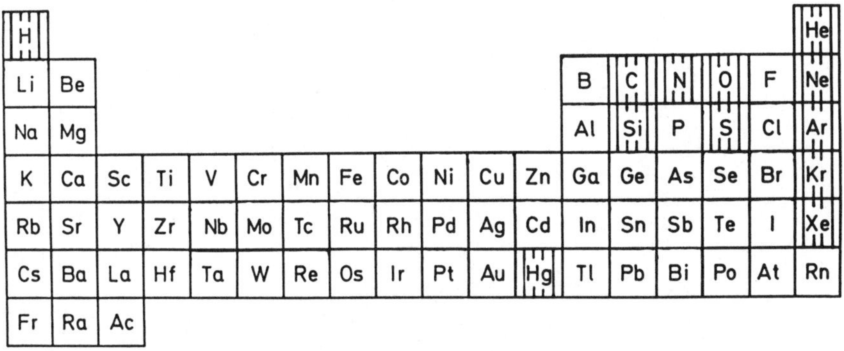

H																	He
Li	Be											B	C	N	O	F	Ne
Na	Mg											Al	Si	P	S	Cl	Ar
K	Ca	Sc	Ti	V	Cr	Mn	Fe	Co	Ni	Cu	Zn	Ga	Ge	As	Se	Br	Kr
Rb	Sr	Y	Zr	Nb	Mo	Tc	Ru	Rh	Pd	Ag	Cd	In	Sn	Sb	Te	I	Xe
Cs	Ba	La	Hf	Ta	W	Re	Os	Ir	Pt	Au	Hg	Tl	Pb	Bi	Po	At	Rn
Fr	Ra	Ac															

Figure 7.13. Elements preferably determined by electron impact ionization IDMS (2).

A preferred chemical form for the determination of silicon is SiF_4 (119). SiF_4 can be produced in the ion source from $BaSiF_6$ samples (120). A problem that can arise with a SiF_4 sample is the formation of aggressive fluorine atoms in the ion source of the mass spectrometer. Because of its high ionization potential and its volatility, mercury is the only metal that is determined by electron impact ionization; however, contaminations in the mass spectrometer are the source of major problems (119). A $Hg(NO_3)_2$ sample can be used in the ion source to form Hg^+ ions after the evaporation of the sample (120). However, there are no known important applications for trace analysis of mercury with IDMS.

A gas inlet system that introduces the above mentioned gases into the ion source of an electron impact mass spectrometer for isotope ratio measurements is shown in Fig. 7.14. A viscous gas flow from the reservoir to the ion source must be used to avoid an isotope shift in the reservoir over time. Therefore, a pressure of approximately 50 mbar is necessary for the sample gas in the reservoir. Because of the molecular gas flow from the ion source toward the pumping system, the measured isotope ratio must be corrected by a factor $\sqrt{m_1/m_2}$ (m_1 and m_2 are the masses of the light and heavy isotopic molecules, respectively) for this isotopic fractionation effect. Since a symmetrical dual inlet system, as shown in

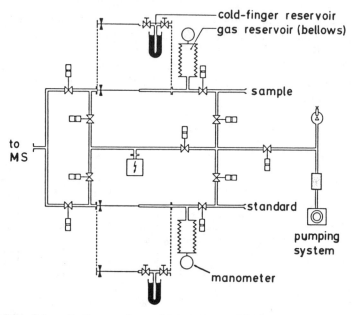

Figure 7.14. Schematic diagram of a gas inlet system used for isotope ratio measurements.

Fig. 7.14, is used and the isotope ratio of the sample is measured by direct comparison with a standard gas of known isotopic composition, the correction factor can be compensated for by determining the $\delta(\%_o)$ value:

$$\delta(\%_o) = \frac{\text{isotope ratio (sample)} - \text{isotope ratio (standard)}}{\text{isotope ratio (standard)}} \times 10^3 \quad (12)$$

To balance the ion currents of sample and standard, the volume of the reservoir can be varied by means of a membrane bellows. For small sample sizes, a cold-finger reservoir is used. Inlet systems of this type have been constructed for high-precision isotope ratio measurements with which relative standard deviations of 0.001% can be achieved under optimum conditions. Although the isotope dilution technique does not demand such high precision, these systems can be readily used for IDMS. Parameters that must be taken into account for precise isotope ratio measurements of gases are discussed in the literature (e.g., 121, 122).

GC-MS together with electron impact ionization can also be used for the analysis of some elements by IDMS. In this way, volatile organometallic compounds (123) or elements that form stable chelates can be determined after a gas chromatographic separation. Selenium is one of the few elements that have been successfully determined (124). Calcium chelates, for example, of trifluoroacetylacetone, have been used to determine the isotope composition of calcium (125). The deviations between the measured and the calculated calcium isotope compositions were in the range of 0.1–2.7%. Nickel and chromium have been determined with GC-IDMS using derivates of chromium acetylacetonate and those of nickel dithiocarbamate (126). When analyzing test solutions, deviations for chromium and nickel of up to 8% from the reference value have been found at the nanogram level.

3.5. Field Desorption Mass Spectrometry

Field desorption mass spectrometry (FD-MS), which is usually used for the analysis of organic compounds, is also applicable to the determination of trace metals, as has been shown by Schulten and coworkers in particular during the last few years (127–133). The principles of field desorption mass spectrometry are described in several review articles (134, 135). The preparation of the field emitter for metal analysis and the deposition of samples are similar to those for organic compounds. A few microliters of the solution containing the element to be determined is deposited on the emitter surface, as shown in Fig. 7.15 (133). The solution is then evap-

Figure 7.15. Sample deposition on the surface of a field emitter by a microliter pipette (133).

orated to dryness. Alternatively, one can use electrodeposition of the metal at the emitter surface.

The desorption of ions is carried out in a high electric field of the ion source by heating the emitter with an electric current. Temperatures of up to 1400°C can be used without influencing the emitter activity. If higher temperatures are necessary for a sufficiently high ion beam of the analyte, a laser-assisted field desorption system must be used (136, 137). Figure 7.16 shows all elements that have been detected by FD-MS through simple electrical heating of the emitter or through laser-assisted heating (133). Up to now, ions have been detected in FD-MS for all the indicated elements, but isotope dilution techniques with this ionization method have been applied for only a few, such as the alkalis, the alkaline earths, and thallium (127–133). The relative standard deviation of isotope ratio determinations with FD-MS is usually less than 1% for isotopes with high abundance; for those with low abundance it lies in the range of 1–10%, which is below the precision that can be achieved by thermal ionization mass spectrometry. Furthermore, high thermal ion intensities can be obtained for most of the elements where laser-assisted heating is necessary.

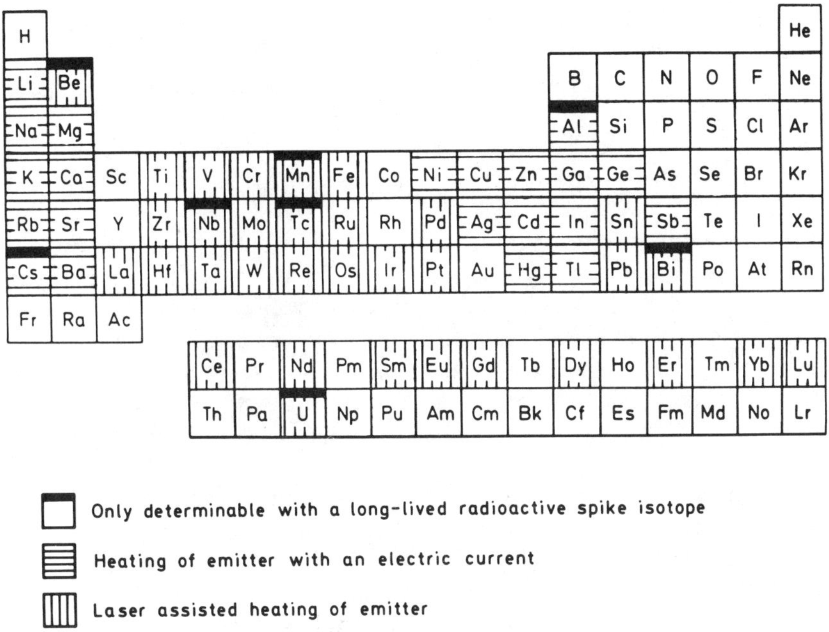

Only determinable with a long-lived radioactive spike isotope

Heating of emitter with an electric current

Laser assisted heating of emitter

Figure 7.16. Metals determinable by field desorption IDMS (2, 133).

Therefore, a thermal ionization mass spectrometer is usually preferred in determinations done by IDMS. However, one particular advantage of FD-MS is that it allows successive analyses of organic compounds and metal traces. For example, chlorophyll can be detected by using a low laser density, and then when the laser power is increased, magnesium becomes detectable (133).

4. INSTRUMENTATION

4.1. Mass Analyzer Systems

In principle, there is no reason to use any mass analyzer systems for the isotope dilution technique other than those that have been described for the particular ionization methods (see Chapters 2–6). In thermal ionization and electron impact mass spectrometry, a magnetic sector field is the only commercial mass analyzer that has been used for precise isotope ratio measurements in the past. This is due to the fact that flat isotope peaks, which can be obtained by a magnetic analyzer, are necessary for the precise determination of isotope ratios. However, measurements of isotope ratios by noncommercial equipment with a thermal ion source combined with a quadrupole mass analyzer have shown that this type of mass spectrometer should also be applicable in isotope dilution analyses (138–140). Indeed, long experience with IDMS has shown that the limiting factor with respect to precision is usually not the mass spectrometer, and therefore not the isotope ratio measurement, but rather the chemical treatment of the sample (see Section 5.2), the blank problem (see Section 2.3), and sample inhomogeneity (23, 34, 77). This means that the use of magnetic mass spectrometers for IDMS often goes far beyond the needs of the isotope dilution technique with regard to instrument performance and complexity. Therefore, in search of a cost-efficient solution, a compact thermal ionization mass spectrometer with a quadrupole analyzer was recently constructed as a commercial instrument in cooperation with Heumann and coworkers (48).

The schematic diagram of a thermal ionization quadrupole mass spectrometer is shown in Fig. 7.17 (141). A turret magazine will take up to 13 samples to be automatically analyzed, which is very important for the application of an analytical method in routine analysis. The absence of a magnetic sector analyzer and the associated high accelerating voltage

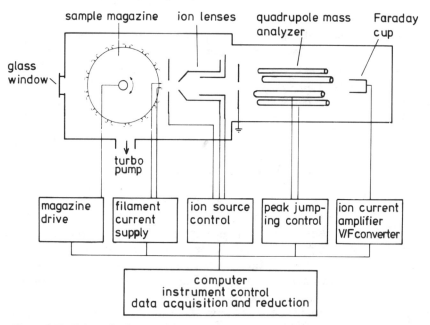

Figure 7.17. Schematic diagram of a thermal ionization quadrupole mass spectrometer.

allow very fast switching between mass peaks, fast scanning of mass spectra (for control of impurities), and a convenient choice of positive and negative thermal ionization. Therefore, it is possible to determine metals as well as nonmetals with this instrument. In contrast to magnetic mass spectrometers, it is relatively easy to transport a thermal ionization quadrupole instrument, so that flexibility is realized. The application of a quadrupole mass analyzer in glow discharge and ICP-MS (see Chapters 4 and 6) for multielement determinations is also shown by the recent trend to the use of simpler and more cost-efficient instruments in inorganic mass spectrometry.

The relative standard deviation for isotope ratio measurements with a thermal ionization quadrupole mass spectrometer usually ranges between 0.05 and 0.5% (23, 48, 142). With this precision the results of trace analyses carried out by IDMS should be comparable for both types of instrumentation—quadrupole and magnetic sector field mass spectrometers. This was shown by a direct comparison of analysis results, for example, for boron, iodine, calcium, and lead (7, 48), as in Table 7.12. Identical results are obtained within the given standard deviations for both types of mass spectrometers. Although the results with the quadrupole mass spectrometer are slightly less precise in some cases, they are as accurate as those obtained using the high-precision magnetic sector field instrument.

4.2. Detector Systems

The principal ion detectors used for isotope ratio measurements are the photoplate and the electrical detection system with a Faraday cup or an electron multiplier. For multielement determinations, a simultaneous detection of all ions registered by a photoplate is useful, as in spark source IDMS. Nevertheless, photoplate detection is usually less precise than electrical detection. This can be seen from the results given in Table 7.13, where rare earth elements have been determined in a standard rock (BCR-1) by spark source IDMS using photoplate and electrical detection (112). The overall relative standard deviation (s_{rel}) of the mean for five replicate determinations is 6.5% for the photoplate and 2.7% for electrical detection. The higher standard deviation for photoplate detection is due mainly to densitometer errors (143) and depends primarily on the measured value of the isotope ratio (144). The data obtained with photoplate detection fall within ± 10% of the "best estimated" values; with electrical detection, within ± 2.7%. That means that the analyses done by electrical detection are also more accurate.

The ion currents for isotope ratio measurements on the detector side

Table 7.12. Comparison of Trace Element Determinations by Thermal Ionization IDMS Using Magnetic Sector Field and Quadrupole Mass Spectrometer

| Sample | Element | Concentration (µg/g) | |
		Magnetic MS	Quadrupole MS
Spinach (NBS 1570)[a]	B	30.3 ± 0.1	30.2 ± 0.2
Aluminum	B	2.77 ± 0.05	2.75 ± 0.02
Milk powder (BCR 151)[a]	I	5.34 ± 0.01	5.44 ± 0.12
Bovine liver (NBS 1577)[a]	I	0.28 ± 0.01	0.31 ± 0.05
Human serum	Ca	92.4 ± 0.01	92.6 ± 0.8
Twice-distilled water	Pb	$(0.23 ± 0.05) \times 10^{-3}$	$(0.23 ± 0.03) \times 10^{-3}$

Source: Refs. 7 and 48.

[a] Standard reference material not certified for the analyzed elements.

Table 7.13. Determination of Rare Earth Elements in Standard Rock BCR-1 by Spark Source IDMS Using Photoplate and Electrical Detection

Element	Photoplate Detection		Electrical Detection		Best Estimated Value
	Conc. (µg/g)	Rel. Std. Dev. (%)	Conc. (µg/g)	Rel. Std. Dev. (%)	
Ce	58.7	13.4	53.6	3.7	53.7
Nd	30.0	6.1	27.8	2.2	28.5
Sm	6.73	5.8	6.79	1.6	6.70
Eu	2.13	6.7	1.97	2.4	1.95
Gd	6.17	5.3	6.38	2.7	6.55
Dy	6.00	5.3	6.40	3.6	6.39
Er	3.33	3.6	3.80	2.4	3.70
Yb	3.32	5.5	3.49	2.7	3.48

Source: Ref. 112.

of the mass spectrometer lie in the range of 10^{-8}–10^{-18} A. If low ion currents ($<10^{-15}$ A) must be registered, an electron multiplier has to be applied. In this case the mass discrimination effect of the electron multiplier has to be corrected for accurate isotope ratio determinations. The secondary electron emission from the first dynode of the multiplier depends on the ion velocity (145). Because the ions have the same average energy, the velocity of the lighter isotopic ions is higher than that of the heavier isotopic ions, leading to a larger amplification. The correction factor can be determined by a comparison of the same ion beam detected with a Faraday cup and an electron multiplier. On the other hand, an electron multiplier together with a multichannel analyzer can be used for single-ion detection. This is a detection mode that is used in ICP-MS (see Chapter 7) and is very often applied in FD-MS (135).

Ion currents in thermal and electron impact ionization mass spectrometry are usually so high that a Faraday cup measurement can be applied. No mass discrimination effect takes place with this system. With a single-collector system of this type, one must follow the ion current of one isotope as a function of the measuring time in order to correct its ion intensity for those times when the other isotopes are registered. In this case, calculations with linear and exponential corrections are always only more or less exact approximations. Nevertheless, the accuracy is usually

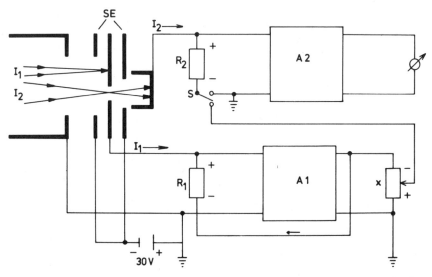

Figure 7.18. Schematic diagram of a double-collector system for isotope ratio determinations ($I_{1,2}$ = ion current of isotopes 1 and 2; R = resistance; SE = secondary electron circular; A = amplifier; x = fraction of potential difference 1 used for compensation).

high enough for IDMS. If the analytical problem to be solved is always the same—that is, if the reference and spike isotopes to be determined do not change very often—the use of a double-collector system is much more convenient. Such multicollector systems, up to nine Faraday cups, are applied especially in high-precision isotope ratio measurements. A schematic diagram of a double-collector system is shown in Fig. 7.18. The use of such compensation switching allows a direct measurement of the isotope ratio to be carried out.

5. APPLICATIONS

5.1. General Aspects

The application of the isotope dilution technique is especially useful in carrying out precise and accurate micro and trace analyses. This is why IDMS has become a routine method in geochronology and in nuclear technology. In these fields accuracy of the results has always been the main analytical interest. IDMS also plays a major role in the certification of standard reference materials. During the last few years, the isotope dilution technique has become important in the analysis of environmental, biological, and medical samples as well.

The detection limit of IDMS essentially depends on the element to be determined, on the matrix in which trace elements have to be analyzed, and on the blank during the sample preparation. The sensitivity of ionization is different from element to element (see Section 3). Therefore, the optimum ionization method for each element to be determined must be selected. The chemical process for the isolation of elements depends strongly on the type of element and on the matrix. From this it follows that sophisticated chemical isolation processes have to be developed in order to obtain high sensitivities. The determination of traces of elements that are abundant in the environment will be mainly influenced by the blank. However, even under unfavorable conditions it is possible to determine trace amounts at the nanogram level. The determination of rare trace elements should be possible down to the femtogram level in a simple matrix such as water.

Thermal ionization, electron impact, field desorption, and spark source mass spectrometry have been mainly applied in IDMS, so the examples given in the following sections are related to these ionization methods. One copper determination in a biological standard reference material was carried out with MIP-MS using the isotope dilution technique (146). The result was less accurate than that obtained by the usual calibration methods in MIP-MS. A standard reference water sample has been analyzed by isotope dilution ICP-MS for copper, strontium, barium, and lead (147). The concentrations determined in this analysis deviated by between 5 and 25% from the certified values. No significant difference was found between single-element and multielement analyses. So far, isotope dilution analyses have been applied in only a few cases with MIP-MS and ICP-MS. These results are not significantly better than those obtained by common calibration methods used in connection with these plasma-exciting instruments, for example, by the standard addition method. Therefore, future investigations will show whether or not the potential for accurate results with IDMS can be transferred to MIP-MS and ICP-MS. There is a good chance for successful development. The application of the isotope dilution technique in laser and secondary ion mass spectrometry does not play an important role at the moment.

5.2. Sample Treatment

For the ionization methods used in IDMS (see Sections 3 and 5.1), sample treatment varies with the three major types of analysis:

1. Multielement determinations (spark source MS)
2. Single element and oligoelement determinations (thermal ionization and field desorption MS)
3. Gas determinations (electron impact MS)

The only processes that must be carried out for spark source mass spectrometers are a homogeneous mixing of the spike and the sample followed by the production of a conducting electrode of this mixture. The principles of this treatment are shown in Fig. 7.12. If interferences in the mass spectrum are caused by the matrix or other elements, the interfering elements can be separated by one of the methods discussed for the second type of analysis. Because a multispike addition must be applied in the multielement analysis, one has to take into account that the stock spike solution of one element can contribute impurities with respect to another element to be determined.

For the second type of analysis, the element(s) to be determined usually have to be separated from the matrix and from other elements in order to prevent undesired effects during the evaporation and ionization process. The addition of the spike to the sample and the mixing of the two should be carried out at the very beginning of sample treatment in order to make full use of one of the major advantages of IDMS, namely that the analytical results are not influenced by a number of sources of error once the isotope dilution has taken place (see Section 2.1). In principle, the following separation techniques can be applied to the elements after the sample has been dissolved:

Cation and anion exchange
Electrolytic deposition
Extraction
Precipitation
Element-specific methods, such as evaporation techniques

The order in which the separation methods are listed above corresponds to the frequency with which these techniques are applied. Separation by an ion exchanger results mostly in a more selective separation of a single element, whereas electrolytic deposition, extraction, and precipitation are usually preferred for the isolation of a group of chemically similar elements. For example, the heavy metals lead, cadmium, and thallium can be electrodeposited simultaneously (24, 103), whereas the halides Cl^-, Br^-, and I^- can be precipitated as silver halides (89, 105) under the same conditions.

Figures 7.19–7.23 show the principles of sample treatments for different matrices that can be used for thermal ionization and field desorption IDMS. All of these examples have important analytical aspects, and the sample treatments listed can be transferred to many other analytical investigations. The chemical treatment of biological and other organic samples for trace metal determinations is shown in Fig. 7.19 (7). After the preparation of a homogeneous starting material, which is one of the conditions for reproducible analysis, a known quantity of the spike solution(s) is (are) added. The sample can then be decomposed either by dry-ashing (148–150) or by wet-ashing with oxidizing acids (151–155). If the sample contains silicate, the silicate portion must be decomposed by hydrofluoric acid in order to determine the total metal content of the sample. The separation of the metals to be determined can be done by ion exchange chromatography, by electrolytic deposition, or by extraction. Electrolytic deposition is preferred if an oligoelement analysis has to be carried out, and the metals to be determined can be deposited at one electrode under the same conditions of electrolysis. The final step is always the isotope ratio measurement in the mass spectrometer.

If trace metals have to be analyzed in inorganic matrices, the decomposition steps of the sample treatment used for biological materials cannot be applied. Geological samples, such as rocks, are usually decomposed

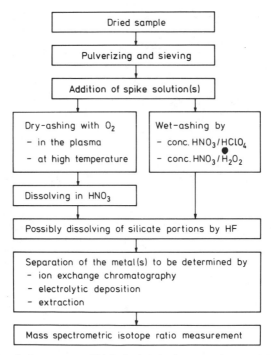

Figure 7.19. Chemical treatment of biological and other organic materials for trace metal determinations with PTI-IDMS (7).

by hydrofluoric acid or by mixtures of hydrofluoric acid and perchloric acid or sulfuric acid (32, 84, 90, 152). The greatest efficiency for this type of decomposition process is obtained by using a Teflon bomb. Alternatively, one can also achieve decomposition with alkaline fusion, for example, with Na_2CO_3 (84, 90, 152). However, alkaline decomposition usually results in a higher blank value than treatment with acids, which can be purified by sub-boiling methods. In water samples the sample treatment can be simplified in contrast to solid analyses. The chemical process for trace and ultratrace determinations of lead, cadmium, and thallium with IDMS is shown in Fig. 7.20 as an example (24). Natural water samples with normal concentrations of dissolved substances can be analyzed down to lead levels of approximately 200 pg/g, cadmium levels of 20 pg/g, and thallium levels of 10 pg/g. If lower concentrations of these elements have to be determined, the filtration and electrodeposition steps must be omitted in order to reduce the influence of the blank. This can be done with high-purity water samples. Then the detection limit for the above-mentioned heavy metals is approximately two orders of magnitude lower.

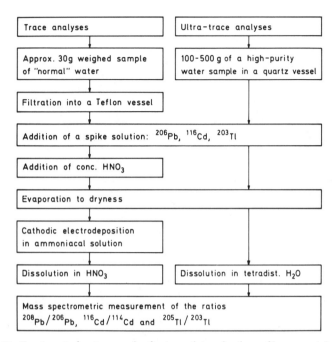

Figure 7.20. Treatment of water samples for trace determinations of heavy metals with PTI-IDMS (24).

Medical samples, for example, cell tissues and physiological liquids, can be homogenized together with a spike solution. In principle, a direct determination of metal in the homogenized material is possible with FD-MS as long as the emitter surface is not influenced by the substance and if isotope dilution has taken place (133). Protein-containing samples can disturb the direct analysis by FD-MS. Therefore, in these samples, the protein must be precipitated before the mass spectrometric analysis is carried out. Possible treatments for biomedical samples are presented in Fig. 7.21.

One important example of the determination of nonmetals in food samples is the determination of traces of iodine. The corresponding sample treatment, which is typical for an extractive isolation of the element to be determined, is shown in Fig. 7.22 (88). After adding the long-lived radioactive ^{129}I as a spike, the decomposition of the sample is carried out with a mixture of chloric, perchloric, and nitric acids. Then iodate is converted into elementary iodine, which can be extracted by carbon tetrachloride. Afterwards, iodine is reextracted into an aqueous solution, reduced to iodide, and precipitated as silver iodide. The last step is the

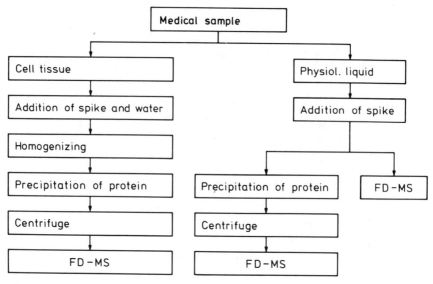

Figure 7.21. Treatment of medical samples for trace metal determinations with FD-IDMS (133).

$^{129}I/^{127}I$ ratio measurement using NTI-MS. The oxidizing decomposition by $HClO_3$ and $HClO_4$ or a mixture of the two is a common method in the sample treatment of organic substances (152, 153).

The last type of sample treatment discussed in connection with thermal ionization IDMS is the simultaneous isolation of the halides Cl^-, Br^-, and I^- from geological samples. This sample processing stands for a selective isolation of elements by precipitation and is shown in Fig. 7.23 (105). After the addition of a mixed spike solution, the sample is decomposed with hydrofluoric acid in a Teflon bomb. Using the closed Teflon bomb for the decomposition of the sample has the advantage that hydrofluoric acid can be used instead of an alkaline fusion. During the decomposition of the sample, the spike and sample isotopes of the halogens are mixed thoroughly. After this mixture, loss of substance has no influence on the analysis result, which means that the Teflon bomb can be opened even if there is a partial loss of volatile hydrogen halides. After centrifugation, a fractionated precipitation of the silver halides is carried out in the acid solution. First, silver iodide and silver bomide are precipitated using a substoichiometric amount of silver nitrate solution. The silver nitrate amount is fixed according to the known amount of the added bromide and iodide spike. Afterwards, silver chloride is precipitated in a second step. The fractionated isolation of the silver halides is necessary

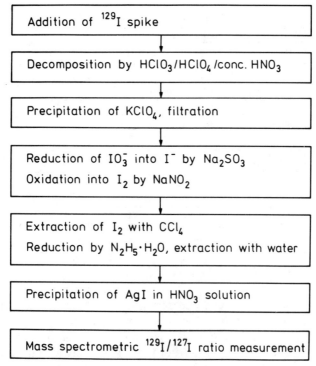

```
┌─────────────────────────────────────────────────┐
│  Addition of ¹²⁹I spike                         │
└─────────────────────────────────────────────────┘
                        ↓
┌─────────────────────────────────────────────────┐
│  Decomposition by HClO₃/HClO₄/conc. HNO₃        │
└─────────────────────────────────────────────────┘
                        ↓
┌─────────────────────────────────────────────────┐
│  Precipitation of KClO₄, filtration             │
└─────────────────────────────────────────────────┘
                        ↓
┌─────────────────────────────────────────────────┐
│  Reduction of IO₃⁻ into I⁻ by Na₂SO₃            │
│  Oxidation into I₂ by NaNO₂                      │
└─────────────────────────────────────────────────┘
                        ↓
┌─────────────────────────────────────────────────┐
│  Extraction of I₂ with CCl₄                      │
│  Reduction by N₂H₅·H₂O, extraction with water   │
└─────────────────────────────────────────────────┘
                        ↓
┌─────────────────────────────────────────────────┐
│  Precipitation of AgI in HNO₃ solution          │
└─────────────────────────────────────────────────┘
                        ↓
┌─────────────────────────────────────────────────┐
│  Mass spectrometric ¹²⁹I/¹²⁷I ratio measurement │
└─────────────────────────────────────────────────┘
```

Figure 7.22. Treatment of food samples for iodine determinations at trace levels with NTI-IDMS (88).

because of the fact that a great excess of chloride can influence the mass spectrometric measurement of bromide and iodide. Using this fractionated precipitation process, it is possible to determine chloride, bromide, and iodide within one sample treatment even if the halide ratios Cl^-/Br^- and Cl^-/I^- in the geological sample are in excess of 10^4 (90, 105).

The third type of trace analysis to be discussed here is gas determination using electron impact mass spectrometry. In solid samples, the extraction of the gas and the purification of the extracted gas are the major steps of sample treatment that must be carried out before the mass spectrometric measurement takes place. Geochronology is one important field where very low argon concentrations in rocks are determined by IDMS (115, 156). Figure 7.24 shows the principles of such a sample treatment. The extraction of argon from the rock sample is carried out in an extraction furnace. First the system is evacuated, and then the sample is inductively heated in a molybdenum crucible to approximately 2000°C.

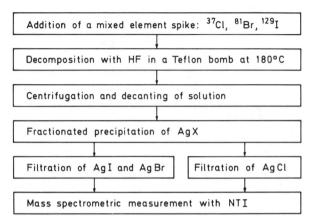

Figure 7.23. Treatment of geological samples for trace determinations of halides with NTI-IDMS (105).

An exact dosage of the ^{38}Ar spike can be achieved by using the tube volume between valves F and G. During the extraction, the gases are adsorbed at the cooled active carbon finger 1. Afterwards, purification of the noble gases from the other gases is carried out in a titanium sublimation pump. As a last step, the isotope-diluted argon is adsorbed at the active carbon finger 3, and from there the gas is introduced into the electron impact ion source of the mass spectrometer. If only small sample sizes are available, the mass spectrometer is run statically. With this procedure, radiogenic ^{40}Ar (for K/Ar dating) and atmospheric argon can

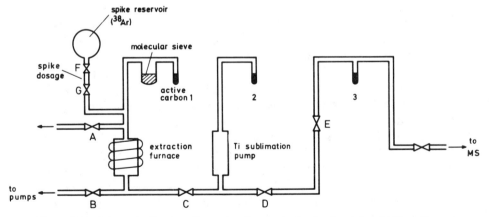

Figure 7.24. Schematic diagram of an extraction apparatus for the determination of argon in rocks by IDMS.

be determined, using ^{36}Ar as an indicator for atmospheric argon. This type of analysis carried out with IDMS can also be followed for other gases.

Electron impact IDMS can also be used for element chelates, which have been separated by gas chromatography, as was mentioned in Section 3.4. After the isotope dilution process has taken place, one must convert the element to be determined into a volatile yet thermally stable chelate. Whereas the precision of isotope ratio measurements is better than 1% when atomic ions are used for the evaluation, the relative standard deviation is in the range of 1–10% when organic compounds are used. This is due to the fact that interferences—for example, from fragment ions and isotopic peaks of the organic compounds—are more frequent in the electron impact ionization of organic compounds than in an ionization process where only atomic ions are produced. Because the precision of the determination is significantly influenced by the precision of the isotope ratio measurement of the isotope-diluted sample [see Eq. (5), Section 2.2], one has to select the element chelates using the criterion discussed above in order to obtain acceptable analysis results.

5.3. Possible Sources of Error

Although IDMS is called a definitive method, one has to take into consideration the possible sources of error, which can affect the result in some particular cases:

1. Isotopic fractionation in the ion source
2. Isotopic variations in nature
3. Isotopic separations during sample treatment
4. Nonequilibration of the isotope dilution process
5. Nonrelevant blank corrections

With respect to the first of these, isotope fractionations in an electron impact ion source can be best corrected by successive measurements of the sample and of a standard with known isotopic composition (see Section 3.4). In thermal ionization mass spectrometry the measured isotope ratio depends on the temperature of the ionization filament, on the mass fractionation effect by evaporation of the sample, and on the chemical compound of the element to be analyzed (80, 157, 158). If necessary, the temperature of the ionization filament can be controlled by an optical pyrometer. As mentioned in Section 3.2, the isotopic fractionation carried out by evaporation of the sample is smaller when a double-filament ion source is used than when a single-filament source is used. The possible effects of different chemical compounds of the element to be determined can be excluded when the process for sample isolation always results in an identical chemical compound.

The effect of variations in natural isotopic compositions (15) is discussed in Section 2.1. This effect is usually negligible for IDMS with a few exceptions such as lead. However, the isotopic composition has to be determined in all analyses where the sample does not have a natural isotopic composition.

Isotopic fractionation effects created by chemical processes are very small, ranging somewhere within a few parts per thousand (159). From the results of different experiments and from theory it follows that isotopic fractionation increases with increasing relative mass difference (160). This means that isotopic fractionations have to be expected only for the lighter elements. From the separation processes for element isolation, discussed in Section 5.2, ion exchange chromatography has produced significant isotopic fractionation, as has been shown, for example, for calcium (161) and chloride ions (162). Nevertheless, the isotopic effects are usually so small that IDMS is not affected. Ion exchanger resins with crown ethers and cryptands as complex-forming anchor groups are systems in which isotopic fractionations of metal ions have been found to be large enough that they could influence results obtained by IDMS if only a fraction of the total element to be determined is isolated (163).

Complete isotope exchange between the sample isotopes of the element to be determined and the spike isotopes must be preserved in order to obtain accurate results. When no isotope exchange takes place between the different species, either the compounds must be decomposed for the determination of the total element content or species analyses must be carried out by using spike solutions that contain these species labeled with enriched isotopes (see Section 5.5). Errors can occur if a partial loss of either the spike or the sample takes place before the mixing of the isotopes. A similar effect results in incomplete decomposition of the sample. It is possible that these types of errors are responsible for the difference in lead results in two sewage sludges shown in Table 7.14, which were obtained by three laboratories with IDMS. The samples are standard reference materials of the Community Bureau of References (BCR) of the European Communities. Whereas laboratories 1 and 2 carried out their analyses during the certification round (164), laboratory 3 analyzed the lead content after the final certification round (2). There is a significant difference between the results of laboratory 1, on the one hand, and laboratory 2 and the certified value, on the other hand. As can be seen, the results of laboratories 2 and 3 agree well with each other and with the

Table 7.14. Systematic Errors in IDMS during Sample Treatment of Lead Determination of Two Sewage Sludges

Source	Decomposition Method after Spiking	Pb Concentration (µg/g)		
		SRM 144[a]	SRM 145[a]	
Laboratory 1	Dry ashing; digestion with HF; dissolution in HNO_3	573 ± 59	419 ± 32	
Laboratory 2	Digestion with HF, $HNO_3/HClO_4$; dissolution in HCl	485.2 ± 2.0	346.3 ± 1.6	
Laboratory 3	Digestion with HNO_3, HNO_3/HF, HNO_3/HCl; dissolution in HCl	483.7 ± 4.8	356.9 ± 7.7	
Certified value	Different methods and instrumentations	495 ± 19	349 ± 15	

Source: Refs. 2 and 164.

[a] BCR standard reference material.

certified value within the limits of error, whereas the results of laboratory 1 deviate significantly from these values. The main difference between laboratory 1 and the other two laboratories, which used the IDMS method, lies in the treatment of the sample. Laboratory 1 applied the dry-ashing method for the decomposition of the sewage sludges, and the other two laboratories used a wet decomposition carried out with different acids. Obviously, a complete mixing of the spike and the sample isotopes did not take place in the case of the dry-ashing method. The abnormally high standard deviation confirms this assumption. The loss of parts of the spike during the sample treatment can possibly explain the difference in the results, because the values analyzed by laboratory 1 are higher than the certified values.

The last source of error to be discussed with respect to IDMS are nonrelevant blank corrections. The blank is usually determined by the same analytical process that is used in the analysis of the sample, but without any of the sample components present. If the absence of the sample results in a different blank amount compared with the situation obtained with a sample, a systematic error occurs. An error can also occur in the blank if the isolation yield of the element to be determined varies greatly from analysis to analysis. For example, the same amount of contamination changes the isotope ratio of the isotope-diluted sample much more in the case of a 50% isolation yield than in the case of a substance with an approximately 100% isolation.

5.4. Analysis of Metals

Spark source mass spectrometry in connection with the isotope-dilution technique is used for multielement analysis in geological and cosmological samples in particular (10–12, 111, 112, 144, 165, 166). The determination of rare earth elements with spark source IDMS in a standard rock using photoplate and electrical detection (112) was discussed in Section 4.2 (Table 7.13). This method can also be applied to metal samples (107, 108, 167). The results of the determination of copper, molybdenum, and tungsten with spark source IDMS in the steel standard NBS 1161 are shown in Table 7.15 (111). The sample treatment corresponds to the procedure shown in Fig. 7.12 (Section 3.3). The steel is first spiked and then dissolved with acid and evaporated to dryness. The resulting residue is mixed with ultrapure graphite powder and briquetted into rod-shaped electrodes. The improvement in precision and accuracy obtained using the isotope dilution technique as compared to the direct method, where relative sensitivity factors are applied for quantitation, is obvious in the data given in Table 7.15. The relative standard deviations for the determination of the three

Table 7.15. Determination of Cu, Mo, and W in Steel Standard NBS 1161 with Spark Source Mass Spectrometry Using Isotope Dilution Technique and Direct Method

Element	IDMS Conc. (wt %)	s_{rel} (%)	Direct Method Conc. (wt %)	s_{rel} (%)	Certified Value
Cu	0.327 ± 0.005	1.5	0.33 ± 0.02	6.1	0.34
Mo	0.313 ± 0.005	1.6	0.29 ± 0.04	13.8	0.30
W	0.0119 ± 0.0003	2.5	0.013 ± 0.002	15.4	0.012

Source: Ref. 111.

elements, which have been analyzed with IDMS, are ≤2.5%. Multi- and oligoelement analyses have also been carried out with spark source IDMS in coal, fly ash, and environmental samples (109, 168–170) and in seawater (171).

Field desorption IDMS has been particularly useful for the determination of alkalis and alkaline earths (128, 131, 132) as well as for thallium (130, 172–174) in medical samples. In this case, the sample treatment shown in Fig. 7.21 (Section 5.2) is applied. The second process given in Fig. 7.21 was used for the determination of lithium in blood serum; the alternative sample treatment was used for the analysis of saliva and urine (175). The results for three individuals with normal lithium levels in human body fluids are presented in Table 7.16. From the data given it follows that the lithium concentration in plasma, saliva, and urine can be determined with a relative standard deviation ranging between 2 and 10%. Without any sample treatment except the evaporation of the solvent, traces of lithium have also been determined in mineral waters, wines, and organic solvents with FD-IDMS (127).

Thermal ionization mass spectrometry is the method most frequently used in isotope dilution analyses. This ionization technique is ideal for

Table 7.16. Determination of Normal Lithium Levels in Human Body Fluids with FD-IDMS

Person No.	Concentration (μmol/L) Blood Plasma	Saliva	Urine
1	2.43 ± 0.14	1.24 ± 0.13	4.69 ± 0.22
2	1.44 + 0.07	0.84 + 0.08	4.29 ± 0.14
3	1.30 ± 0.11	1.30 ± 0.03	7.54 ± 0.17

Source: Ref. 175.

Table 7.17. Determination of Rare Earth Elements in Standard Rock (BCR-1) by Thermal Ionization IDMS

Element	Conc. (μg/g)	s_{rel} [a] (%)
Ce	54.3	0.5
Nd	29.4	0.6
Sm	6.80	1.1
Eu	2.00	1.3
Gd	6.79	1.1
Dy	6.55	0.7
Er	3.74	0.8
Yb	3.50	0.6

Source: Ref. 176.

[a] 95% confidence level.

the determination of rare earth elements of which a second stable isotope is available as a spike (see Table 7.5, Section 2.4). In this case, measurements are carried out at various temperatures of the ionization filament (see Section 3.2). Therefore, the same elements could also be determined with thermal ionization IDMS in the BCR-1 standard rock as was done with spark source IDMS (see Section 4.2). The results are summarized in Table 7.17 (176) and agree well with the best estimated values for the standard rock (see Table 7.13). The overall relative standard deviation is 0.8%, which is somewhat better than the corresponding value of 2.7% for spark source IDMS with electrical detection (112). Highly accurate determinations of rare earth elements in small quantities of lunar samples have been carried out by Nguyen and coworkers (177) and in inorganic matrices by Heumann and coworkers (97).

Special attention has been focused on the analysis of alkalis and alkaline earth elements with PTI-IDMS. For example, the concentrations of lithium, potassium, magnesium, and calcium have been determined in blood sera by Garner and coworkers (79) as summarized in Table 7.18. The samples were decomposed by $HNO_3/HClO_4$ after the spiking process, followed by chromatographic separation of the analyte elements. This procedure is discussed in Section 5.2 (see Fig. 7.19). The relative standard deviation is 0.5% for Li, 0.2% for K, 0.5–0.8% for Mg, and 0.2–0.3% for Ca. For calcium the standard deviation agrees well with results in blood sera obtained by Moore and Machlan (178) and by Heumann and coworkers (7); for potassium, with those of Gramlich and coworkers (179). This precision obtained with PTI-IDMS is better than that achieved with FD-IDMS (see Table 7.16). The results are also more precise than those

Table 7.18. Determination of Alkalis and Alkaline Earths in Blood Sera with PTI-IDMS

	Concentration (mmol/L)		
Element	Lot 1	Lot 2	Lot 3
Li	1.004 ± 0.005	1.968 ± 0.010	2.954 ± 0.015
K	2.794 ± 0.006	4.765 ± 0.010	6.867 ± 0.014
Mg	0.359 ± 0.003	1.032 ± 0.005	1.738 ± 0.009
Ca	1.282 ± 0.004	2.253 ± 0.007	3.264 ± 0.007

Source: Ref. 79.

achieved with atomic spectroscopy methods (7). To point out the fact that PTI-IDMS of alkalis and alkaline earths can also be used with other matrices, the following determinations are mentioned: caesium in pilot-plant effluents (180), alkaline earths in chemicals (29, 181, 182), and calcium in geological samples and meteorites (22, 183) and in polar snow (184, 185).

The elements lead, cadmium, thallium, copper, and zinc are the heavy metals most frequently determined with PTI-IDMS. This is due to the extreme significance of these elements in environmental, medical, and food samples. The determination of these heavy metals can be carried out by means of either a monoelement analysis or, under certain conditions (see Section 3.2), a oligoelement analysis.

The high quality of PTI-IDMS in analysis of heavy metals can be seen in Table 7.19, which shows a selection of lead determinations in eight

Table 7.19. Determination of Traces of Lead in Standard Reference Materials with PTI-IDMS

	Pb Concentration (μg/g)	
Sample	IDMS	Reference Value
Fuel (NBS 1636 A)	11.2 ± 0.3	11.2 ± 0.2
Fly ash (NBS 1633 A)	71.8 ± 0.6	72.4 ± 0.4
Pine needles (NBS 1575)	10.6 ± 0.3	10.8 ± 0.5
Orchard leaves (NBS 1571)	47.0 ± 0.5	45 ± 3
Tomato leaves (NBS 1573)	6.03 ± 0.15	6.3 ± 0.3
Sea plant (IAEA SP-M-1)	15.9 ± 0.2	16 ± 7
Lake sediment (IAEA SL-1)	39.5 ± 0.1	40 ± 14
Glass (NBS 611)	427 ± 1	426

Source: Ref. 186.

Table 7.20. Oligoelement Analysis of Standard
Reference Material "Orchard Leaves" (NBS 1571)
with PTI-IDMS

	Concentration (μg/g)	
Element	IDMS	Reference Value
Tl	0.0467 + 0.0003	Not certified
Cd	0.120 ± 0.002	0.11 ± 0.01
Cu	12.85 ± 0.09	12 ± 1
Pb	45.1 ± 0.4	45 ± 3

Source: Ref. 103.

standard reference materials carried out by Broekman and van Raaphorst
(186). The samples of biological origin were treated as shown in Fig. 7.19
(see Section 5.2), and they were decomposed with HNO_3 and $HClO_4$. HF
was also added to samples containing silica. The separation of lead was
done by electrolytic deposition. The silica gel technique (see Section 3.2)
was applied in the mass spectrometric measurement. In all cases, the
IDMS values are identical to the reference values within the limits of
error. It is obvious that extremely accurate results can be obtained with
this type of IDMS.

Table 7.20 shows results of thallium, cadmium, lead, and copper determinations of the NBS standard reference material "Orchard Leaves"
obtained by Hilpert and coworkers with PTI-IDMS (103). The elements
rubidium, strontium, barium, and chromium were also measured. It was
possible to successively measure the different elements in the sample by
increasing the temperature of the filament (see Section 3.2) using the silica
gel technique. Again, the results are identical to the reference values
within the limits of error. Other oligoelement analyses of more than three
heavy metals have been carried out, for example, in sewage sludges and
soils by Götz and Heumann (187) and in biological samples, seawater,
and chemicals by Murozumi and coworkers (53, 188–190).

Using monoelement PTI-IDMS, the heavy metals lead, thallium, cadmium, and silver have been assessed in biological and medical samples
(34, 54, 60, 191–198); silver, tellurium, palladium, cadmium, and zinc in
geochemical and cosmological samples (58, 59, 195, 199); lead, uranium,
thorium, and thallium in glass (200); and lead, cadmium, thallium, and
copper in water samples, seawater, and polar ice samples (24, 201–204).
Boutron and Patterson were able to determine lead traces in the range of
a few picograms per gram in Antarctic ice samples (204). An interesting
new development in mass spectrometry is the resonance ionization tech-

Table 7.21. Determination of Traces of Technetium with NTI-IDMS

Sample	Tc Concentration	Spike Used for IDMS
Aqueous solution 1	37.2 ± 0.1 ppb ^{97}Tc	^{99}Tc
Aqueous solution 2	653 ± 2 ppb ^{97}Tc	^{99}Tc
Irradiated Al foil		
With ^{98}Mo	0.19 ± 0.01 μg ^{99}Tc	^{97}Tc
With ^{99}Tc	12.64 ± 0.08 μg ^{99}Tc	^{97}Tc

Source: Ref. 207.

nique (205), which exploits lasers to selectively and efficiently ionize gas-phase atomic species. The application of resonance ionization mass spectrometry (RIMS) is especially useful for elements where thermal ionization is particularly insensitive. This was the case in the past, for example, for iron. Therefore, Fassett and coworkers have developed a method to determine iron traces with resonance ionization IDMS. They applied this method in the iron analysis of human serum and water samples (206).

Picogram amounts of ^{99}Tc in environmental samples with PTI-IDMS were measured by Anderson and Walker (67) using the resin bead technique (see Section 3.2). As an alternative, Kastenmayer and Heumann applied the formation of negative thermal ions to determine this radioactive element (66, 86, 207). Technetium is one of the transition metals that can be detected with the NTI technique as a TcO_4^- ion (see Fig. 7.7 and Table 7.10, Section 3.2). The major advantage to measuring negative thermal ions of technetium is that there is no interference from molybdenum isotopes, the most oxygen-rich negative thermal ion of molybdenum being MoO_3^- (see Table 7.10, Section 3.2), which is found in the positive thermal ionization mode, because of the formation of the atomic ions Tc^+ and Mo^+, respectively. Table 7.21 shows the results of the determination of two aqueous ^{97}Tc solutions using a ^{99}Tc spike. Solutions 1 and 2 were obtained after separating technetium from ruthenium, which was irradiated in a nuclear reactor for 1 and 12 months, respectively. Using a ^{97}Tc spike, the amount of ^{99}Tc was determined in two samples of irradiated aluminium foil that were spiked with ^{98}Mo and ^{99}Tc, respectively, before the irradiation took place. In the case of the first aluminium foil, the ^{99}Tc that was analyzed had been produced by nuclear reactions of ^{98}Mo with thermal neutrons. The second analysis shows the result of the unirradiated ^{99}Tc. These results are also shown in Table 7.21.

As already mentioned, the determination of actinides, especially of uranium and plutonium, with PTI-IDMS, is a well-known technique and is the most common method used in nuclear technology today for these

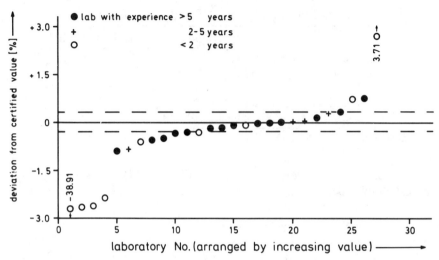

Figure 7.25. Interlaboratory study of an uranium solution (certified value = 1.7154 ± 0.0052 mg/g) with PTI-IDMS (2, 210)

elements (104, 208–211). The determination of thorium and uranium in geological samples using PTI-IDMS is also an established analytical method (212, 213). The results presented in Fig. 7.25 for an uranium solution were obtained in an interlaboratory study with PTI-IDMS (210, 211). In this figure the deviation of one of the investigated solutions from the certified value is plotted versus the number of the laboratory, where laboratory numbers were assigned in order of increasing deviation values. The results of most of the laboratories lie within the uncertainty of the certified value of ±0.3%, indicated by the dashed lines. However, accuracy is also dependent on the degree of experience of the participating laboratory. Five of the results show deviations of more than 1.5%; these all come from laboratories with less than two years experience. Similar results were obtained for other uranium and plutonium solutions. The results in Fig. 7.25 show that PTI-IDMS is normally an accurate method for determining uranium traces.

5.5. Nonmetal and Species Analyses

The nonmetals that can be measured with NTI-MS are included in Fig. 7.7 and Table 7.10 (Section 3.2). Under certain conditions it is possible to determine the concentration of not only the nonmetal itself but also of its species with NTI-IDMS. The most important species of nonmetals in inorganic chemistry are the anions. IDMS procedures developed to de-

Table 7.22. Anions of Nonmetals Determined with
NTI-IDMS

Element	Analyzed Anion	Spike Compound	Refs.
C	CN^-	$^{13}CN^-$	74
N	NO_2^-, NO_3^-	$^{15}NO_2^-$, $^{15}NO_3^-$	76,77
Se	SeO_3^{2-}, SeO_4^{2-}	$^{82}SeO_3^{2-}$, $^{82}SeO_4^{2-}$	40
Cl	Cl^-	$^{35}Cl^-$	32,81,82
Br	Br^-	$^{81}Br^-$	84,85
I	I^-, IO_3^-	$^{129}I^-$, $^{129}IO_3^-$	87–90

Source: Ref. 48.

termine these anions are listed in Table 7.22. One of the conditions in species-specific determinations is that no isotopic exchange is allowed to take place between the different species. For example, there is no isotopic exchange of nitrogen between NO_2^- and NO_3^- for pH values ≥ 4 (77). On the other hand, a chemical separation of different species must be carried out, because the same negative thermal ions measured in the mass spectrometer can usually be formed by different elemental species. For example, NO_2^- ions are formed by both nitrate and nitrite samples (75). During the analysis of boron, nitrite, nitrate, and cyanide, no atomic ions of elements other than the negative thermal ions BO_2^-, NO_2^-, and CN^- are measured in the mass spectrometer (see Table 7.10, Section 3.2). This means that the isotope abundances h_S and h_{Sp} in Eqs. (1)–(4) (Section 2.1) must be replaced by the measured natural and spike abundances of the ion intensities of mass numbers 26 and 27 for cyanide, 42 and 43 for boron, and 46 and 47 for nitrite and nitrate.

Both direct and indirect analytical procedures can be applied in species determinations. For example, to determine NO_2^- and NO_3^-, either one can add a $^{15}NO_2^-$ and $^{15}NO_3^-$ spike to the sample and then separate nitrite from nitrate, or one can first separate them after spiking with $^{15}NO_3^-$ and, then, in a second analysis, oxidize NO_2^- to NO_3^- and analyze the sum of both by adding only the $^{15}NO_3^-$ spike to the sample. The first possibility is shown in Fig. 7.26 as an example of the determination of nitrate and nitrite in food samples. Using the principles of the sample treatment shown in Fig. 7.26, Unger and Heumann obtained the results listed in Table 7.23 (77). As one can see, it is possible to determine nitrate and nitrite in plant materials and food samples with NTI-IDMS, in which the nitrate content can exceed the nitrite concentration by a factor of 1000. The standard deviations are in the range of 0.2–6%. The difficulties in obtaining accurate nitrate analyses of food samples were shown by the

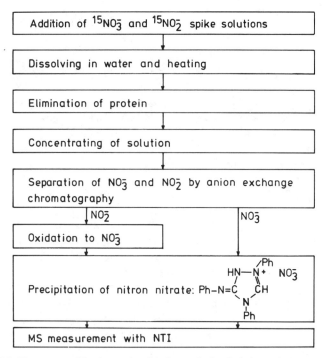

Figure 7.26. Treatment of food samples for the analysis of nitrite and nitrate species with NTI-IDMS (77)

Table 7.23. Nitrate and Nitrite Species Analysis with NTI-IDMS

	Concentration (μg/g)	
Sample	NO_3^-	NO_2^-
Spinach (NBS 1570)[a]	$(16.08 \pm 0.04) \times 10^3$	39.4 ± 0.9
Aquatic plant (BCR 61)[a]	$(4.41 \pm 0.05) \times 10^3$	1.86 ± 0.11
Smoked cut of lean pork	165 ± 4	18.0 ± 0.5
Raw ham	61 ± 3	28.0 ± 0.6
Creek water	39.2 ± 0.2	0.65 ± 0.03

Source: Ref. 77.

[a] Standard reference material not certified for NO_3^- and NO_2^-.

Community Bureau of Reference (BCR) of the European Communities, which organized an interlaboratory study to analyze nitrate traces in three milk powder samples (214). The results of the participating laboratories using various analytical methods differ by up to 45% at the 150 μg/g level, up to 75% at the 30 μg/g level, and up to a few hundred percent at the 3 μg/g level (2, 7). There are a number of arguments that support the opinion that NTI-IDMS has been the most accurate method in this interlaboratory study. NTI-IDMS seems to be a successful analytical method for the calibration of nitrate traces in food samples, although this cannot be verified by other methods at the moment.

Using the sample treatment shown in Fig. 7.23, Weiss and Heumann were able to determine the halides Cl^-, Br^-, and I^- simultaneously in geological samples with NTI-IDMS (105, 215). NTI-IDMS results for some geochemical and cosmological standard reference materials are compared in Table 7.24 with the available results of other methods such as X-ray fluorescence spectroscopy, photometric methods, ion-selective electrodes, and spark source MS (216–221). It is possible with NTI-IDMS to simultaneously determine traces of chloride in geological samples down to the low microgram per gram level, and bromide and iodide down to some nanograms per gram. The problem of accurate halide analyses in geological samples using other methods is obvious in the great variation of the results—a factor of 18 between the minimum and the maximum value for chloride analyses in granite GS-N and a factor of 760 for iodide in granite GH. Up to now, chloride, bromide, and iodide could not be certified in a standard rock, but NTI-IDMS offers a good prospective for such a certification.

The boron concentration in metallic materials considerably influences the physical and chemical properties of these substances. Therefore, traces of boron must be determined in a great variety of metals and alloys. The results of an interlaboratory study organized by the Community Bureau of Reference of the European Communities are presented in Table 7.25. Cyclotron proton activation analysis (CPAA) with the nuclear reaction indicated in the table, ICP atomic emission spectroscopy, a photometric method, NTI-IDMS, and a BF_4^--selective electrode are the methods that were applied in the boron analysis of an aluminium–magnesium alloy and aluminium (23, 222). Zeininger and Heumann determined the boron concentration with NTI-IDMS using a decomposition of the metals by hydrochloric acid and a separation of boron by the volatile boric acid trimethylester (23). The very high accuracy of the determination with NTI-IDMS is obvious, as the results are in good agreement with those of the other methods. The individual result of a single method only deviated up to 4% from the average value of all methods. Boron analyses

Table 7.24. Halide Trace Analyses in Geochemical Standard Reference Materials of Centre de Recherches Pétrographiques et Géochimiques, Nancy, with NTI-IDMS

Sample	Chlorine (µg/g)		Bromine (µg/g)		Iodine (µg/g)	
	IDMS	Other Methods	IDMS	Other Methods	IDMS	Other Methods
Granite GS-N	442 ± 3	47–830	2.18 ± 0.02	3	0.038 ± 0.004	—
Granite GH	60.5 ± 0.6	58–503	0.094 ± 0.003	2	0.020 ± 0.001	0.15–114
Disthene DT-N	27.0 ± 0.5	30–140	0.058 ± 0.012	—	0.063 ± 0.002	—
Bauxite BX-N	27.7 ± 0.3	45	0.983 ± 0.014	1	2.52 ± 0.05	—

Source: Refs. 105.

Table 7.25. Boron Interlaboratory Study on Aluminium and Aluminium/Magnesium Alloy with NTI-IDMS and Other Methods

Method	Concentration (μg/g)	
	Al/Mg	Al
CPAA: $^{10}B(p, \alpha)^7Be$		1.16 ± 0.10
ICP atomic emission	31.8 ± 1.5	1.20 ± 0.03
Photometry	30.5 ± 1.1	$1.27 + 0.08$
NTI-IDMS	29.0 ± 0.3	1.25 ± 0.05
Ion-selective electrode	29.7 ± 1.0	
Average	30.3 ± 1.2	1.22 ± 0.05

Source: Refs. 23 and 222.

with NTI-IDMS are also possible with biological materials (see Table 7.12, Section 4.1) as well as with seawater and fresh water samples (92). In the past, boron determinations with IDMS were only carried out with PTI using the $Na_2BO_2^+$ ion emitted from a borax sample. With this ionization method, Duchateau and de Bièvre determined the boron content in some chemicals (73), and Luo and coworkers in a water sample (223). The major advantages of NTI-IDMS over PTI-IDMS are the absence of possible interferences in the mass spectrum, the higher sensitivity, and a much more stable ion current.

5.6. Gas Analyses

As mentioned in Section 5.2, electron impact IDMS was most frequently used in the past for the determination of noble gas traces, especially for the analysis of argon in geochronology. A schematic diagram of an extraction apparatus for the determination of argon in rocks with electron impact IDMS for K/Ar and for $^{40}Ar/^{39}Ar$ dating is shown in Fig. 7.24. A detailed description of the sample treatment, the isotope dilution process, the mass spectrometric measurements, possible sources of error, and the evaluation of the data to determine ages of geological samples is given in the fundamental literature of geochronology (e.g., 115, 156, 224, 225). Additionally, $^{40}Ar/^{38}Ar$ determinations can be used to determine exposure of meteorites to cosmic rays (226). Measuring the concentration and isotopic composition of the noble gases helium, neon, argon, krypton, and xenon in Antarctic meteorites, information about spallogenic isotopes that were produced by cosmic rays; about radiogenic ^4He and ^{40}Ar from natural decay of potassium, uranium, and thorium; as well as about trapped primordial or solar gas could be obtained (227, 228). To determine ter-

Table 7.26. Determination of Hydrogen in Titanium, Zirconium, and Zircaloy Standard Reference Materials with Electron Impact IDMS

	Concentration (μg/g)	
Sample	IDMS	Certified Value
Zirconium NBS 358	107.2 ± 0.6	107 ± 3
Titanium NBS 1087	56.1 ± 0.4	57.5 ± 2.5
Zircaloy JAERI Z-10	9.3 ± 0.4	8.2 ± 1.1[a]
Zircaloy JAERI Z-17	15.8 ± 0.5	Not certified

Source: Ref. 231.

[a] Mean value of nine laboratories

restrial ages of Antarctic meteorites, the measurement of ^{81}Kr has been used by Schultz and coworkers (229).

Recently, the hydrogen concentration in titanium and in zirconium and its alloys was determined with electron impact IDMS by Watanabe and coworkers (230, 231). In this analysis the sample was equilibrated with a known quantity of deuterium at 850°C. Table 7.26 shows a selection of results for four NBS and JAERI standard reference materials. The IDMS results show very good agreement with the certified values. The relative standard deviations are in the range of 0.5–1% for concentrations above 50 μg/g and in the range of 2–10% for hydrogen contents of 7–20 μg/g.

Only a few alternative analytical methods are available for accurate determinations of trace gases, but there is a prospect of expanding electron impact IDMS in this field. Also, it is possible to determine element concentrations in inorganic or organic compounds with the gas IDMS method. For example, nitrogen from an organic compound can be converted into gaseous N_2, and after spiking with ^{15}N-enriched N_2 the nitrogen content can be determined by IDMS. Using optical emission spectroscopy for the isotope ratio measurement of nitrogen, this type of isotope dilution technique was applied by Middelboe (232).

6. CONCLUSION

The most accurate results in mass spectrometry are obtained if the isotope dilution technique is applied. Accuracies in trace analyses can be achieved with relative standard deviations of less than 1% under certain conditions. Therefore, the application of IDMS is especially recommended for the calibration of other analytical data, for the certification of standard reference materials, and in all cases where the accuracy of the result is the major analytical interest. The greatest amount of experience for mono- and oligoelement analyses using IDMS exists with thermal ionization and electron impact ionization, and for multielement determinations, with spark source mass spectrometry.

Future trends in IDMS might be in species-specific determinations, as the first examples presented in Section 5.5 have shown. In multielement analyses an expansion of the isotope dilution technique to other ionization methods, for example, to ICP-MS and glow discharge mass spectrometry, can be expected. Also, the construction of simpler and more cost-efficient instruments can lead to more frequent application of IDMS; this has been shown by the construction of a compact quadrupole thermal ionization mass spectrometer.

REFERENCES

1. J. Büttner, R. Borth, J. H. Boutwell, M. G. Broughton, and R. C. Bowyer, *J. Clin. Chem. Clin. Biochem.*, **18**, 69 (1980).

2. K. G. Heumann, *Fresenius Z. Anal. Chem.*, **324**, 601 (1986).

3. P. de Bièvre, *Adv. Mass Spectrom.*, **7A**, 395 (1978).

4. K. G. Heumann, *Toxicol. Environ. Chem. Rev.*, **3**, 111 (1980).

5. K. G. Heumann, *Trends Anal. Chem.*, **1**, 357 (1982).

6. K. G. Heumann, *Int. J. Mass Spectrom. Ion Phys.*, **45**, 87 (1982).

7. K. G. Heumann, *Biomed. Mass Spectrom.*, **12**, 477 (1985).

8. L. Siekmann and H. Breuer, *J. Clin. Chem. Clin. Biochem.*, **20**, 883 (1982).

9. I. Björkhem, A. Bergman, O. Falk, A. Kallner, O. Lantto, L. Svensson, E. Åkerlöf, and R. Blomstrand, *Clin. Chem.*, **27**, 733 (1981).

10. K. P. Jochum and M. Seufert, *Geol. Rundschau*, **69**, 997 (1980).

11. K. P. Jochum, M. Seufert, and H. J. Knab, *Fresenius Z. Anal. Chem.*, **309**, 285 (1981).

12. K. P. Jochum, A. W. Hofmann, E. Ito, M. Seufert, and W. M. White, *Nature*, **306**, 431 (1981).

13. F. Basolo and R. G. Pearson, *Mechanismen in der anorganischen Chemie*, 2nd ed., Georg Thieme, Stuttgart, 1973.

14. R. G. Wilkins, *The Study of Kinetics and Mechanism of Transition Metal Complexes*, Allyn and Bacon, Boston, 1974.

15. N. E. Holden, R. L. Martin, and I. L. Barnes, *Pure Appl. Chem.*, **56**, 675 (1984).

16. J. Hoefs, *Stable Isotope Geochemistry*, Springer, New York, 1973.

17. A. Götz and K. G. Heumann, unpublished, 1986.

18. T. J. Chow, C. B. Snyder, and J. L. Earl, *Stable Isotopes in the Life Sciences*, Proc. Technical Committee Meeting in Leipzig, IAEA, Vienna 1977, pp. 95–108.

19. J. D. Schladot, K. Hilpert, and H. W. Nürnberg, *Adv. Mass Spectrom.*, **8A**, 325 (1980).

20. M. Keinonen and T. Jaakkola, *Advances in Mass Spectrometry 1985, Part B*, J. F. J. Todd, Ed., Wiley, New York, 1986, p. 1069.

21. J. Trettenbach and K. G. Heumann, unpublished, 1984.

22. K. G. Heumann, E. Kubassek, and W. Schwabenbauer, *Fresenius Z. Anal. Chem.*, **287**, 121 (1977).

23. K. G. Heumann and H. Zeininger, *Int. J. Mass Spectrom. Ion Processes*, **67**, 237 (1985).

24. J. Trettenbach and K. G. Heumann, *Fresenius Z. Anal. Chem.*, **322**, 306 (1985).

25. A. Eberhardt, R. Delwiche, and J. Geiss, *Z. Naturforsch.*, **19a**, 736 (1964).

26. L. J. Moore and E. F. Heald, *Adv. Mass Spectrom.*, **7A**, 448 (1978).

27. K. Habfast, *Int. J. Mass Spectrom. Ion Phys.*, **51**, 165 (1983).

28. M. H. Dodson, *J. Sci. Instrum.*, **2**, 490 (1969).

29. W. Riepe and H. Kaiser, *Fresenius Z. Anal. Chem.*, **223**, 321 (1966).

30. P. de Bièvre and G. H. Debus, *Nucl. Instrum. Methods*, **32**, 224 (1965).

31. B. N. Colby, A. E. Rosecrance, and M. E. Colby, *Anal. Chem.*, **53**, 1907 (1981).

32. K. G. Heumann, F. Beer, and R. Kifmann, *Talanta*, **27**, 567 (1979).

33. P. de Bièvre, "Isotope Dilution Mass Spectrometry," in *Trace Metal Analysis in Biological Samples*, Biomedical Publishers, Foster City, CA, 1987.

34. K. G. Heumann, P. Kastenmayer, and H. Zeininger, *Fresenius Z. Anal. Chem.*, **306**, 173 (1981).

35. Cambridge Isotope Laboratories, Woburn, Massachusetts.

36. A. Hempel GmbH, Düsseldorf, Germany.

37. Mound Laboratory, Miamisburg, Ohio.

38. Oak Ridge National Laboratory, Union Carbide Corp., Nuclear Division, Oak Ridge, Tennessee.

39. Oris Stable Isotopes, Gif-sur-Yvette, France.

40. R. Grosser, K. G. Heumann, and W. Schindlmeier, *Advances in Mass Spectrometry 1985, Part B*, J. F. J. Todd, Ed., Wiley, New York, 1986, p. 1057.

41. I. Wendt and W. Stahl, Massenspektrometrische Untersuchung der natürlichen Variationen der Calcium-Isotopenverhältnisse, *Bericht der Bundesanstalt für Bodenforschung*, Hannover, 1968.

42. I. Langmuir and K. H. Kingdon, *Proc. Roy. Soc. (London)*, **107**, 61 (1925).

43. M. Kaminsky, *Atomic and Ionic Phenomena on Metal Surfaces*, Springer-Verlag, New York, 1965.

44. H. Kawano and F. M. Page, *Int. J. Mass Spectrom. Ion Phys.*, **50**, 1 (1983).

45. R. C. Weast, Ed., *Handbook of Chemistry and Physics*, 63rd ed., CRC Press, Boca Raton, FL, 1982.

46. P. B. Papić, M. M. Ciríc, and K. F. Zmbov, *Bull. Soc. Chim. Beograd*, **44**, 195 (1979).

47. R. J. Zollweg, *J. Chem. Phys.*, **50**, 4251 (1969).

48. K. G. Heumann, W. Schindlmeier, H. Zeininger, and M. Schmidt, *Fresenius Z. Anal. Chem.*, **320**, 457 (1985).

49. D. J. Rokop, R. E. Perrin, G. W. Knobeloch, V. M. Armijo, and W. R. Shields, *Anal. Chem.*, **54**, 957 (1982).

50. P. A. Akishin, O. T. Mikitin, and B. M. Panchenkov, *Geokhimiya*, **5**, 425 (1957).

51. A. E. Cameron, D. H. Smith, and R. L. Walker, *Anal. Chem.*, **41**, 525 (1969).

52. I. L. Barnes, T. J. Murphy, J. W. Gramlich, and W. R. Shields, *Anal. Chem.*, **45**, 1881 (1973).

53. M. Murozumi, S. Nakamura, T. Igarashi, and K. Yoshida, *Nippon Kagaku Kaishi*, **1**, 122 (1981).

54. E. Waidmann, K. Hilpert, J. D. Schladot, and M. Stoeppler, *Fresenius Z. Anal. Chem.*, **317**, 273 (1984).

55. A. Broekman and J. G. van Raaphorst, *Fresenius Z. Anal. Chem.*, **318**, 398 (1984).

56. W. R. Kelly, F. Tera, and G. J. Wasserburg, *Anal. Chem.*, **50**, 1279 (1978).

57. K. J. R. Rosman, J. R. de Laeter, and A. Chegwidden, *Talanta*, **29**, 279 (1982).

58. J. R. de Laeter and N. Mermelengas, *Geostand. Newsl.*, **2**, 9 (1978).

59. R. D. Loss, K. J. R. Rosman, and J. R. de Laeter, *Geostand. Newsl.*, **7**, 321 (1983).

60. J. W. Gramlich, L. A. Machlan, T. J. Murphy, and L. T. Moore, in D. D. Hemphill, Ed., *Trace Substances in Environmental Health*, Vol. 11, University of Missouri, 1977, p. 376.

61. J. R. de Laeter, M. T. McCulloch, and K. J. R. Rosman, *Earth Planet. Sci. Lett.*, **22**, 226 (1974).

62. J. Völkening and K. G. Heumann, *Advances in Mass Spectrometry 1985, Part B*, J. F. J. Todd, Ed., Wiley, New York, 1986, p. 1059.

63. D. Tuttas, Finnigan MAT, Application Note No. 46, Bremen, 1981.

64. D. H. Smith, R. L. Walker, L. K. Bertram, J. A. Carter, and J. A. Goleb, *Anal. Lett.*, **12**, 831 (1979).

65. D. H. Smith, W. H. Christie, and R. E. Eby, *Int. J. Mass Spectrom. Ion Phys.*, **36**, 301 (1980).

66. P. Kastenmayer, thesis, University of Regensburg, 1984.

67. T. J. Anderson and R. L. Walker, *Anal. Chem.*, **52**, 709 (1980).

68. M. H. Studier, E. N. Sloth, and L. P. Moore, *J. Phys. Chem.*, **66**, 133 (1962).

69. F. R. Niederer, D. A. Papanastassiou, and G. J. Wasserburg, *Geochim. Cosmochim. Acta*, **45**, 1017 (1981).

70. A. H. Turnbull, "Surface Ionisation Techniques in Mass Spectrometry," AERE-Report 4295, Harwell, UK, 1963.

71. P. de Bièvre and G. H. Debus, *Int. J. Mass Spectrom. Ion Phys.*, **2**, 15 (1969).

72. H. Zeininger and K. G. Heumann, *Int. J. Mass Spectrom. Ion Phys.*, **48**, 377 (1983).

73. N. L. Duchateau and P. de Bièvre, *Int. J. Mass Spectrom. Ion Phys.*, **54**, 289 (1983).

74. K. G. Heumann and H. Hable, *Proceedings, 33rd Ann. Conf. Mass Spectrom. Allied Topics*, San Diego, 1985, p. 166.

75. M. Unger and K. G. Heumann, *Int. J. Mass Spectrom. Ion Phys.*, **48**, 373 (1983).

76. K. G. Heumann and M. Unger, *Fresenius Z. Anal. Chem.*, **315**, 454 (1983).

77. M. Unger and K. G. Heumann, *Fresenius Z. Anal. Chem.*, **320**, 526 (1985).

78. W. R. Shields, T. J. Murphy, E. L. Garner, and V. H. Dibeler, *J. Amer. Chem. Soc.*, **84**, 1519 (1962).

79. E. L. Garner, L. A. Machlan, J. W. Gramlich, L. J. Moore, T. J. Murphy, and I. L. Barnes, *NBS Spec. Publ.*, **422**, 951 (1976).

80. K. G. Heumann and R. Hoffmann, *Adv. Mass Spectrom.*, **7**, 610 (1978).

81. K. G. Heumann, F. Beer, and H. Weiss, *Mikrochim. Acta*, 95 (1983).

82. K. G. Heumann, R. Kifmann, W. Schindlmeier, and M. Unger, *Int. J. Environ. Anal. Chem.*, **10**, 39 (1981).

83. E. J. Catanzaro, T. J. Murphy, E. L. Garner, and W. R. Shields, *J. Res. NBS* (*Phys. Chem.*), **8**, 318 (1964).

84. K. G. Heumann, W. Schrödl, and H. Weiss, *Fresenius Z. Anal. Chem.*, **315**, 213 (1983).

85. K. G. Heumann and W. Schindlmeier, *Fresenius Z. Anal. Chem.*, **306**, 245 (1981).

86. K. G. Heumann, P. Kastenmayer, W. Schindlmeier, M. Unger, and H. Zeininger, *Proceedings, 31st Ann. Conf. Mass Spectrom. Allied Topics*, Boston, 1983, p. 581.

87. K. G. Heumann and W. Schindlmeier, *Fresenius Z. Anal. Chem.*, **312**, 595 (1982).

88. W. Schindlmeier and K. G. Heumann, *Fresenius Z. Anal. Chem.*, **320**, 745 (1985).

89. K. G. Heumann and H. Seewald, *Fresenius Z. Anal. Chem.*, **320**, 494 (1985).
90. K. G. Heumann and H. Weiss, *Fresenius Z. Anal. Chem.*, **323**, 852 (1986).
91. J. E. Delmore, *Int. J. Mass Spectrom. Ion Phys.*, **43**, 273 (1982).
92. H. Zeininger, thesis, University of Regensburg, 1984.
93. E. N. Treher and D. J. Rokop, *Proceedings, 31st Ann. Conf. Mass Spectrom. Allied Topics*, Boston, 1983, p. 828.
94. A. Perksy, E. F. Greene, and A. Kuppermann, *J. Chem. Phys.*, **49**, 2347 (1968).
95. J. D. Waldron, *Advances in Mass Spectrometry*, Pergamon, London, 1959, p. 97.
96. L. D. Nguyen and M. de Saint-Simon, *Int. J. Mass Spectrom. Ion Phys.*, **9**, 299 (1972).
97. K. G. Heumann and J. Trettenbach, *Fresenius Z. Anal. Chem.*, **310**, 146 (1982).
98. K. G. Heumann, *Int. J. Mass Spectrom. Ion Phys.*, **9**, 315 (1972).
99. A. E. Cameron, *Actinides Rev.*, **1**, 299 (1969).
100. M. Lounsbury, *Can. J. Chem.*, **34**, 259 (1956).
101. M. Köppe, Diplomarbeit, University of Regensburg, 1985.
102. L. L. Ames, P. N. Walsh, and D. White, *J. Phys. Chem.*, **71**, 2707 (1971).
103. K. Hilpert, E. Waidmann, and M. Stoeppler, *Advances in Mass Spectrometry, 1985, Part B*, J. F. J. Todd, Ed., Wiley, New York, 1986, p. 1063.
104. P. R. Trincherini, H. Ullah, and S. Facchetti, *Eur. Appl. Res. Rept., Nucl. Sci. Technol.*, **2**, 295 (1980).
105. H. Weiss, thesis, University of Regensburg, 1985.
106. J. Trettenbach, thesis, University of Regensburg, 1984.
107. R. Alvarez, P. J. Paulsen, and D. E. Kelleher, *Anal. Chem.*, **41**, 955 (1969).
108. P. J. Paulsen, R. Alvarez, and C. W. Mueller, *Anal. Chem.*, **42**, 673 (1970).
109. C. Karr, Ed., *Analytical Methods for Coal and Coal Products*, Vol. 1, Academic, New York, 1978, p. 406.
110. H. J. Dietze and I. Opauszky, *Isotopenpraxis*, **15**, 309 (1979).
111. K. P. Jochum, M. Seufert, and S. Best, *Fresenius Z. Anal. Chem.*, **309**, 308 (1981).
112. J. van Puymbroeck and R. Gijbels, *Fresenius Z. Anal. Chem.*, **309**, 312 (1981).
113. H. Kienitz, Ed., *Massenspektrometrie*, Verlag Chemie, Weinheim, 1968.
114. M. R. Litzow and T. R. Spalding, *Mass Spectrometry of Inorganic and Organometallic Compounds*, Elsevier, New York, 1973.
115. G. B. Dalrymple and M. A. Lanphere, *Potassium-Argon Dating*, Freeman, San Francisco, 1969.
116. I. Friedmann, *Geochim. Cosmochim. Acta*, **4**, 89 (1953).

117. H. Craig, *Geochim. Cosmochim. Acta,* **12,** 133 (1957).

118. W. Dansgaard, *Meddel. Grφnl.,* **165,** 1 (1961).

119. H. Birkenfeld, G. Haase, and H. Zahn, *Massenspektrometrische Isotopenanalyse*, VEB Deutscher Verlag der Wissenschaften, Berlin, 1969.

120. E. J. Spitzer and J. R. Sites, "Isotopic Mass Spectrometry of the Elements," ORNL Report 3528, Oak Ridge, TN, 1963.

121. W. G. Mook and P. M. Grootes, *Int. J. Mass Spectrom. Ion Phys.,* **12,** 273 (1973).

122. C. Brunnée, *Chem.-Tech.,* **7,** 111 (1978).

123. T. Nielsen, H. Egsgaard, and E. Larsen, *Anal. Chim. Acta,* **124,** 1 (1981).

124. D. C. Reamer and C. Veillon, *Anal. Chem.,* **53,** 2166 (1981).

125. R. Kownatzki, F. Peters, G. H. Reil, and G. Maass, *Biomed. Mass Spectrom.,* **7,** 540 (1980).

126. W. Schäfer and K. Ballschmiter, *Fresenius Z. Anal. Chem.,* **315,** 475 (1983).

127. H. R. Schulten, U. Bahr, and W. D. Lehmann, *Mikrochim. Acta,* 191 (1979).

128. H. R. Schulten, B. Bohl, U. Bahr, R. Mueller, and R. Palavinskas, *Int. J. Mass Spectrom. Ion Phys.,* **38,** 281 (1981).

129. H. R. Schulten, *Fachzeitschr. Lab.,* **26,** 533 (1982).

130. R. Ziskoven, C. Achenbach, H. R. Schulten, and R. Roll, *Toxicol. Lett.,* **19,** 225 (1983).

131. H. R. Schulten, P. B. Monkhouse, C. Achenbach, and R. Ziskoven, *Experientia,* **39,** 736 (1983).

132. R. Palavinskas, K. Kriesten, and H. R. Schulten, *Comp. Biochem. Physiol.,* **79A,** 77 (1984).

133. H. R. Schulten, U. Bahr, and R. Palavinskas, *Fresenius Z. Anal. Chem.,* **317,** 497 (1984).

134. D. Glick, Ed., *Methods of Biochemical Analysis*, Vol. 24, Wiley, New York, 1977, p. 313.

135. H. R. Schulten, *Int. J. Mass Spectrom. Ion Phys.,* **32,** 97 (1979).

136. H. R. Schulten, R. Müller, and D. Haaks, *Fresenius Z. Anal. Chem.,* **304,** 15 (1980).

137. H. R. Schulten, P. B. Monkhouse, and R. Müller, *Anal. Chem.,* **54,** 654 (1982).

138. A. L. Yergey, N. E. Vieira, and J. W. Hansen, *Anal. Chem.,* **52,** 1811 (1980).

139. J. R. Lloyd and F. H. Field, *Biomed. Mass Spectrom.,* **8,** 19 (1981).

140. H. S. McKown, D. H. Smith, and R. L. Sherman, *Int. J. Mass Spectrom. Ion Phys.,* **51,** 39 (1983).

141. M. Schmidt, "Thermionic Quadrupole Mass Spectrometer THQ for Trace Element Analyses," Tech. Rep. No. 406, Finnigan MAT, Bremen, 1984.

142. M. Schmidt, "Thermoquad THQ, Testing Instrument Performance with the

NBS 987 Strontium Standard," Application Rep. No. 61, Finnigan MAT, Bremen, 1985.

143. J. Franzen, K. H. Maurer, and K. D. Schuy, *Z. Naturforsch.*, **21a**, 37 (1966).

144. K. P. Jochum, in B. Sansoni, Ed., *Instrumentelle Multi-Elementanalyse*, VCH-Verlag, Weinheim, 1985.

145. C. Brunnée, *Z. Phys.*, **147**, 161 (1957).

146. D. J. Douglas, E. S. K. Quan, and R. G. Smith, *Spectrochim. Acta*, **38B**, 39 (1983).

147. H. E. Taylor and J. R. Garbarino, *Abstracts Colloquium Spectroscopicum Internationale XXIV*, Vol. 3, Gesellschaft Deutscher Chemiker, Garmisch-Partenkirchen, 1985, p. 398.

148. E. Scheubeck, A. Nielsen, and G. Iwantscheff, *Fresenius Z. Anal. Chem.*, **294**, 398 (1979).

149. S. E. Raptis, G. Knapp, and A. P. Schalk, *Fresenius Z. Anal. Chem.*, **316**, 482 (1983).

150. G. Kaiser, P. Tschöpel, and G. Tölg, *Fresenius Z. Anal. Chem.*, **253**, 177 (1971).

151. T. T. Gorsuch, *Destruction of Organic Matter*, Pergamon, Oxford, 1970, p. 19.

152. R. Bock, *Aufschlussmethoden der anorganischen und organischen Chemie*, Verlag Chemie, Weinheim, 1972.

153. G. Knapp, B. Sadjadi, and H. Spitzy, *Fresenius Z. Anal. Chem.*, **274**, 274 (1975).

154. L. Kotz, G. Kaiser, P. Tschöpel, and G. Tölg, *Fresenius Z. Anal. Chem.*, **260**, 207 (1972).

155. L. Kotz, G. Henze, G. Kaiser, S. Pahlke, M. Veber, and G. Tölg, *Talanta*, **26**, 681 (1979).

156. O. A. Schaeffner and J. Zähringer, *Potassium Argon Dating*, Springer-Verlag, Berlin, 1966.

157. K. G. Heumann, K. H. Lieser, and H. Elias, *Recent Developments in Mass Spectroscopy*, University of Tokyo Press, Tokyo, 1970, p. 457.

158. K. G. Heumann, E. Kubassek, W. Schwabenbauer, and I. Stadler, *Fresenius Z. Anal. Chem.*, **297**, 35 (1979).

159. P. Krumbiegel, *Isotopieeffekte*, Akademie-Verlag, Berlin, 1970.

160. J. Bigeleisen and M. G. Mayer, *J. Chem. Phys.*, **15**, 261 (1947).

161. K. G. Heumann, F. Gindner, and H. Klöppel, *Angew. Chem. Int. Ed. Engl.*, **16**, 719 (1977).

162. K. G. Heumann and R. Hoffmann, *Angew. Chem. Int. Ed. Engl.*, **15**, 55 (1976).

163. K. G. Heumann, in F. L. Boschhe, Ed., *Topics in Current Chemistry*, Vol. 127, Springer-Verlag, New York, 1985, p. 77.

164. E. Colinet, B. Griepink, and H. Muntau, BCR information EUR 8836 EN and EUR 8837 EN, Brussels, 1983.

165. H. J. Dietze, *Isotopenpraxis,* **15,** 46 (1979).

166. H. J. Knab and H. Hintenberger, *Meteoritics,* **13,** 522 (1978).

167. P. J. Paulsen, R. Alvarez, and D. E. Kelleher, *Spectrochim. Acta,* **24B,** 535 (1969).

168. J. A. Carter, R. L. Walker, and J. R. Sites, *Trace Elements in Fuel (Adv. Chem. Ser.* Vol. 141), Amer. Chem. Soc., Washington, D.C., 1975, p. 74.

169. J. A. Carter, J. C. Franklin, and D. L. Donohue, Amer. Chem. Soc. Symp. Ser., Div. Fuel Chem. 22, Chicago, 1977.

170. D. W. Koppenaal, R. G. Lett, F. R. Brown, and S. E. Manahan, *Anal. Chem.,* **52,** 44 (1980).

171. A. P. Mykytiuk, D. S. Russell, and R. E. Sturgeon, *Anal. Chem.,* **52,** 1281 (1980).

172. C. Achenbach, R. Ziskoven, F. Koehler, U. Bahr, and H. R. Schulten, *Angew. Chem.,* **91,** 944 (1979).

173. R. Ziskoven, C. Achenbach, U. Bahr, and H. R. Schulten, *Z. Naturforsch.,* **35c,** 902 (1980).

174. H. R. Schulten, W. D. Lehmann, and R. Ziskoven, *Z. Naturforsch.,* **33c,** 484 (1978).

175. W. D. Lehmann, U. Bahr, and H. R. Schulten, *Biomed. Mass Spectrom.,* **5,** 536 (1978).

176. V. V. Armstrong and A. A. Verbeek, *S. Afr. J. Chem.,* **33,** 30 (1980).

177. L. D. Nguyen, M. de Saint-Simon, G. Puil, Y. Yokoyama, and F. Arbey, *Adv. Mass Spectrom.,* **6,** 619 (1974).

178. L. J. Moore and L. A. Machlan, *Anal. Chem.,* **44,** 2291 (1972).

179. J. W. Gramlich, L. A. Machlan, K. A. Brletic, and W. R. Kelly, *Clin. Chem.,* **28,** 1309 (1982).

180. P. Chastagner, *22nd Ann. Rocky Mountain Conf.,* Denver, 1980.

181. S. K. Aggarwal, V. D. Kavimandan, H. C. Jain, and C. K. Mathews, *Talanta,* **24,** 701 (1977).

182. Y. M. Miller and M. S. Chupakhin, *Zh. Anal. Khim. Engl. Transl.,* **23,** 987 (1968).

183. M. Shima, M. Imamura, and M. Honda, *Mass Spectrosc. Japan,* **16,** 277 (1968).

184. M. Murozumi and S. Nakamura, *Bunseki Kagaku,* **22,** 1548 (1973).

185. M. Murozumi and S. Nakamura, *Bunseki Kagaku,* **23,** 912 (1974).

186. A. Broekman and J. G. van Raaphorst, *Fresenius Z. Anal. Chem.,* **315,** 30 (1983).

187. A. Götz and K. G. Heumann, *Fresenius Z. Anal. Chem.,* **325,** 24 (1986).

188. M. Murozumi, S. Nakamura, and K. Suga, *Shitsuryo Bunseki,* **29,** 371 (1981).

189. M. Murozumi, *Bunseki Kagaku,* **30,** 19 (1981).

190. M. Murozumi, T. Igarashi, and S. Nakamura, *Nippon Kagaku Kaishi,* **1982,** 54.

191. E. Wenig and P. Zink, *Arch. Toxikol.,* **22,** 255 (1967).

192. L. J. Moore, J. W. Gramlich, and L. A. Machlan, in D. D. Hemphill, Ed., *Trace Substances in Environmental Health,* Vol. 9, University of Missouri, 1975, p. 311.

193. L. A. Machlan, J. W. Gramlich, T. J. Murphy, and I. L. Barnes, *NBS Spec. Publ.,* **422,** 929 (1976).

194. M. Murozumi, S. Nakamura, T. Kato, T. Igarashi, and H. Tsubota, *Nippon Kagaku Kaishi,* **1978,** 226.

195. M. Murozumi, S. Nakamura, and K. Suga, *Nippon Kagaku Kaishi,* **1981,** 385.

196. J. Everson and C. C. Patterson, *Clin. Chem.,* **26,** 1603 (1980).

197. P. R. Trincherini and S. Facchetti, in S. Facchetti, Ed., *Analytical Techniques for Heavy Metals in Biological Fluids,* Elsevier, New York, 1981, p. 255.

198. I. L. Barnes, T. J. Murphy, and E. A. I. Michiels, *J. Assoc. Off. Anal. Chem.,* **65,** 953 (1982).

199. K. J. R. Rosman and J. R. de Laeter, *Geochim. Cosmochim. Acta,* **38,** 1665 (1974).

200. I. L. Barnes, E. L. Garner, J. W. Gramlich, L. J. Moore, T. J. Murphy, L. A. Machlan, and W. R. Shields, *Anal. Chem.,* **45,** 880 (1973).

201. M. Murozumi, S. Nakamura, and M. Yuasa, *Bunseki Kagaku,* **26,** 626 (1977).

202. M. Murozumi, S. Nakamura, T. Igarashi, and H. Tsubota, *Nippon Kagaku Kaishi,* **1978,** 565.

203. L. I. Zhuk, I. Y. Nikolishin, L. A. Kuranova, E. S. Gureev, and A. A. Kist, *Uzb. Khim. Zh.,* **4,** 65 (1980).

204. C. F. Boutron and C. C. Patterson, *Geochim. Cosmochim. Acta,* **47,** 1355 (1983).

205. J. D. Fassett, J. C. Travis, L. J. Moore, and F. E. Lytle, *Anal. Chem.,* **55,** 765 (1983).

206. J. D. Fassett, L. J. Powell, and L. J. Moore, *Anal. Chem.,* **56,** 2228 (1984).

207. P. Kastenmayer and K. G. Heumann, unpublished, 1984.

208. E. E. Filby, G. W. Webb, R. A. Rankin, and W. A. Emel, *Proc. Nucl. Mater. Management, 1981,* p. 95.

209. B. Ganser, M. Wantschik, and L. Koch, *Int. J. Mass Spectrom. Ion Phys.,* **48,** 405 (1983).

210. W. Beyrich, W. Golly, G. Spannagel, P. de Bièvre, and W. Wolters, The IDA-80 Measurement Evaluation Programme on Mass Spectrometric Isotope Dilution Analysis of Uranium and Plutonium, Vol. I: *Design and Results,* KfK 3760, EUR 7990e, Kernforschungszentrum Karlsruhe, 1984.

211. P. de Bièvre, M. Gallet, F. Hendrickx, W. Lycke, W. Wolters, K. R. Eberhardt, J. D. Fassett, J. W. Gramlich, L. A. Machlan, E. Mainka, and H. Wertenbach, The IDA-80 Measurement Evaluation Programme on Mass Spectrometric Isotope Dilution Analysis of Uranium and Plutonium, Vol. II: *Preparation, Characterization and Transport of the Test Samples*, KfK 3761, EUR 7991e, Kernforschungszentrum Karlsruhe, 1984.

212. K. G. Heumann and K. H. Lieser, *Fresenius Z. Anal. Chem.*, **257**, 18 (1971).

213. W. Zheng, W. Lung, and Z. Zhou, *Zhongguo Dizhi Kexueyuan Yichang Dizhi Kuangchan Yanjiuso Sokan*, **8**, 121 (1984).

214. Community Bureau of Reference, Commission of the European Communities, Project 222, Brussels, 1981.

215. H. Weiss and K. G. Heumann, *Geochim. Cosmochim. Acta*, in press (1987).

216. K. Govindaraju, *Geostand. Newsl.*, **1**, 67 (1977).

217. R. Fuge and C. C. Johnson, *Geostand. Newsl.*, **3**, 51 (1979).

218. K. Govindaraju, *Geostand. Newsl.*, **6**, 91 (1982).

219. K. Govindaraju, *Geostand. Newsl.*, **8**, 173 (1984).

220. L. Leoni, M. Menichini, and M. Saitta, *X-Ray Spectrom.*, **11**, 156 (1982).

221. P. J. Aruscavage and E. Y. Campell, *Talanta,* **30**, 745 (1983).

222. Community Bureau of Reference, Commission of the European Communities, ''The Certification of Boron in Primary Ingot Aluminum BCR No. 25,'' Brussels, 1984.

223. S. K. Luo, C. C. Ou-Yang, F. C. Chang, and Y. C. Yeh, *J. Chinese Chem. Soc.,* **30**, 229 (1983).

224. H. Faul, *Ages of Rocks, Planets, and Stars*, McGraw-Hill, New York, 1966.

225. E. Jäger and J. C. Hunziker, Eds., *Isotope Geology*, Springer-Verlag, New York, 1979.

226. P. Lämmerzahl and J. Zähringer, *Geochim. Cosmochim. Acta,* **30**, 1059 (1966).

227. H. W. Weber and L. Schultz, *Z. Naturforsch.,* **35a**, 44 (1980).

228. H. W. Weber, O. Braun, L. Schultz, and F. Begemann, *Z. Naturforsch.,* **38a**, 267 (1983).

229. L. Schultz and M. Freundel, *Meteoritics,* **19**, 310 (1984).

230. K. Watanabe and M. Ouchi, *Bunseki Kagaku,* **34**, 677 (1985).

231. K. Watanabe, M. Ouchi, and K. Gunji, *Fresenius Z. Anal. Chem.,* **323**, 225 (1986).

232. V. Middelboe, *Stable Isotopes in Life Sciences*, Proc. Tech. Comm. Meet. Mod. Trends Biol. Appl. Stable Isotopes, IAEA, Wien, 1977, p. 239.

CHAPTER

8

RECENT TRENDS AND FUTURE PROSPECTS

R. GIJBELS and F. ADAMS

University of Antwerp, Antwerpen-Wilrijk, Belgium

Within the range of analytical methods, mass spectrometry occupies a favored place because of its unique analytical characteristics: its unprecedented sensitivity and detection limits with a large dynamic range of concentration and its many applications due to the variety of measurable phenomena on the isotopic, elemental, and molecular levels. The method has drawn on and influenced many fields, including atomic physics, reaction kinetics, geochronology, and, most of all, chemical analysis.

Advances have been extremely rapid in the last decade, and present-day instrumentation for mass spectrometry is immensely more powerful and of higher performance than the equipment available just a few years ago. At the same time a plethora of new means for ion excitation, isotope separation, and data acquisition and reduction have appeared that in various combinations provide entirely new instruments and continuously extend the range of possible applications. The latitude in methodology and the characteristics of the hardware that demand interaction with the equipment have made mass spectroscopists likely to modify instruments, extend their applicability, and even develop entirely new instruments (1)—

that is, there is a profound interaction between those who design and build new mass spectrometric instrumentation and those who use it, which tends to push the field quickly forwards. This scientific push is at least as important to the rapid expansion of the field as the demand from the applications side.

1. MASS SPECTROMETRY IN ORGANIC ANALYSIS

The methodological developments and the increase in analytical capabilities and potentials are perhaps more dramatically illustrated in the field of organic analysis than in the inorganic analysis of solids, the topic of this book. Although this area is too vast to develop to any extent in this monograph, we will briefly survey some of the highlights.

In the mass spectrometric analysis of organic compounds, ionization procedures have been introduced that allow the determination of ionic and nonvolatile compounds that could not previously be mass spectrometrically studied (2). The history of desorption ionization starts in 1969 with the publication of the mass spectrum of glucose, a compound of low volatility, by Beckey (3), who used field desorption. The range of methods for desorption ionization is summarized in Table 8.1 (2). They are based on desorption ionization, a family of methods that includes fast atom bombardment (FAB) (4), secondary ion excitation as in SIMS, fission fragment (plasma desorption) methods (5), laser desorption, and field ionization (desorption) (6).

These are called "soft" ionization methods because they minimize fragmentation and tend to yield both the molecular weight and structural data. FAB, which became commercially available only in 1981, enjoyed immediate success. To the reader of this book, this may be surprising, as at first glance the ionization mode distinguishes itself from SIMS only in an apparently minor detail, the substitution of a molecular beam for an ion beam. However, the novel aspect of the method resides elsewhere, and, indeed, neutral ion beams instead of charged ones have long been used for both inorganic and organic analysis. A practical factor is that existing sector instruments with their sources at high voltage could be readily adapted to this mode of excitation (2). A more significant advantage is connected with the way the sample is prepared. Using desorption of the ions from a glycerol solution, which by diffusion continuously supplies the analyte to the surface, ensures that the high vacuum requirements and concomitant beam density requirements of static SIMS are eliminated. Hence, the charge state of the bombarding particle is not important; rather, the liquid mat.ix is, and the beam energy is of secondary impor-

Table 8.1. Techniques for Desorption Ionization[a]

Field Desorption (FD), 1969. Samples are placed on microdendrites, usually carbon grown on a fine metal wire. Ions are desorbed by the combined action of heat and the very high fields present in the source.

Plasma Desorption (PD), 1974. Samples are supported on a thin foil and energized by the passage of fission fragments from ^{252}Cf or ions from a particle accelerator. Mass analysis is performed by time-of-flight measurements.

Secondary Ion Mass Spectrometry (SIMS), 1977. Solid samples or samples mixed with a solid matrix are energized by ions with kiloelectronvolt energy.

Electrohydrodynamic Ionization (EHMS), 1978. Samples are dissolved in glycerol containing an electrolyte. Desorption takes place directly from solution under the influence of high fields without the application of heat.

Laser Desorption (LD), 1978. Ions are desorbed from thin or thick samples by a pulsed laser in transmission or reflection geometry. Commercial systems include a time-of-flight mass analyzer.

Thermal Desorption, 1979. Samples are introduced into the source on a direct probe. Heating of the probe desorbs ions and neutrals. No ionization filament is used.

Fast Atom Bombardment (FAB), 1981. Samples, usually in glycerol solution, are energized by atoms of kiloelectronvolt energy. Fluxes are higher than in SIMS. Add-on sources of several types are available for existing mass spectrometers.

[a]The year corresponds with first use; after Busch and Cooks (2).

tance. Hence, Burlingame et al. (7) are right in proposing to call FAB "liquid SIMS" or perhaps "atom SIMS."

Another important area of progress is the instrumentation itself. Modern mass spectrometers are constructed by adapting building blocks that approach state-of-the-art units of solid-state electronics, vacuum systems, magnet design, precision machining, and computerized data acquisition and processing (8). The fundamental aim of molecular mass spectrometry is to successfully form ions from the molecules of interest without undue degradation of their structure. The ion source is the most critical part of the setup in this respect, but other parts of the system must also meet rather unique requirements. Thus, mass analyzers must be constructed that allow the mass spectrometric analysis of high masses up to several hundred thousand atomic mass units.

The magnetic analyzer is the most important mass spectrometer (actually it is a momentum analyzer), and significant advances have recently been made in new magnetic and electrostatic analyzer combinations. Quadrupole instruments are now available with characteristics that are nearly as good as conventional magnetic sector instruments but with the advantage of being smaller, considerably less expensive, and simpler to operate under computer control. A significant and promising new method in the analysis of organic compounds is the Fourier transform mass spectrometer (FT-MS) (9). This method illustrates very well the speed at which instrumental developments are rapidly transforming mass spectrometry. Indeed, the method arose from ion cyclotron resonance, which itself derives from the principle of mass-to-charge ratio analysis in the cyclotron. As an ion cyclotron resonance mass spectrometer, the technique originated with the Omegatron instrument built at the National Bureau of Standards in 1948. The unprecedented resolution in excess of 10^8 at m/z = 18 (10) allows numerous new applications and, moreover, fundamental work in the areas of ion–molecule gas-phase reaction chemistry and gas-phase metal ion chemistry. The ion trap promises to be significant in making mass spectrometry available at low cost, in particular as a detector for gas chromatography.

Mass spectrometry of large fragile and nonvolatile molecules opens up possibilities for unprecedented applications in biology and medicine (1). These possibilities allow metabolic analyses with immediate diagnostic value, for example, in the detection of particular genetic diseases, or they may be applied in structure elucidations as in sequencing peptides and proteins (11).

At the same time the possibilities for the analysis of increasingly complex mixtures increased dramatically, allowing "needle-in-the-haystack" analytical problems to be tackled (12). These depend on the successful integration of separation techniques with mass spectrometry as in gas chromatography–mass spectrometry (GC-MS) or in liquid chromatography–mass spectrometry (LC-MS). Tandem mass spectrometry or mass spectrometry–mass spectrometry (MS-MS), with its potential for separating complex samples rapidly, often with minimal, if any, sample cleanup (13), has several advantages over both GC-MS and LC-MS as it is a virtually instantaneous separation method that does not suffer from the drawbacks of the chromatographic separations. A tandem mass spectrometer consists of one or more ion sources, two mass analyzers separated by a fragmentation region, and an ion detector. The entire setup aims at increasing the selectivity of the measurement. The principle is straightforward and can be compared to that of conventional GC-MS. A mixture is introduced into the ion source, where ionization produces ions characteristic of the individual components. The separation of the analyte is achieved by the mass selection of the first mass analyzer, replacing the chromatographic separation of GC or LC. The ion then undergoes collisionally activated dissociation through collisions with neutral gas molecules in the fragmentation region, and mass analysis of the daughter ions by the second mass analyzer permits the specific identification of the components. Despite the novelty of the methodology (which was introduced in 1970), there now exist several more or less complex commercially available instruments that can be exploited in various operation modes such as daughter scan, parent scan, neutral loss scan, and selected reaction monitoring. We refer to a book (14) and to recent review articles (13, 15) for the theory, instrumentation, and applications. It was recently determined that the two sector instruments may not provide adequate selectivity for some applications, and more complex instrumentation is now evolving.

All this shows the enormous progress of the potential of mass spectrometry in fields outside the direct scope of this book. They should not, however, detract from the developments that have occurred in the field of inorganic analysis during the same period. Indeed, as is shown in the preceding chapters, the expansion of the methodology in inorganic analysis has also been very rapid. Moreover, the applications in molecular, elemental inorganic, and isotopic analysis seem to become increasingly intertwined and interrelated. In what follows we will briefly review some topics within the realm of this book that have not been treated extensively and then make some comments on the future prospects of the methodology as it appears now.

2. SENSITIVE MASS SPECTROMETRY OF ISOTOPES

2.1. Measurement of Stable Isotope Abundance

Measurements of the abundances of stable isotopes were the first applications of mass spectrometry after the discovery of isotopes. Therefore they have been treated extensively in this book, especially in regard to the isotope dilution method. One major challenge within the field is the accurate and sensitive measurement of the isotopic composition of micrometer-size samples. An illustration of the capabilities in this respect is the determination of the isotopic composition of magnesium in interplanetary dust particles that are typically 10 μm in size and contain on the order of 10^{-10} g of magnesium, which could be determined with a precision of better than 0.1% by conventional isotope mass spectrometry (16). It is clear that direct analysis by ion microscopy would be of tremendous help in such areas of research.

Hart et al. (17) used ion microprobe techniques to determine the variation in lead isotope ratios within a single octahedral crystal of galena (PbS). The crystal showed uniform concentric zones of lead isotope ratios with a total variation of ~3–4%. The results were accurate to 0.1–0.2% in comparison with conventional mass spectrometric analysis performed on microsamples of the same crystal. The ion probe was thus shown to hold promise for establishing a lead isotope "chronostratigraphy" for ore-bearing solutions from which galena is formed. Microanalysis of galena by SIMS was described by Pimminger et al. (18) for the determination of the sulfur isotope ratio. Even at high mass resolution ($m/\Delta m \geq 5000$) a precision and accuracy of 0.2–0.3% for the $^{34}S/^{32}S$ ratio could be obtained in an analyzed area of 8 μm diameter. Hence, in practical applications a variation in the sulfur isotope ratio in zonal growth structures of single galena octahedrons could be observed.

The ability to accurately measure in situ isotope ratios at high spatial resolution is obviously of use in many areas of research and in various technological problems such as in the study of the diffusion of isotopically enriched tracers. Such measurements require great care and optimization, as discussed in Refs. 17 and 18, but are possible on commercially available instruments such as the Cameca IMS-3F and 4F ion microscope/microprobe. The precise measurement of isotope abundance ratios is usually accomplished by sequential scanning of a section of the mass spectrum at high enough mass resolution to eliminate any mass spectral interferences. The magnetic peak-switching mode does not provide sufficient precision, especially if high mass resolution is required, and a fast electrostatic peak-switching unit was therefore developed by Slodzian and coworkers (19) for the Cameca IMS-3F ion microanalyzer. This provides a convenient and flexible means for isotope measurements with a precision on the order of $1-2 \times 10^{-3}$. More precise measurements would require simultaneous detection capability. Of course, the accuracy of the measurements depends on various artifacts such as fractionation during the sputtering and ionization processes (20), the effect of the energy passband of the secondary ions, and other instrumental parameters.

2.2. Geochronology

Single-grain $^{207}Pb/^{206}Pb$ and U/Pb age determinations have been carried out by, among others, Hinthorne et al. (21) and Vander Wood and Clayton

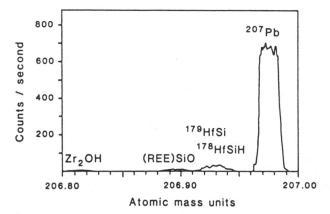

Figure 8.1. Scan over a 200 milli-mass-unit interval below ^{207}Pb in a lunar zircon (^{207}Pb concentration ~140 ppm). Mass resolution defined at 1% of peak height is 7200 and is limited by the choice of a wide collector slit to obtain flat-topped peaks. ^{207}Pb is resolved from major interferences including Zr$_2$OH, HfSi, HfSiH, and species probably composed of SiO plus rare earth elements (24).

(22) with a 10-μm spatial resolution using an ion microprobe mass analyzer. Because of the limited mass resolution of the instrument ($M/\Delta M$ = 300), a peak-stripping technique was necessary to correct for the unresolved mass spectral interferences from a variety of complex polyatomic cluster ion species in the mass range of the lead and uranium isotopes.

The design of a large and sensitive high-resolution ion microprobe identified by the designers as SHRIMP (23) made it possible to eliminate such interferences, as was demonstrated by the in situ U/Pb age determination in zircon crystals in thin sections of lunar breccia (24). SHRIMP (sensitive high-resolution ion microprobe), which was developed and constructed at the Australian National University, incorporates a secondary ion mass analyzer based on a design of Matsuda (25) that attempts minimal second-order aberration coefficients in energy and directional focusing. It includes a cylindrical 85° electrostatic sector, an electrostatic quadrupole lens, and a homogeneous 72.5° magnetic field with non-normal entry. The magnet has a large turning radius of 100 cm to allow the use of a wide source slit (40 μm) for high sensitivity at a mass resolution of 10,000. About 45% of the secondary ions are passed by the entrance slit, and about 90% of these are transmitted to the collector slit. Figure 8.1 shows a scan over a 200 milli-mass-unit interval below ^{207}Pb in a zircon (the ^{207}Pb concentration is about 140 ppm). Mass resolution defined at 1% of the peak height is 7200 and is limited by the choice of a wide collector slit to obtain flat-topped peaks. ^{207}Pb is resolved from major molecular interferences including Zr_2OH, HfSi, HfSiH, and species probably composed of SiO plus rare earth elements. This type of instrumentation is not yet commercially available.

2.3. Accelerator-Based Mass Spectrometry

New techniques of mass spectrometry employ tandem accelerators at energies of millions of electronvolts and work with multiply charged ions and charge-exchange processes. Such mass spectrometers have attained isotope abundance sensitivities of up to 10^{16}, many orders of magnitude larger than the present limits of conventional instruments operated at energies of thousands of electronvolts ($\sim 10^{10}$ keV) (26–28). This very high sensitivity is emphasized by the term "ultrasensitive mass spectrometry;" on the basis of the instrumentation it would be better to call it "accelerator mass spectrometry" (AMS).

This new field within mass spectrometry has been applied to measurements of radioisotopes that are useful as tracers or chronometers such as ^{10}Be, ^{14}C, ^{26}Al, ^{36}Cl, and ^{129}I (29).

One of the important applications is to the measurement of radiocarbon (^{14}C) for determination of the age of archeological and geological samples. Although the beta decay counting used conventionally for dating has been a great success, it suffers from serious drawbacks that sometimes greatly limit its applicability. The most serious drawback is that gram amounts of sample and a few days of measurement are required to achieve a precision of the order of 1%. With AMS, a milligram amount of sample and a few hours of experimentation are sufficient (30). With $^{14}C/^{12}C$ ratios of 10^{15}, age determinations of $\sim 40,000$ years can be achieved, making AMS vastly superior to the conventional counting approach.

This technique, which has also been called mass and charge spectrometry (MACS), promises to become an analytical tool with unprecedented atomic sensitivity. The layout of the instrument developed by Purser et al. (26) is shown as an example in Fig. 8.2. It consists of a SIMS system with a primary cesium ion beam of 30 keV. The negative secondary ions are accelerated to an energy of 20 keV before mass analysis at low resolution. After this initial mass analysis, the ions are accelerated again to 3 MeV, electrons are removed by gas collisions in a stripper, and the positive ions are produced in a high charge state, for example, 3^+ for carbon. The electrostatic 3-MeV potential is then used again, and the ions arriving at ground potential have an energy of up to 12 MeV depending on the charge state. Finally, the ions are again mass analyzed in a low-resolution electrostatic deflection system ($E/\Delta E = 100$) and a magnetic

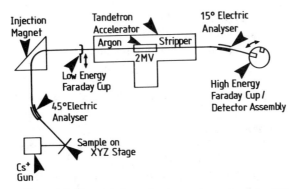

Figure 8.2. Layout of accelerator mass spectrometer (32).

deflector ($M/\Delta M = 200$). The entire setup provides tremendous specificity ($<1:10^{15}$) and sensitivity.

This "SIMS at high energy" can, of course, be used, as already mentioned above, for trace element determinations (31); trace element abundances as low as 10^{-11} have been reported (32). A high sensitivity requires that Cs^+ bombardment produce an intense beam of negative ions of the element of interest. This cannot be achieved, for instance, for iron, and therefore Blattner et al. (32) were unable to detect iron in even a highly doped silicon sample (10^{19} atoms/cm³). Alternatively, a positive ion source combined with an alkali metal charge-exchange canal allows the direct production of Fe^-.

High specificity implies that interfering species such as polyatomic cluster ions can be completely destroyed during the production of multiply charged positive ions. Becker and Dietze (33) found that the intensity ratio M^{3+}/M^+ in a radiofrequency spark could still be as high as 10^{-1} for some aromatic species such as coronene. Also, for simpler molecules and for those unable to delocalize the electronic charge in π electron orbitals, the Coulomb explosion of highly charged states might not always be as complete as expected (34).

The difficulty of studying natural, unprepared materials, such as minerals, was shown by Rucklidge et al. (35), who used AMS to detect ^{194}Pt in a nickel sulfide mineral. Despite remaining interferences, it was estimated that platinum can be detected below the 1 ppb level in this type of sample, when increasing the charge of platinum from +3 to +10 at the high-energy end, and when applying subsequent magnetic filtering.

Blattner et al. (32) evaluated accelerator-based SIMS for the ultratrace elemental characterization of bulk silicon. Boron, arsenic, and antimony were restricted to a useful detection limit of about 5×10^{14} atoms/cm³

(~10 ppb) in these experiments because of the background. No useful data could be obtained for phosphorus because of an insufficient sensitivity in the low-energy mass spectrometer and/or unresolvable interferences from $^{30}SiH^-$ dissociation in the charge-stripper canal. Results for phosphorus in silicon have, however, been reported by a group from Texas Instruments using the AMS setup at the University of Arizona (36).

A lot of research remains to be done to outline the methodology and optimal instrumentation and the potential for applications of accelerator-based mass spectrometry. Detection of the superheavy elements would require methods with high isotope abundance sensitivity in excess of 10^{15} and very low detection limits. The main problem here is the need to avoid atomic and especially molecular isobaric interferences. Dietze (37) suggested resonance ionization mass spectrometry (RIMS) as a potential solution to these experimental difficulties.

3. SECONDARY ION MASS SPECTROMETRY—ION MICROSCOPICAL ANALYSIS

This area, which was extensively treated in Chapter 6, deserves some emphasis in these concluding remarks, in view of the rapid instrumental and methodological progress. Indeed, refinements in the methodology that considerably increase its potential are appearing in several areas.

3.1. Image Analysis

Intrinsically, ion microscopy is capable of three-dimensional characterization by acquiring a series of ion images as a function of sputtered depth. The realization of this potential has been slow due to technological limitations in storage and manipulation of the large sets of data involved.

Digital image-processing systems for both quantitation and image enhancement and data management are now approaching the commercial stage. A general advance in microcomputer technology and software now permits the exciting possibilities of true three-dimensional characterization. With each point of sputtered layer of material corresponding to a separate sample associated with a given volume, different types of analysis are possible, as argued below. This microvolume corresponds to a "voxel" with these coordinates x, y, and z that is independently registered, stored, and processed (38). The computing power remains the primary obstacle for the time being, considering that the total set of data may consist of a stack of pictures of a number of elements each with up to 1024×1024 pixel points and with a large dynamic intensity range.

Once all this information is properly under computer control, transaxial imaging (x–z, y–z, or any plane of interest) and many hitherto unexplored probing applications should become feasible together with any demonstrated advantages of digital manipulation of images, including the use of various standard image processing and enhancement algorithms such as signal averaging, smoothing, image overlay, and pseudo-color-coding and display.

The evolution of digital image acquisition in ion microscopy up to 1983 was reviewed by Rüdenauer (39), who was the first to set up online digital image acquisition for a scanning ion microprobe (40). Digitized images (ion counts) were first stored on magnetic disk and then digitally processed and transferred to digital video for visualization on a TV monitor. On-line digital image acquisition on a high-sensitivity video camera was first achieved on the ion microscope by Furman and Morrison (41). Video camera recording suffers from the drawback that local sensitivity variations of the ion detector as well as the video camera have to be accounted for. On the other hand, the video is a very cheap, versatile, and virtually limitless data storage medium. Brenna et al. (42) reviewed the artifacts introduced, including noise, degradation of spectral resolution, and linearity of response, and came to the conclusion that quantitative analysis is achievable by standardization with images of known intensity distribution. A different approach to a digital system for ion microscopes is the resistive anode encoder (RAE) system described by Odom et al. (43). The RAE is a pulse-counting, position-computing device that provides a direct means of both localizing ion signals in the image plane of the detector and providing pulse-counting capabilities of the ion intensity in this image a quantitative image detector for the Cameca IMS and there it is being commercialized as 3F and 4F ion microscopes by Charles Evans & Associates (Redwood City, CA).

3.2. Focused Primary Ion Beams—Improvements in Spatial Resolution

Significant improvements in lateral resolution are feasible only with the scanning ion probe principle (44). This implies the development of primary ion beams with very small diameters. The development of very high brightness focused liquid metal ion sources (LMIS) of Ga^+ and In^+ based on field ionization has made it possible to obtain chemical maps with a lateral resolution for 40–60 keV Ga^+ at 20 nm at a few picoamperes of probe current (45). This resolution corresponds to the fundamental limit set by the size of the collision cascades initiated by the incident ions.

Table 8.2. Comparison of Working Modes of the Ion Microscope and Ion Microprobe for the Cameca IMS-4F

	Ion Microscope	Ion Microprobe
Analyzed area	$\geq 10 \times 10 \ \mu m$	$\leq 10 \times 10 \ \mu m$
Beam diameter	$50 \ \mu m$	$\leq 1 \ \mu m$
Ion image	Direct ion image, resolution $\sim 1 \ \mu m$	Raster scan image, resolution $\sim 0.2 \ \mu m$
Dynamic transfer system[a]	Usable for depth profile analysis; increases sensitivity	Usable for depth profile analysis and ion image
Line scan	Analyzed area limited by aperture	Analyzed area corresponds to beam diameter

[a] The field of view is rastered over the analyzed area.

3.3. Increases in Capabilities

The Cameca IMS-4F instrument, which has been commercially available since 1986, is a refined version of the IMS-3F described in detail in Chapter 6. It combines capabilities of the ion microscope and the ion microprobe. The advantages of both modes are well documented in Chapter 6 and will be only briefly summarized here.

The ion microscope gives the best sensitivity and hence the fastest ion image-acquisition rate when areas larger than several tens of micrometers have to be analyzed. On the other hand, the ion microprobe offers better sensitivity when the dimensions of the analyzed area are small (below a few micrometers). This implies that the primary beam diameter can be finely focused with a sufficiently high current density. Within the IMS-4F instrument, a cesium ion source of a novel design (no liquid metal ion source) produces an intense and stable primary beam with a diameter as small as $0.2 \ \mu m$ that can be used for the ion microprobe operation mode.

Table 8.2 gives typical characteristics and parameters for the two operation modes as supplied by the manufacturer.

3.4. Methods of Postionization in SIMS

With SIMS, the determination of true concentrations is frequently difficult because of matrix effects. Also, the low ionization efficiency has an adverse effect on detection sensitivity. Historically, electron beam postionization was first employed in 1958 for enhancing the positive secondary ion yield (46). In secondary neutral mass spectrometry (SNMS), the emission process and the ionization processes are decoupled and can be optimized separately. Oechsner and coworkers demonstrated the potential of SNMS as a tool for quantitative mass spectrometric analysis of thin films and bulk solids (47). The ionization fraction may be high and constant to within a factor of 4 for all elements. We will focus on laser postionization. Other methods rely on glow discharge (see Chapter 2) and on thermal processes and are discussed in a review paper by Reuter (48).

4. LASER-ASSISTED MASS SPECTROMETRY

Of growing interest is the use of high-intensity laser light beams in the formation of ions prior to mass spectrometric analysis. The use of lasers in combination with mass spectrometry opens entirely new and potentially very attractive possibilities for enhancing the specificity and/or the sensitivity of the analyses.

Resonance ionization spectroscopy (RIS) is a photoionization process in which atoms in the gas phase are ionized by the absorption of photons that energetically match quantum-selected states of these atoms (49, 50). The simplest scheme for RIS involves a single-color, two-photon process that first excites the atom to an intermediate excitation state that is energetically more than halfway to the ionization potential of that element. Absorption of a second photon then leads to ionization. Lasers are used as a source for photons so that sufficient power is available to promote a bound-state electron to a free state under saturated conditions. The result of RIS is an ion and an electron. Detection of the electron by a proportional counter has been shown to be an extremely sensitive analytical tool, resulting in a single-atom detection capability (51). There is also analytical potential in the ions generated when they are extracted and then separated and detected in a mass spectrometer. Such a combination of RIS and MS (RIMS) could lead to several interesting applications such as selective elemental ionization (49, 50) that would minimize isobaric interferences encountered in more conventional mass spectrometric analysis (52). For a more comprehensive discussion, reference is made to a review by Fassett et al. (53).

Laser ionization mass spectrometry has been reported for some elements, using a variety of vaporization and atomization sources, including thermal vaporization, glow discharge, laser ablation, and ion beam sputtering. Ion beam sputtering is an attractive method for producing the atom reservoir for laser ionization, because the process of atom formation is highly controlled physically and spatially. Provided the laser beam is powerful enough to saturate the ionization, the overall ionization efficiency becomes unity regardless of the ionization potential of the element. Not only does one gain in sensitivity but, perhaps more significantly, methods can be developed that will make SIMS less element-dependent, thus counterbalancing the lack of fundamental understanding of the secondary ion emission process.

Laser photon postionization with tunable dye lasers (multiphoton resonance ionization, MPRI) could increase the specificity of measurement of a particular species even at the isotopic level. The method is presently in the demonstration stage, but several groups have shown interesting aspects of the analytical potential with an ion microprobe (54). Nonresonant laser excitation with a pulsed laser beam and a time-of-flight mass spectrometer is exactly the reverse situation, since it attempts to minimize selectivity (55).

5. MASS SPECTROMETRY IN SPACE

The package of instrumentation for the Galileo mission to Jupiter includes, together with spectrometers and a helium abundance detector based on measurement of the refractive index, a neutrals mass spectrometer. It will be used to determine the abundances and isotope ratios of the major constituents of the Jovian atmosphere. As with all instruments designed for space applications, the mass spectrometer's components, electronics, electron impact ion source, quadrupole mass analyzer, and sputter-ion pump vacuum system were chosen for ruggedness and adaptability. This is just one example of the use of mass spectrometry in the analysis of extraterrestrial space; indeed dozens of specially designed mass spectrometers have already flown on U.S. and U.S.S.R. space probes for atmospheric studies of Earth, Mars, and Venus and to test for the possible presence of organic molecules in the surface soil of Mars. These experiments prove that mass spectrometric equipment can be designed to operate in extremely hostile environments and at the same time illustrate the wide applicability of the technique.

6. CONCLUSIONS

All the examples given here illustrate the rapid development of the methodology and the fact that inorganic mass spectrometry is a fast-moving field with gradual refinements in the instrumentation that increase its attractiveness for various applications. As a field, mass spectrometry is rich in history and in potential. It has drawn from and influenced many fields, including atomic physics, reaction kinetics, geochronology, chemical analysis, and, most recently, biomedicine. Advances in all these fields have depended heavily on this instrumental method.

Specificity, accuracy, sensitivity, and speed are the key performance characteristics that must justify the cost of any analytical method. Mass spectrometry is a method that is competitive within the analytical arsenal despite its complexity and cost. It is often not only the method of choice, but the only one that could provide the required answers to many of the questions raised in science and technology. Areas of interest include materials science, semiconductor research and development, and general-purpose development and quality control in research and industry. A considerable fraction of the analytical work requires only qualitative interpretation of the results, but quantitative work is definitely possible in many areas. Indeed, lack of understanding of the underlying physical or chemical principles sometimes prohibits the use of quantification methods on a theoretical basis, but empirical approaches then often give analytical results with a total relative uncertainty of the order of 10%, provided suitable standards are available and systematic uncertainties are properly circumvented.

REFERENCES

1. R. G. Cooks, K. L. Busch, and G. L. Glish, *Science,* **222,** 273 (1983).

2. K. L. Busch and R. G. Cooks, *Science,* **218,** 247 (1982).

3. H. D. Beckey, *Int. J. Mass Spectrom. Ion Phys.,* **2,** 500 (1969).

4. K. L. Rinehart, Jr., *Science,* **218,** 254 (1982).

5. D. F. Torgerson, R. P. Skowronski, and R. D. Macfarlane, *Biochem. Biophys. Res. Commun.,* **60,** 616 (1974).

6. H. R. Schulten, U. Bahr, and R. Palavinskas, *Fresenius Z. Anal. Chem.,* **317,** 497 (1983).

7. A. L. Burlingame, T. A. Baillie, and P. J. Derrick, *Anal. Chem.,* **58,** 165R (1986).

8. W. V. Ligon, Jr., *Science,* **205,** 151 (1979).

9. M. B. Comisarow and A. G. Marshall, *Chem. Phys. Lett.,* **25,** 282 (1974).

10. M. Allemann, H. P. Kellerhals, and K. P. Wanczek, *Int. J. Mass Spectrom. Ion Processes,* **46,** 139 (1983).

11. K. Biemann, *Anal. Chem.,* **58,** 1288A (1986).

12. F. W. McLafferty, *Science,* **214,** 280 (1981).

13. J. V. Johnson and R. A. Yost, *Anal. Chem.,* **57,** 758A (1985).

14. F. W. McLafferty, Ed., *Tandem Mass Spectrometry*, Wiley, New York, 1983.

15. R. A. Yost and D. D. Fetterolf, *Mass Spectrom. Rev.*, **2**, 1 (1983).

16. T. M. Esat, D. E. Brownlee, D. A. Papanassiou, and G. J. Wasserburg, *Science*, **206**, 190 (1979).

17. S. R. Hart, N. Shimizu, and D. A. Sverjensky, *Econ. Geol.*, **76**, 1873 (1981).

18. M. Pimminger, M. Grasserbauer, E. Schroll, and I. Cerny, *Anal. Chem.*, **56**, 407 (1984).

19. G. Slodzian, J. G. Lorin, R. Dennebouy, and A. Havette, in A. Benninghoven, J. Okano, R. Shimizu, and H. W. Werner, Eds., *Secondary Ion Mass Spectrometry (SIMS) IV*, Springer-Verlag, Berlin, 1983, pp. 153–157.

20. N. Shimizu and S. R. Hart, *J. Appl. Phys.*, **53**, 1303 (1982).

21. J. R. Hinthorne, C. A. Andersen, R. L. Conrad, and J. F. Lovering, *Chem. Geol.*, **25**, 271 (1979).

22. T. Vander Wood and R. N. Clayton, *J. Geol.*, **93**, 251 (1985).

23. S. W. Clement, W. Compston, and G. Newhead, in A. Benninghoven, Ed., *Proceedings of the International SIMS Conference.* Münster, Springer-Verlag, Berlin, 1977.

24. W. Compston and I. S. Williams, Proceedings of the 14th Lunar and Planetary Conference, part 2, *J. Geophys. Res.*, **89** Suppl., B525 (1984).

25. H. Matsuda, *Int. J. Mass Spectrom. Ion Phys.*, **14**, 219 (1974).

26. K. H. Purser, A. E. Litherland, and H. E. Gove, *Nucl. Instrum. Methods*, **162**, 637 (1979).

27. G. N. Flerov and G. M. Ter-Akopian, *Pure Appl. Chem.*, **53**, 909 (1981).

28. H. J. Dietze and S. Becker, *ZFI-Mitt.*, **57**, 1 (1982).

29. A. E. Litherland, K. P. Bouleons, L. R. Kilius, J. C. Rucklidge, H. E. Gove, E. Elmore, and K. H. Purser, *Nucl. Instrum. Methods*, **186**, 463 (1981).

30. T. Nakamura, H. Yamashita, N. Nakai, T. Sakase, S. Sato, and A. Sakai, in A. Benninghoven, J. Okans, R. Shimizu, and H. W. Werner, Eds., *Secondary Ion Mass Spectrometry (SIMS) IV*, Springer-Verlag, Berlin, 1983, pp.175–177.

31. A. E. Litherland, J. C. Rucklidge, G. C. Wilson, W. E. Kieser, L. R. Kilius, and R. P. Beukens, in A. Benninghoven, J. Okans, R. Shimizu, and H. W. Werner, Eds., *Secondary Ion Mass Spectrometry (SIMS) IV*, Springer-Verlag, Berlin, 1983, pp. 170–174.

32. R. J. Blattner, J. C. Huneke, M. D. Strathman, R. S. Hockett, W. E. Kieser, L. R. Kilius, J. C. Rucklidge, G. C. Wilson, and A. E. Litherland, in A. Benninghoven, R. J. Colton, D. S. Simons, and H. W. Werner, Eds., *SIMS V*, Springer-Verlag, Berlin, 1986, pp. 192–194.

33. S. Becker and H. J. Dietze, *Int. J. Mass Spectrom. Ion Processes*, **54**, 337 (1983).

34. L. Morray and I. Cornides, *Int. J. Mass Spectrom. Ion Processes*, **62**, 263 (1984).

35. J. C. Rucklidge, M. P. Gorton, G. C. Wilson, L. R. Kilius, A. E. Litherland, D. Elmore, and H. E. Gove, *Can. Mineral.*, **20**, 111 (1982).

36. D. J. Donahue, *Bull. Am. Phys. Soc.*, **28**, 991 (1983).

37. H. J. Dietze, *ZFI-Mitt.*, **101**, 73 (1985).

38. J. A. McHugh, in A. W. Czanderna, Ed., *Methods of Surface Analysis*, Elsevier, Amsterdam, 1975, p. 226.

39. F. G. Rüdenauer, *Surf. Interface Anal.*, **6**, 132 (1984).

40. H. Liebl, *J. Phys.*, **E8**, 797 (1975).

41. B. K. Furman and G. H. Morrison, *Anal. Chem.*, **52**, 2305 (1980).

42. J. T. Brenna and G. H. Morrison, *Anal. Chem.*, **58**, 428 (1986).

43. R. W. Odom, B. K. Furman, C. A. Evans, Jr., C. E. Bryson, W. A. Petersen, M. A. Kelly, and D. H. Wayne, *Anal. Chem.*, **55**, 578 (1983).

44. H. Liebl, *Vacuum*, **33**, 525 (1983).

45. R. Levi-Setti, G. Crow, and Y. L. Wang, *Scanning Electron Microsc.*, **2**, 535 (1985).

46. R. E. Honig, *J. Appl. Phys.*, **29**, 549 (1958).

47. H. Oechsner and W. Gerhard, *Phys. Lett.*, **A40**, 211 (1972).

48. W. Reuter, in A. Benninghoven, R. J. Colton, D. S. Simons, and H. W. Werner, Eds., *SIMS V*, Springer-Verlag, Berlin, 1985, pp. 94–102.

49. J. P. Young, G. S. Hurst, S. D. Kramer, and M. G. Payne, *Anal. Chem.*, **51**, 1050 (1979).

50. G. S. Hurst, M. G. Payne, S. D. Kramer, and J. P. Young, *Rev. Mod. Phys.*, **51**, 767 (1979).

51. G. S. Hurst, M. H. Nayfeh, and J. P. Young, *Appl. Phys. Lett.*, **30**, 229 (1977).

52. D. L. Donahue, J. P. Young, and D. H. Smith, *Int. J. Mass Spectrom. Ion Processes*, **43**, 293 (1982).

53. J. D. Fassett, L. J. Moore, J. C. Travis, and J. R. De Voe, *Science*, **230**, 262 (1985).

54. D. L. Donahue, W. H. Christie, D. E. Goeringer, and H. S. McKown, *Anal.Chem.*, **57**, 1193 (1985).

55. C. H. Becker and K. T. Gillen, *J. Vac. Sci. Technol.*, **A3**, 1347 (1985).

INDEX